TO THE DIGITAL AGE

JOHNS HOPKINS STUDIES IN THE HISTORY OF TECHNOLOGY   MERRITT ROE SMITH, SERIES EDITOR

ROSS KNOX BASSETT

TO THE DIGITAL AGE

RESEARCH LABS, START-UP COMPANIES, AND THE
RISE OF MOS TECHNOLOGY

THE JOHNS HOPKINS UNIVERSITY PRESS
BALTIMORE AND LONDON

Johns Hopkins Paperbacks edition, 2007

9   8   7   6   5   4   3   2   1

The Johns Hopkins University Press

2715 North Charles Street

Baltimore, Maryland 21218-4363

www.press.jhu.edu

The Library of Congress has cataloged the hardcover edition of this book as follows:

Bassett, Ross Knox, 1959–

To the digital age : research labs, start-up companies, and the rise of MOS technology /
Ross Knox Bassett.

    p.   cm. — (Johns Hopkins studies in the history of technology)

Includes bibliographical references and index.

ISBN 0-8018-6809-2 (hardcover : alk. paper)

1. Metal oxide semiconductors—History. 2. Electronics—Social aspects.

I. Title. II. Series.

TK7871.99.M44 B373 2002

621.39′732′09—dc21   2001001861

ISBN 10: 0-8018-8639-2 (pbk. : alk. paper)

ISBN 13: 978-0-8018-8639-3

A catalog record for this book is available from the British Library.

To my mother, Alvira H. Bassett,

and in memory of my father,

A. Knox Bassett (1919–2000)—my favorite historian

CONTENTS

vii

ACKNOWLEDGMENTS

As I pen the last words of a project that has been on my mind for the better part of two decades, I take great joy in acknowledging the many people and institutions that have enabled me to reach this point. This book had its formal origins as a doctoral dissertation in the Program in History of Science at Princeton University. My advisor, Michael Mahoney, has been a strong supporter of this project and a wise counselor at every step along the way. His comments on various drafts greatly improved both the dissertation and this book. His own work has set the highest standard for me, in both the quality of its scholarship and the grace and clarity of its writing. I have learned much about the study of history through him. I thank Norton Wise for serving on my dissertation committee and for his counsel throughout my graduate career at Princeton. Bill Leslie of Johns Hopkins University, who served as the outside reader of the dissertation, gave me comments on later revisions and much-needed advice thereafter. The members of the program seminar in history of science at Princeton provided a challenging yet congenial atmosphere to try out new work. I would like to thank my compatriots from graduate school, Ann Johnson, Leo Slater, and David Brock, for their contributions to this work, both during graduate school and thereafter.

This book has also benefitted from the wider history and history of technology communities. Several sections were first presented at meetings of the Society for the History of Technology, where I received encouragement and helpful comments. Other material was presented at seminars at Rutgers University and the Chemical Heritage Foundation. William Aspray, Craig Casey, Steven Usselman, Ron Kline, Paul Ceruzzi, Alex Roland, and Henry Lowood have provided valuable comments on sections of this work. I thank David Hounshell for a number of helpful discussions on this project. I would also like to acknowledge my fellow historians of semiconductor technology, Lillian Hoddeson, Michael Riordan, Robert Arns, Daniel Holbrook, and Christophe Lécuyer, for help and encouragement. My colleagues at North Carolina State University, Holly Brewer, Jonathan Ocko, and David Zonderman, gave me useful comments on an early version of the introduction. John David Smith has taken a strong interest in this project and given me his sage counsel on editorial matters. William Kimler gave me the benefit of his biological expertise in straightening me out on a biological analogy. My department heads at North Carolina State, Edith Sylla and Tony Lavopa, have provided invaluable support for my work.

I would also like to thank the various archives and libraries that granted me access to their materials, and the numerous archivists and librarians who made it possible for me to use those materials. Sheldon Hochheiser of the AT&T Archives helped me navigate its massive collections. Jodelle French and Rachel Stewart of the Intel Museum gave me access to the Intel Museum's collection of oral histories and other materials. Marge McNinch and other librarians at the Hagley Museum and Library made my trips there very productive and enjoyable. The late Phyllis Smith, first of RCA and then the Sarnoff Corporation, deserves special mention for her efforts to preserve RCA's history. She enthusiastically gave me access to Sarnoff's library and its collection of laboratory notebooks. The management of the Sarnoff Corporation graciously permitted me to use its material. Alex Magoun of the David Sarnoff Library has provided me with both insight into the history of RCA and additional materials. Emerson Pugh and the other members of IBM's Technical History Project have performed a great service by preserving documents relating to IBM's history. I thank Emerson Pugh for generously serving as my liaison to the IBM Archives, and I thank Robert Godfrey for granting me access to the Technical History Project collection as well as other materials in the IBM Archives. Dawn Stanford of the IBM Archives and Tom Way and Jeff Couture of IBM Microelectronics helped me obtain photos of IBM's semiconductor technologies. Henry Lowood of Stanford University introduced me to Stanford's rich collections, and Maggie Kimball and the other members of the Department of Special Collections and University Archives at the Stanford University Libraries helped me to use them. I would also like to acknowledge Nance Briscoe at the National Museum of American History for her work gathering materials on semiconductor history and her assistance in using the museum's collections. William Van Der Vort of the IEEE Electron Device Society gave me access to its records.

In the course of this project I have interviewed many participants in this story. While I cannot thank each one individually, this work is much richer for their support. A number of engineers and scientists involved in the development of the MOS transistor further assisted me by providing me with materials, commenting on drafts of my manuscript, or entering into an extended dialogue on the history of the MOS transistor. I want to thank A. E. Anderson, Lee Boysel, Paul Castrucci, George Cheroff, Bob Cole, Dale Critchlow, Bruce Deal, Rico DiPietro, Alan Fowler, Jim Kelley, Glen Madland, Robert Meade, Gordon Moore, Tom Rowe, and Merlin Smith. While I am keenly aware that my work will never satisfy any of the participants fully, I deeply appreciate their help and their willingness to entrust a portion of their own history to me. I would particularly like to acknowledge the support of Rolf Landauer. When I, as a graduate student, approached him, one of the senior members of the IBM research community, he warmly invited me up to see him. And while he clearly thought IBM's MOS work reflected well on IBM, his was not the glossy history of public relations brochures but a history that was willing to see the past in all its complexity. I deeply regret that he did not live to see the finished product. All of the people and institutions that have given me access to materials have done so with absolutely no conditions attached.

The National Science Foundation awarded me a dissertation improvement grant (Grant No. SBR94–21889), which enabled me to do essential archival research. Stanford University Libraries awarded me its SDF Project Fellowship, and the IEEE Center for Electrical History awarded me its Fellowship in Electrical History. Generous fellowship support from Princeton University also enabled me to finish my graduate training in a timely fashion.

At Johns Hopkins University Press, Bob Brugger took an interest in this project from a very early stage. I thank an anonymous reviewer for helpful comments. Melody Herr provided essential support in getting this work into print. Elisabeth Dahl displayed infinite patience and saved me from numerous errors with her meticulous copyediting. I thank Merritt Roe Smith for including this work in his series in the history of technology.

I thank Lee Boysel, Bruce Deal, Robert Donovan, Merlin Smith, the AT&T Archives, the David Sarnoff Library, Lucent Technologies, Penton Media, the

IBM Archives, the IBM Corporation, the *IBM Journal of Research and Development*, the Stanford University Libraries Department of University Archives and Special Collections, the Heinz Nixdorf MuseumsForum, and the MIT Press for permission to reproduce figures. I also thank Fairchild Semiconductor, the Intel Museum, Oswald Viva, Bob Norman, Robert Meade, and John Kendall for contributing pictures that could not be included for production reasons. I thank Harwood Academic, a part of the Taylor and Francis Group, for permission to include portions of my article "When Is a Microprocessor Not a Microprocessor? The Industrial Construction of a Semiconductor Innovation," which first appeared in *Exposing Electronics*, Artefacts: Studies in the History of Science and Technology, ed. Bernard Finn, Robert Bud, and Helmuth Trischler (Amsterdam, the Netherlands: Harwood Academic, 2000), 115–34.

Some of my debts on this project are so old, they were doubtless written off many years ago. I would like to thank Paul Matter, my manager at IBM fifteen years ago, for permitting an engineer to take some history courses at Vassar College. At Vassar, James Merrell and Clyde Griffen introduced a frustrated engineer to the joys of the academic study of history. At Cornell, R. Laurence Moore gently steered me in the direction of the history of technology, while Ronald Kline introduced me to the history of technology community.

My wife, Debbie, provided constant support and love through my long and at times agonizing transformation from an engineer to a historian. I want to thank my wife for graciously tolerating the state of my study, which never would have passed muster in one of Andrew Grove's "Mr. Clean" inspections. I thank my daughter, Anna, for her love and her frequent prayers that I would finish my book carefully and not rip any of the pages. My father, A. Knox Bassett, instilled in me a love of history. Although he had no formal training in history, he was a great historian of our family, Pittsburgh sports, and the Dravo Corporation. His example convinced me that any worthwhile history had to be more than just a discussion between academics. I dedicate this book to my mother and to the memory of my father for their years of devoted support and selfless love. While many have contributed, the final responsibility for what is written here is mine alone.

TO THE DIGITAL AGE

Introduction

The MOS (metal-oxide-semiconductor) transistor, the fundamental element in digital electronics, is the base technology of late-twentieth-century and early-twenty-first-century America. Through it digital electronics have entered almost every area of American life, first through the calculator, then through the digital watch, and finally through the microprocessor. The rise of the MOS transistor has made what was once ludicrous commonplace. In Woody Allen's 1969 movie, *Take the Money and Run,* his character responds to the question "Have you ever had any experience in running a high-speed digital electronic computer?" by replying that his aunt has one—an absurd assertion given the state of computing in 1969. Where would his aunt get the $5 million or so necessary to buy one? Would she have it in a cavernous basement with raised floors and special cooling equipment? Would she oversee the teams of programmers and technicians required to use it? Would she use it for calculating missile trajectories, computing payrolls for thousands of employees, or storing her recipes? MOS technology has ruined a good joke. The adjectives that in the 1960s served as reminders of computers' foreignness to people's everyday experience are gone, tautological today. Of course a computer is high-speed, digital, and electronic. What else would it be? Today's image of a computer is of a personal computer, vastly more powerful than the largely forgotten 1969 behemoth, easily affordable, and relatively simple to use. Even if the aunt of today were too timid to buy one, she would likely own dozens of computers simply by having a car, a camera, a sewing machine, or almost any other household appliance of recent vintage.

MOS technology is central to understanding this transformation of the computer and digital electronics. The density of MOS circuitry led to low-cost semiconductor memory and the microprocessor, making a personal computer possible. From the first popular personal computer, the Altair 8800 in 1975, MOS technology development has continued apace, making possible not only more and more powerful personal computers, but also the ever cheaper microcontrollers that have made digital electronics ubiquitous. In 1999, the semiconductor industry sold more than 5 billion microcontrollers and 328 million microprocessors.[1]

Although the term *revolution* is used promiscuously in regard to semiconductor electronics, with the precise nature of the revolution left to the reader's imagination, the MOS transistor has been revolutionary in a way that can be well defined: it overthrew the previously dominant bipolar transistor. The bipolar transistor, invented by William Shockley in 1948, was the first technologically important transistor and for roughly two decades was the default meaning of the word "transistor." The first transistor radio, the first transistorized television, the first integrated circuit, and the computer Woody Allen had in mind in 1969 all used bipolar transistors. In the 1960s, after some in the industry had begun work on the MOS transistor, a debate raged over whether it would find a place, and if it did, what the division of labor between the two transistors would be. While partisans on each side predicted the demise of the other technology, a sensible moderate view was that there would be room for both. In 1997, Gordon Moore, the head of research and development (R&D) at Fairchild Semiconductor in the 1960s and the cofounder of Intel, noted that contrary to his expectations in the 1960s, the final split was over 99 percent MOS.[2]

As Moore suggests, it has been a most improbable triumph. If one were to travel back through time to the late 1950s and try to explain to semiconductor engineers the course their field would take, one would be hard-pressed to know which they would find more incredible, the growth of semiconductor electronics or the fact that the MOS transistor was the vehicle of that growth. A 1950s engineer with a sense of history would have recognized the MOS transistor as a close relative of several of Bell Labs' missteps on its way to inventing the transistor in 1947 Everything that happened in the intervening years reinforced the view that the MOS transistor was a failure. One could certainly build a transistor that way, but it would be much slower and much less reliable. An MOS transistor did not represent progress. It was not the wave of the future. The future would belong to new semiconductor materials, such as gallium arsenide, that promised faster speeds—or entirely new electronic devices based on new scientific principles discovered at industrial research labs.

But the future has belonged to the MOS transistor, a triumph closely tied to the rise of integrated circuits. While in the 1960s the MOS transistor was much slower than the bipolar transistor, it offered the offsetting benefit of simplicity, which allowed one to put more transistors on an integrated circuit. The MOS transistor had the further advantage that it scaled; one could improve its performance by shrinking its dimensions by a constant factor.[3] MOS transistors could also be combined in circuit configurations using almost no power (complementary MOS or CMOS), hardly an important issue in the 1960s, but absolutely essential if millions of transistors were ever to operate on an integrated circuit without melting the chip. In 1960 engineers at RCA built one of the first MOS transistors ever, with its gate (a key parameter affecting both the size and speed of the transistor) having a length of 300 microns; in 2000 Intel produced transistors having gate lengths of 0.18 microns. This relentless push towards smaller and smaller sizes has made it possible to put more and more transistors on an integrated circuit and it has made those transistors faster, driving the bipolar transistor, which did not scale and lacked an ultra-low-power configuration, to the brink of extinction and creating a hostile environment for would-be competitors.[4]

Like all revolutions, the MOS revolution has produced winners and losers. Perhaps the biggest winner has been Intel, the leader in the first major market for MOS products, semiconductor memory, and the pioneer of another important MOS product, the microprocessor. Although Intel started with both bipolar and MOS technologies, it quickly allowed its bipolar work to wither. Since 1992, Intel has been the world's largest producer of integrated circuits.

Intel's public face has changed over time; first it was a memory company, then it was a microprocessor company, and later it called itself an Internet company. Fundamentally, however, since its second year of existence, Intel has been an MOS company. Although Intel sold random access memories, erasable programmable read-only memories, and microprocessors, among other products, it made MOS transistors. Those specific products were a way for Intel to arrange those transistors into parts that customers would buy. Intel's fundamental skill was the ability to fabricate chips containing first thousands and then millions of MOS transistors. With that base secure, all else was possible; without it, nothing was.[5]

Andrew Grove's slogan "only the paranoid survive" has defined a company ever vigilant for new threats or opportunities, willing to transform itself in any way necessary to meet changing conditions. One can easily imagine Intel's managers tossing and turning through many a sleepless night, never able to let go of the challenges facing them. Intel has displayed no sentimental attachment to technologies or products, abandoning dynamic random access memory and being willing to give up the $x86$ architecture at a point when it seemed to have run its course. But even a paranoid has to trust something. Based on Intel's history one might imagine another picture that would also convey an important truth about the company: Intel's senior managers yawning in boredom as they heard about exotic new semiconductor technologies that had attracted attention in some companies and laboratories in the 1970s and 1980s as the possible electronics technologies of the future. For Intel, MOS technology has always been the technology of both the present and the future, and it has never considered life without it.[6]

Of course there have been losers in this revolution. The MOS transistor exposed the managerial weaknesses at Fairchild Semiconductor, the most important semiconductor firm of the 1960s. Despite numerous efforts over two decades, Fairchild was unable to compete in MOS technology and ultimately disappeared as a separate corporation.[7] The unprecedented losses IBM suffered in the early 1990s, $15 billion over three years, were at root due to the MOS transistor's undermining the role of the mainframe computer as the foundation of the company. In the retrenchment that followed, IBM decided it could no longer afford to support both bipolar and MOS technologies, and it halted its bipolar work, by this time largely confined to high-performance mainframe computers. Among the tens of thousands of IBMers who lost their jobs were thousands displaced when IBM closed its East Fishkill semiconductor manufacturing facility, which had been the center of bipolar technology work and once one of the premier semiconductor plants in the world. IBM East Fishkill, named one of America's best-managed factories by *Fortune* in 1984, joined the Lowell mills and the Homestead steel works as exemplars of technological leadership that subsequently became symbols of industrial decline.[8]

The MOS Transistor and Technological Trajectories

Why is it that technological change has been so traumatic for corporations, even ones themselves committed to technological change? Clayton Christensen has provided a useful framework using the concept of a technological trajectory. A technological trajectory is a pattern of development along specific axes that consumers and producers of the technology find relevant. Gordon Moore's 1965 observation that the number of transistors it had been possible to put on an integrated circuit had doubled over each of the previous five years, now enshrined as Moore's Law, represents semiconductor technology's path along one axis. But any technology has many attributes, and a fully defined technological trajectory considers the development along each axis that producers or consumers find relevant. From the invention of the transistor in 1947 to the present, relevant axes for semiconductor technology would have included circuit speed, power consumption, reliability, and cost. Different customers may value different attributes, so a technology developed for one user may follow a different trajectory than that developed for another user. A technological trajectory is not a simple mechanical phenomenon, as the name implies, but a result of the acts of individuals and organizations.[9]

A trajectory sustained in a given direction over time, such as Moore's Law, suggests a balance between the producers and the consumers of the technology, where producers move the technology along particular axes and consumers value those improvements enough to pay for them. A trajectory acts as a focusing mechanism, defining progress in specific terms. Engineers and managers, who know both the possibilities of the technology and the expectations of their customers, understand those axes that customers value most and have ideas on how to make advances along those lines. By highlighting the areas deemed relevant, the trajectory implicitly defines those areas that are irrelevant and can be safely ignored.

Christensen defines technologies based on their relation to existing technological trajectories. Sustaining technologies have benefits along a well-established (and well-valued) axis. They make existing products better. On the other hand, a disruptive technology leads to poorer performance based on traditional trajectories but has advantages along an axis deemed irrelevant by existing customers. A sustaining technology is typically commercialized quickly by established companies because they understand the technology's historic trajectory and the demands of consumers, but a disruptive technology may languish for lack of interest. Some entity, often a new firm, must find new products and new markets that exploit the disruptive technology's distinctive attributes.[10]

The MOS transistor was a disruptive technology compared to the bipolar transistor. When it was first conceived in the late 1950s, the MOS transistor did not constitute an advance based on the existing trajectory of transistor technology. Its speed was much slower than existing bipolar junction transistors', and its reliability was so poor, it essentially ceded all the progress Bell had made since the transistor's invention. Not surprisingly, it was an outsider to semiconductor technology who first proposed it, and not surprisingly Bell Labs management rejected it.

Almost simultaneously with the invention of the MOS transistor came an invention that had the potential to redefine what constituted a good transistor. Integrated circuits put several transistors and other electrical components on a single piece of silicon. Some who looked at the possibilities of integrated circuits saw that if the goal was to put many transistors on a single chip, a transistor that was easy to build might have significant advantages. The MOS transistor's simplicity, an attribute that had little saliency prior to integrated circuits, spurred its advocates on in the face of daunting problems.

Initially, the MOS transistor could not be used in most applications that used bipolar transistors; it was too slow. The 1960s were a period of uncertainty about whether MOS would ever prove to be a viable technology. Would it ever achieve adequate reliability? Would its advantages in ease of fabrication ever translate into products that were unattainable with bipolar technology? No one could know. After an extended search, proponents of MOS technology finally were able to prove its advantages in markets where speed was not critical but density and cost were. These included pocket calculators, computer memories, and later the microprocessor.

Those most closely tied to existing technologies often find disruptive technologies unattractive because of their inferiority by traditional standards of measure. The MOS transistor was most fervently supported by individuals and organizations that had little stake in the existing bipolar technology. These outsiders included a mechanical engineer with previous background in electromechanical relays who had just come to Bell Labs' semiconductor group, a newly hired inventor at Fairchild R&D, a group of engineers at IBM Research who had previously worked on unsuccessful projects, and a young engineer working at Douglas Aircraft. The involved organizations included a research group at IBM that had failed to make substantial contributions to the company; a development laboratory in Boeblingen, West Germany, whose possibilities were circumscribed by its position outside the United States; and several small start-ups on the San Francisco Peninsula.

Bipolar and MOS technologies had an ambivalent relationship to one another. Although MOS technology took root most securely where there was no existing bipolar work, MOS depended on the infrastructure that had been built up to support the bipolar transistor. It was made from silicon and used the planar process, the common material and fabrication process throughout the industry in the 1960s. As the established bipolar technology advanced and new equipment improved bipolar transistors, in many cases MOS transistors improved as well. Instead of trying to catch up with the accelerating bipolar technology from a standing start, the MOS transistor rode piggyback.

At the same time, even though one could describe the steps needed to make an MOS transistor as a subset of the steps used to produce a bipolar transistor, engineers in the 1960s found that attempts to produce MOS transistors this way invariably led to disaster. Many of the fabrication processes required subtle but significant modifications to yield good MOS transistors. MOS engineers developed their own special techniques for applying metal, growing oxides, rinsing a wafer, or cleaning a diffusion tube. In the late 1960s the practitioners of MOS technology had a rule that diffusion furnaces used to make bipolar transistors could not be used to make MOS transistors; contaminants from the bipolar transistors would compromise the MOS transistors. Gordon Moore said that "MOS was a religion as well as a technology" and that to be a participant "one was required to practice rites for reasons bipolar people could not readily accept."[11]

Intel perhaps best represents the ambivalent relation between MOS technology and bipolar technology. It was founded in 1968 by Robert Noyce and Gordon Moore, two of the leading figures in the industry, whose success, both corporately and personally, was closely associated with the bipolar transistor. Intel's formation came after Fairchild had been unable to create a successful MOS business, a failure for which Moore, as the head of R&D, bore some responsibility. Two of Intel's original employees, Andrew Grove and Les Vadasz, had been among the industry's most prominent MOS researchers, laboring on it for half a decade. Intel began work on both technologies, but by abandoning all existing products, it put bipolar and MOS technology on an equal footing. Although Intel was one of the few places where the strictures against making MOS and bipolar on the same equipment were successfully violated, within a short time Intel had established separate MOS and bipolar manufacturing lines. Shortly after that, Intel concentrated its efforts on MOS, with bipolar technology relegated to a supporting role. Ultimately, Intel resolved the ambivalence.

East Coast/West Coast:
MOS Technology and Vertical Integration

This book covers the period between the end of World War II and 1975. Nineteen forty-five marked the origins of the research that led to the transistor at Bell Labs. By 1975, MOS technology was well established, with successful products (semiconductor memories and microprocessors, to name only two) and a clear trajectory (ever smaller dimensions and higher densities). By the late 1970s, the dollar value of MOS digital integrated circuits produced in the United States surpassed that of bipolar digital integrated circuits.[12]

This is a book, not the encyclopedia that would be required to offer an exhaustive history of the MOS transistor. Scores of companies had MOS programs in the late 1960s and early 1970s, and a number of them made significant contributions to the technology. (Most no longer exist.) The first chapter deals with the origins of MOS transistor work at Bell Labs, RCA, and Fairchild. While the MOS transistor originated at Bell Labs, RCA and Fairchild were the first organizations to see its possibilities and bring it to the attention of the wider community.

The primary focus is on three firms: Fairchild Semiconductor, Intel, and IBM. This selection can be justified by their sheer importance: Fairchild and IBM were at the forefront in MOS research, while Intel and IBM were two of the leading MOS producers by 1975. But these firms are interesting for other reasons. The leaders of Intel came from Fairchild; a study of both allows one to investigate how the managers of a firm that had failed to create a successful MOS product line at one company transcended those problems at another.

But most intriguingly, IBM and Intel represent two polar opposite approaches to semiconductor technology, which might be symbolized by their locations: East and West. Many have noted that semiconductor technology was born in the East and moved to the West, and the MOS transistor repeats this pattern. Why did this happen? Why were the possibilities of semiconductor electronics so much more successfully exploited by firms on the San Francisco Peninsula, even when much of the research was done in the laboratories of large East Coast firms? Each East Coast company's difficulties requires a separate explanation. AT&T was hindered by its position as a government-regulated monopoly that did not sell semiconductors on the open market. RCA's managers were preoccupied with other lines of work, diversifying their electronics company into rent-a-cars, carpets, and frozen foods just at the time semiconductor electronics and the MOS transistor were coming of age, in the apparent belief that synergies would come from being a maker of both TVs and TV dinners.

At first glance, IBM represents the great exception to the Western orientation of semiconductor technology. Throughout most of the 1960s, 1970s, and 1980s its facilities in New York, Vermont, and overseas made it the world's largest producer of semiconductors for digital applications. In the early 1980s, IBM began referring to New York's mid-Hudson Valley, the site of its East Fishkill semiconductor facility, as "Silicon Valley East," calling attention to its often ignored position as the leading producer of semiconductors.[13] The name implied an essential similarity between what went on in East Fishkill and what went on in the semiconductor firms of Mountain View, Santa Clara, and Sunnyvale, California. At one level there was a similarity. Both made silicon integrated circuits using the same basic principles of semiconductor physics. And if the criteria for the title were the amount of silicon processed or the dollar value of transistors made, IBM's one enormous plant might have given the Hudson Valley a valid claim to the name. But while journalist Don Hoefler coined the term *Silicon Valley* in 1971 in reference to the proliferation of semiconductor start-up companies on the San Francisco Peninsula, ironically, silicon is not the defining characteristic of Silicon Valley. As the wide range of start-up companies Silicon Valley has supported over the last thirty years has made clear, Silicon Valley is a technology region whose ecology is marked by a technological dynamism, a sensitivity to the possibilities of making money combined with a widespread availability of resources to support entrepreneurship. On that score the Hudson Valley was no Silicon Valley.[14]

IBM and Intel, fundamentally different organizations, developed fundamentally different MOS technologies. IBM represents the classic case of an oligopolistic vertically integrated corporation, with many functions internalized so that they could be removed from the vagaries of the market and actively controlled by managers. Because IBM's semiconductors were not sold directly but went into IBM's computers, they were immune to market forces. IBM's managers had essentially unlimited resources to work with and the freedom to ignore the semiconductor industry to develop the MOS technology they thought best for IBM. By the early 1970s IBM's MOS program had resulted in essentially one product, random access memory, a very important technology for mainframe computers.

On the other hand, Intel, a small start-up during this period, was an example of a vertically disintegrated corporation, able to take advantage of the capabilities of other companies to minimize its capital expenditures. Intel was much more exposed to market forces and lived in constant peril of being destroyed by competitors or changes in conditions. In spite of this frailty, Intel's structure gave it significant advantages in conducting economic experiments with this new MOS technology. Intel developed a wide range of MOS products, including random access memories, shift registers, watch circuits, and microprocessors.

While Intel was highly successful in exploiting the possibilities of MOS technology, the networked nature of Silicon Valley and the semiconductor industry as a whole meant that the development of MOS technology was not the sole achievement of any single company. The complexity of semiconductor technology made for a situation where no one company could be totally self-sufficient. Each one made use of advances that occurred elsewhere. Some firms that existed for only a few years before failing played an important role in the establishment of MOS technology. Practitioners established formal and informal networks of communication between firms, which were crucial for advancing the new technology.

The story of the MOS transistor includes some of the best-known people in the industry, such as Gordon Moore and Andrew Grove. At the same time, it includes many characters who are not well known, even to those in the field. The industry and the profession have standards for recognizing work that often present a distorted view of the past. Those who were recognized by the profession were often those who published papers and worked for large, successful companies that could mount a campaign on their behalf. Wise managers knew that these were not the only ones who contributed. Many engineers and technicians who never wrote papers were nonetheless members of the MOS community and played crucial roles in perfecting and spreading techniques that were important for building MOS transistors.

While the development of the MOS transistor stands as one of the most important chapters in the history of semiconductor technology, it is also one of the least known. Corporations typically tell the history of technology from their own perspective and have little incentive to emphasize the contributions of others, even in an industry, such as the semiconductor industry, where such contributions were crucial. IBM, making semiconductors for internal use, had little need to publicize its work, and its actions were hidden even from insiders in the semiconductor industry. The disruptive nature of MOS technology has also wreaked havoc with historical understanding. While Bell Labs made every effort to publicize the original transistor work of Bardeen, Brattain, and Shockley, when the MOS transistor was first conceived, most saw it as pathological, and there was no glory to be gained by being the inventor of a pathological technology. By the time it had achieved significant commercial success, a decade had passed, new companies were in positions of technological leadership, many of the early figures had left the field, and the early history had been forgotten. And so the path the MOS transistor took from being a bad idea to being the technological foundation of American society has been obscured. This book is an attempt to illuminate that path.

How a Bad Idea Became Good (to Some)

THE EMERGENCE OF THE MOS TRANSISTOR, 1945–1963

The story of the origins of the MOS transistor is the story of how a bad idea became good (to some). Even though the MOS transistor showed striking resemblances to structures Bell Labs scientists had proposed in their investigations that led to the invention of the transistor in 1947, the formal announcement of the MOS transistor's existence as a potentially useful technology did not occur until 1963. In the intervening years semiconductor technology had matured, but it had done so technically by repudiating the principles behind Bell's early semiconductor work, which had tried to construct a semiconductor amplifying device that operated on the surface of the semiconductor.

Instead, Bell and the industry it created followed another path, which was blazed by William Shockley. Shockley invented the bipolar junction transistor, whose genius was that it bypassed the seemingly intractable problems of the semiconductor surface—problems that had plagued Bell's initial work. By the late 1950s, his transistor and variations on it dominated semiconductor technology. Bell Labs' concentrated efforts led to advances that made Shockley's transistor faster, more reliable, and able to handle higher power levels—some of the key measures by which engineers judged a transistor.

When researchers first described what is now known as the MOS transistor at a professional conference, it provoked very little interest from other semiconductor researchers. This response was hardly surprising, for in significant ways the MOS transistor was a return to the bad old days when the semiconductor surface controlled the performance of a transistor. Furthermore, by conventional measures, the MOS transistor seemed to lack any offsetting advantages. The MOS transistor was substantially slower than the bipolar transistor and much less reliable, making it unusable in the most critical bipolar transistor applications.

Two new developments made something like an MOS transistor seem promising to some in the early 1960s, when it had seemed pathological before. First, the use of silicon dioxide to passivate silicon surfaces gave hope that the problems of the semiconductor surface could be mitigated. Second, the advent of integrated circuits brought a new measure by which to judge transistors. Speed was still important, and the MOS transistor was slow, but the difficulties of trying to build multiple components on a single chip made the simplicity of the MOS transistor attractive to some. They believed that if the problems of the semiconductor surface could be solved, it would be possible to put many more MOS transistors than bipolar transistors on a chip.

This vision was not compelling to everyone. The MOS transistor was the work of outsiders. The researchers who conceived of it and worked on it at Bell Labs, RCA, and Fairchild were by and large newcomers to semiconductor technology with little stake in Shockley's bipolar junction transistor. Although the MOS transistor was initially conceived at Bell Labs, the organizational commitment to the older bipolar transistor was too great for the new technology to take root, and management quickly halted work on the MOS transistor, dismissing it as unpromising. However, the expansive nature of the R&D environment at RCA and Fairchild provided a space where the new technology could flourish. By 1963, even though the MOS transistor was still not ready to be sold as a product and many doubted that it would ever be successful, RCA, Fairchild, and other companies had seen its promise clearly enough to devote substantial R&D resources to it.

Bell Labs and the Bipolar Junction Transistor: Commitment to a Technology

Bell Labs invented the transistor in 1947, and by 1955 its development effort had progressed to the point where it had made a major commitment to one particular type of transistor technology, the silicon diffused bipolar junction transistor. This commitment made sense at a number of different levels. Economically it was a good fit with the needs of the Bell System, Bell Labs' prime customer, as it met the requirements of many different systems and offered low costs. Technically it had a number of favorable attributes. It avoided the worst problems of the semiconductor surface that had plagued the point-contact transistor. It was susceptible to detailed mathematical analysis, which was appealing to the many mathematically oriented physicists and electrical engineers at Bell Labs and was essential to Bell Labs' transistor development methodology. It represented a mature technology that could be expected to remain viable for many years into the future, and researchers had many ideas on how to extend the performance of the technology in the more distant future. This commitment and the reasons for it are the essential context for understanding the initial work on what is now called the MOS transistor and the response to that work at Bell Labs.

In the late 1930s Bell Labs officials, energized in part by the development of the quantum mechanical theory of solids, considered the possibility of a solid-state amplifying device, whose reliability would be much superior to that of the electromechanical relays then being used to switch telephone calls. Although World War II interrupted any direct work on such a device, it led to a large body of work in solid-state physics at various institutions aimed at producing improved radar detectors. Near the end of the war Bell Labs assembled an interdisciplinary team at its Murray Hill, New Jersey, site to search for a semiconductor amplifier, which it discovered in December 1947.[1]

Several aspects of this original transistor effort have a bearing on the history of what would become the MOS transistor. William Shockley, one of the leaders of the Bell Labs effort, was originally thinking in terms of a surface field-effect device, where one would apply an electric field to a semiconductor as a way to vary its surface conductivity. However, a structure designed to test the soundness of the concept failed to show the anticipated effects (figure 1.1). Shockley

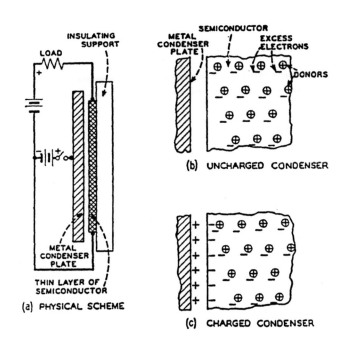

FIGURE 1.1 · WILLIAM SHOCKLEY'S INITIAL TEST STRUCTURE. Shockley's idea was that by applying a positive voltage to the metal condenser (capacitor in today's terms) plate, an increase in the number of electrons on the surface of the semiconductor would result (c). One thousand volts was applied to the metal plate, which was less than a millimeter from the semiconductor surface, but no current was observed in the semiconductor. The experiment is described in Michael Riordan and Lillian Hoddeson, *Crystal Fire: The Birth of the Information Age* (New York: W.W. Norton, 1997), 112–13. The drawing itself is from William Shockley, *Electrons and Holes in Semiconductors* (Princeton, N.J.: D. Van Nostrand, 1950), 30. © 1950 D. Van Nostrand, © 1977 William Shockley.

calculated that the inability to measure any change in resistance due to varying the voltage meant that the effect must be fifteen hundred times smaller than expected by his theory. Shockley showed these results to John Bardeen, a theoretical physicist Bell Labs had hired to strengthen its solid-state effort, and in March 1946 Bardeen proposed that the failure of the experiment was due to surface states—traps on the semiconductor surface that held electrons immobile and shielded the rest of the semiconductor from the effects of the applied electric field.[2]

In response to Bardeen's explanation, a number of scientists in the solid-state group at Bell Labs—with the notable exception of Shockley—focused their

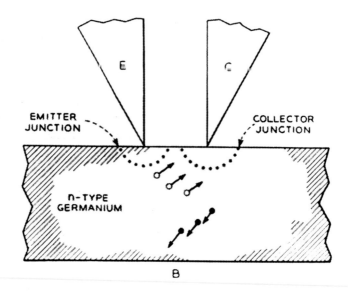

FIGURE 1.2 · SCHEMATIC DIAGRAM OF THE POINT-CONTACT TRANSISTOR. From J. A. Morton, "Present Status of Transistor Development," *Proceedings of the IRE* 40 (November 1952): 1315. © 1952 IRE.

attention on understanding surface states. They still aimed to produce a surface field-effect device where a metal contact was separated from the surface by a thin oxide, but these efforts were thwarted when Walter Brattain, the experimental physicist working with Shockley and Bardeen, inadvertently removed the water-soluble germanium oxide, leading in short order to the discovery of the point-contact transistor by Brattain and Bardeen, which Bell Labs announced to the public in June 1948.[3]

It would be difficult to consider the point-contact transistor a triumph of the quantum theory of solid-state physics. The device consisted of two metal point contacts pressed against a block of semiconductor in close proximity to one another (figure 1.2). As such it looked much like the descendant of the cat's whisker crystal rectifier that it was. Furthermore, the device's operation was critically dependent on two poorly understood processing steps affecting the semiconductor surface. One was a surface chemical treatment, while the other, called *forming,* was an empirical process in which large currents were passed through the metal point contact in the reverse direction until the device displayed the desired characteristics. Bardeen and Brattain were unable to offer any quantitative explanation for how the device worked. Researchers who attempted to use the device in circuits faced problems, as the device was prone to wide swings in its operating characteristics or sudden failure.[4]

In 1948 Shockley, driven in his work partly by frustration over his lack of contribution to the point-contact transistor, proposed the junction transistor, now known as the bipolar junction transistor. It does not appear to have been a conscious attempt to circumvent the shortcomings of the point-contact transistor, but that was its long-range effect. In a 1949 paper, "The Theory of p-n Junctions in Semiconductors and p-n Junction Transistors," Shockley presented a detailed mathematical theory for the *p-n* junction. *P-n* junctions, junctions of *p-* and *n*-type semiconducting material, act as rectifiers or diodes, conducting current in one direction only. The critical point of action is the *p-n* junction, which is predominantly inside the semiconductor rather than on the surface. Shockley's junction transistor consisted of two *p-n* junctions placed back to back, which he was able to analyze in detail (figure 1.3). In his article introducing it, Shockley commented that the p-n junction transistor "has the interesting feature of being calculable to a high degree."[5]

While the point-contact transistor represented an experimental discovery, utterly dependent on the current state of materials and experimental techniques, Shockley's device was a theoretical entity, conceived out of a detailed

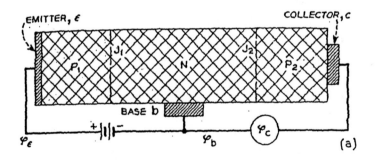

FIGURE 1.3 • SCHEMATIC DIAGRAM OF WILLIAM SHOCKLEY'S JUNCTION TRANSISTOR. The critical points of action are J1 and J2, which are largely inside the semiconductor, rather than on the surface. From William Shockley, "The Theory of p-n Junctions in Semiconductors and p-n Junction Transistors," *Bell System Technical Journal* 28 (July 1949): 470. © 1949 Lucent Technologies Inc. All rights reserved. Reprinted with permission.

knowledge of how electrons move in a semiconductor crystal and little concern about what would be required to translate that knowledge into a physical device. Shockley's junction transistor existed in theory only until 1950, when Gordon Teal and Morgan Sparks, chemists at Bell Labs, produced the first device. Its realization depended on the development of techniques for growing germanium in single crystal form, work in which Shockley played no direct part.[6]

Bell Labs was a massive effort to control the chaotic and uncertain process of inventing and developing new technology. One way it tried to do this was through a very clear and firm organizational distinction between research and development. Research, where the transistor's inventors worked, consisted mainly of scientists who used their superior knowledge of physical laws to produce novelty—"new materials, new techniques, and new knowledge," in the words of one Bell Labs manager.[7] Once a new device, such as the transistor, had been created, research quickly transferred the work of making it into a viable product over to a development group. Development was essentially an engineering function—concerned with economic and technological considerations. Managers at Bell Labs believed that if researchers worked on these types of problems they would lose their scientific edge.[8]

Shortly after the public announcement of the transistor in June 1948, Mervin J. Kelly, executive vice president of Bell Labs, established a development group headed by Jack Morton to convert the work of Brattain and Bardeen into a device that could be used in Bell System applications. Morton had received his master's degree in electrical engineering from the University of Michigan in 1936 and joined Bell Labs that same year. In the 1940s Morton led the development of a microwave triode, a device that represented a major engineering accomplishment and a significant advance over previous triodes. Morton, a strong-willed, driving leader, was a dominant figure in the developing semiconductor industry in the 1950s and in Bell Labs' transistor development efforts until his tragic death in 1971.[9]

One of Morton's first actions in this role was to write a memo describing the research and development process in structured terms in which he called Bell Labs a "teleological organism." Morton asserted that the two main functions in solving engineering problems were analysis and synthesis. For Morton, analysis meant the formulation of design theories that would "relate the performance properties of the device to its structure and material properties."[10] Synthesis was using the results of these design theories to determine the fabrication methods necessary to achieve the required performance. While Shockley's initial paper had provided some elements of a design theory for the junction transistor, Bell Labs engineers soon filled in other parts of it. On the other hand, Bell Labs researchers were unable to develop a quantitative design theory for the point-contact transistor, which relied on poorly understood surface effects.[11]

The diffusion process, one of the key breakthroughs in junction transistor technology, followed Morton's model of analysis leading to synthesis and work moving from research to development. The junction transistor design theory had shown that as the base of the transistor was made narrower, the transistor would operate at higher frequencies. In 1953 Shockley and other members of his research organization proposed a new method of fabricating transistors that would produce narrower base widths. In the diffusion process, the impurities that controlled the semiconductor's electrical characteristics were introduced onto the surface of the semiconductor. As the semiconductor was heated they then diffused into it in a manner that could be tightly controlled as a function of the time the sample was exposed and the temperature. Shockley's group produced the first diffused transistor in 1954. This process allowed a physical control of the transistor's characteristics that approached the mathematical control that Shockley's design theory had demonstrated.[12]

While Bell Labs' research and development organization made impressive progress on the technical problems surrounding the transistor, a number of economic problems related to the unique position of the Bell System put strong constraints on Bell's efforts to use transistors in its equipment. Although the transistor was a radical device, the Bell System was a profoundly conservative organization, as a regulated monopoly with a $20 billion investment in an existing plant providing adequate service. For other firms, transistors were items to be sold, either by themselves or in products, but for Bell, transistors were capital goods that would go into complex systems having a forty-year life span. A continuous revolution in semiconductor technology, with radical innovations rendering previous work obsolete, was not in the interest of the Bell System.[13]

For Bell, transistors needed to offer concrete economic advantages before their usage on a large scale could be justified. As part of this, their reliability had to be proven so that Bell would know the expected costs of transistor failures over the life of the equipment. Bell's emphasis on reliability pushed out the usage of transistors in two ways, first by requiring work to address the reliability problems and then by requiring long periods of testing—on the order of years— to prove the device's reliability. Furthermore, Bell had to devise a way to manufacture transistors economically. Between 1948 and 1958 it had forty-five systems under development that used transistors. Prior to 1955 if each system had used the transistor that was optimal for that system, a wide range of transistor types and manufacturing technologies would have been employed. This situation threatened to keep all transistor costs high and jeopardize the future of the systems under development.[14]

The complexity of the systems using transistors was another factor delaying the introduction of the new technology. The research that had led to the transistor had been inspired by a vision of using semiconductors as the foundation of a system to switch telephone calls. The Electronic Central Office, later called ESS No. 1 (Electronic Switching System), was one of Bell's largest and highest-profile projects using transistors. In 1951 Bell Labs began systems studies for the Electronic Central office, and by 1960 Bell had a trial system in operation. A system such as ESS No. 1, which finally entered commercial service in 1965, took many years to develop, under the expectation that it would be long-lived. As late as 1964 Bell System managers expected that the process of phasing out their mechanical switching gear and implementing the transistorized system would not be complete until 2000![15]

These factors should be seen behind the decision made by Jack Morton in March 1955 to reorient his development efforts to concentrate on silicon diffused junction transistors. His decision to focus on silicon was significant because germanium was better understood and the dominant material used for transistors at the time. His decision also reflected a subtle analysis of a wide range of factors, for if the decision had been made purely on the basis of which material produced faster transistors, then germanium would have been the clear choice.[16]

In light of the fact that silicon has been the dominant material used for producing semiconductors up to the present time, Morton's decision can be analyzed in two ways. One could stress how insightful Morton was, with the success of silicon technology proving that he made the correct decision. Without doubting that silicon technology has properties that have contributed to the growth of the semiconductor industry, the decision marked the shifting of

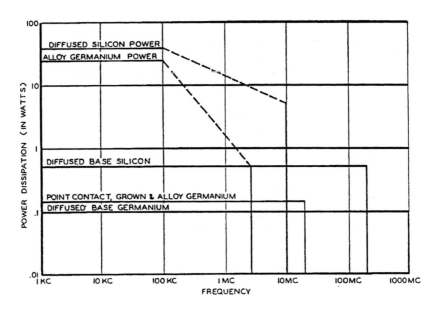

FIGURE 1.4 · THE POWER AND FREQUENCY CAPABILITY OF VARIOUS TRANSISTOR TECH-
NOLOGIES. Morton used this figure to show that the silicon diffused technology was capable of
meeting the requirements of the vast majority of transistor applications. The only area not
bounded by diffused silicon technology was the lower right corner (high speed, low power). This
chart was presented by Morton at the internal Bell Hershey conference, then in an article in the
Proceedings of the IRE in 1958, and by A. E. Anderson in the *Western Electric Engineer* in 1959. From J.
A. Morton and W. J. Pietenpol, "The Technological Impact of Transistors," *Proceedings of the IRE* 46
(June 1958): 957. © 1958 IRE.

resources from germanium to silicon, a self-reinforcing process that helped
ensure silicon's continued dominance.

Morton moved much of the effort to develop more mature technologies out
of the Murray Hill laboratories, and he brought three men from the research
organization into key management positions in his development group to lead
the shift to diffused devices. As Morton reported in 1958 to a conference of the
Bell System presidents, his organization's two main objectives were "to provide
all diffused silicon devices for all but the very highest frequency applications"
and "to make possible large economies of manufacture through the use of a
common material, structure and diffusion techniques."[17] Morton's expectation
was that silicon diffused devices would meet the needs of almost all applications, as shown in figure 1.4, a figure that Morton and his managers presented

both inside Bell and to the larger technical community. Given the long lead time to develop systems and their long life expectancy, it was to Bell's advantage to push the technology to a point of stability, where it could meet the needs of its systems with a mature, long-lived technology. Bell's early commitment to diffusion and silicon can be seen as an effort to speed up the drive to that plateau, while paying scant attention to points along the way.[18]

Bell's unique position provides insight into one of the first cases where a major transistor innovation came from outside Bell Labs. In 1954 Texas Instruments produced the first silicon bipolar transistors using the grown junction technique, a primitive method yielding slow-speed transistors. Bell Labs focused its efforts on the more difficult silicon diffused transistor, and Morton "had refused to commit development work to silicon grown junction transistors."[19] Bell Labs, developing transistors that would be used in capital goods, had to look many years ahead, while Texas Instruments looked for something that would make money right away.

The Problem of Semiconductor Surfaces and the Silicon–Silicon Dioxide Surface Device

As Bell Labs worked to ensure that the transistor would provide many years of reliable service, one of the major problem areas that plagued the junction transistor was the semiconductor surface, for while the move to *p-n* junctions had bypassed the worst problems of the semiconductor surface by moving to the interior of the semiconductor, *p-n* junction devices still had surfaces. Bell Labs' 1958 volume *Transistor Technology* included in its "Final Surface Treatment" chapter a heading entitled "Folklore and Witchcraft," which stated, "the surface protection techniques now used represent an interesting blend of folklore origin, theoretical direction, and experimental justification on a satisfactorily large and representative scale."[20] As late as 1962, in a fiftieth-anniversary issue of the *Proceedings of the Institute of Radio Engineers,* semiconductor engineers from both Bell Labs and Texas Instruments reviewing the state of the art noted that the semiconductor surface remained a major problem area.[21]

Scientists and engineers at Bell Labs directed a great deal of effort toward understanding the problems of the semiconductor surface and lessening their effect on junction transistors. One of the most significant efforts in the development group was led by M. M. (John) Atalla, who joined Bell Labs' transistor effort in 1956 as a supervisor. Atalla, an Egyptian who had earned his Ph.D. in mechanical engineering from Purdue, joined Bell Labs' New York City operations in 1950, where he worked on problems related to the reliability of electromechanical relays. His move to Murray Hill seems to have been a recognition of both his ability and the impending obsolescence of the relay with the invention of the transistor. Those working under Atalla included some of his group that had been working on relays as well as recently hired scientists and engineers.[22]

Atalla's initial work focused on studying the surface of semiconductors. In 1955 Carl Frosch and L. Derick, while working on diffused transistors, accidentally grew a layer of silicon dioxide over the silicon wafer. Further work showed that the silicon dioxide layer could serve as a mask preventing the dopants from diffusing into the silicon wafer. By opening and closing holes in the oxide they were able to control the placement of dopants with great precision. Atalla picked up this work on oxidation, attempting to passivate the surface of silicon through the formation of a layer of silicon dioxide over it. In the course of its work Atalla's group took structures much like those Shockley had proposed in his 1945 attempt to detect a field effect and used them as a way of measuring the effectiveness of silicon dioxide in removing surface states. Atalla demonstrated that silicon dioxide was extremely effective in eliminating one important class of surface states.[23]

Both Bell Labs and the industry at large saw the potential significance of Atalla's work. Speaking at an internal Bell conference in 1958, Morton asserted that Atalla's method of surface oxidation seemed to provide a surface that was "insensitive to its environment."[24] Morton claimed that if this were so, it could replace more costly means of protecting the surface of transistors, such as vacuum sealing, and thereby lower the cost of transistors by 25 percent. Atalla presented his work at the Institute of Radio Engineers' Semiconductor Device Research Conference in 1958, where researchers from RCA judged it to be one of the three most important non-RCA papers presented. They asserted that his results seemed "like a milestone in the surface field."[25]

From this work a new device emerged, and although it lacked a name, physically it had the structure of the modern MOS transistor. Although the precise path Atalla and his group followed to its conception is unclear, it seems to have

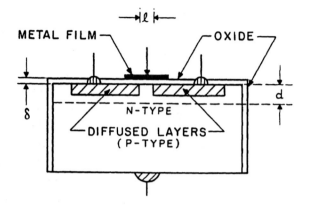

FIGURE 1.5 • ATALLA AND KAHNG'S SILICON–SILICON DIOXIDE SURFACE DEVICE—NOW KNOWN AS THE MOS TRANSISTOR. The oxide forms the dielectric of a capacitor between the metal film (now known as the gate) and the semiconductor (labeled "N-type" in the drawing). This is the reason for its eventual name, metal-oxide-semiconductor transistor. As a negative charge (negative voltage) is applied to the metal film, it produces a positive charge at the surface of the *n*-type semiconductor, which forms a conducting path (channel) between the two *p*-type diffused layers. The device shown here would now be called *p*-channel because the applied voltage creates a *p* channel at the surface of the *n*-type semiconductor material. To make an *n*-channel device, a *p*-type semiconductor is used and *n*-type diffused layers are created. To make an *n*-channel device, a *p*-type to be applied to the metal film. From D. Kahng, "Silicon-Silicon Dioxide Surface Device," Technical Memorandum MM-61-2821-1, 16 January 1961, AT&T Archives, Warren, New Jersey. Property of AT&T Archives. Reprinted with permission of AT&T.

been an extension of their work oxidizing silicon surfaces. Atalla appears to have conceived it, but it was an invention in a different sense than the transistors of Bardeen and Brattain and Shockley. The invention of both the point-contact transistor and the junction transistor involved novel effects. The principles that Atalla's device used were well known; veterans in the field would have recognized them as ones that had been tried without success by Bardeen, Brattain, and Shockley. Atalla recycled these principles using the advanced fabrication techniques that Bell Labs had developed to make diffused bipolar junction transistors. In some sense Atalla's biggest breakthrough was an intellectual one, thinking that such a device was worth making at all.[26]

In November 1959 Atalla put Dawon Kahng, a newly hired Ph.D. electrical engineer from Ohio State, to work on building such a structure. Atalla and Kahng presented their device at the 1960 Solid-State Device Research Conference, and in January 1961 Kahng published an internal technical memorandum titled "Silicon–Silicon Dioxide Surface Device" describing it. The device they described was much like the structure used in measuring surface states, consisting of metal evaporated on top of a thin oxide layer grown on the surface of silicon. As for the silicon itself, two p-type regions were separated by an n-type region beneath the metal electrode (figure 1.5). The device operated by the formation of a p-type inversion layer on the surface of the n-type semiconductor when a negative voltage was applied to the metal film (now called the gate). Kahng's technical memorandum was circumspect in the claims that it made for the new device. He developed a theory for the device's operation but then noted the wide disparity between the theory and the device's actual behavior, which he attributed to surface states. While he acknowledged that the presence of surface states would limit the device's applicability, he did claim several advantages for his device, most notably "ease of fabrication and the possibility of application in integrated circuits."[27]

Two Inventions: The Responses within Bell Labs and the Larger Community

Atalla and Kahng's presentation of their work at the 1960 Solid-State Device Research Conference at Carnegie Institute of Technology in Pittsburgh reveals the larger community of semiconductor researchers and their response to this new device. At this conference Bell Labs presented another significant paper, concerning the epitaxial transistor. And while Atalla and Kahng's work was to have far greater long-term importance, it was the epitaxial transistor that captured the attention of both conference participants and Bell Labs management.

The semiconductor industry as it developed after Bell Labs' announcement of the transistor was characterized by the free flow of information between companies. This pattern was established by Bell Labs under the strong encouragement of the military. Bell Labs licensed other companies to make transistors, sponsored several symposia detailing its transistor knowledge, and published a wide assortment of literature teaching others how to make and use transistors. As the field grew, technical conferences sponsored by professional societies were one of the most important vehicles for the transfer of information. By the late 1950s the Institute of Radio Engineers (IRE) sponsored three annual conferences that attracted many semiconductor researchers: the Solid-State Device

Research Conference, the Electron Device Conference, and the Solid-State Circuits Conference.[28]

The Solid-State Device Research Conference (SSDRC) evolved out of the IRE Electron Tube Conference, which was begun in 1938 for the informal presentation of ideas among a small group of workers in the field. In 1952 semiconductor researchers split to form their own conference, which, after going through an assortment of names in the preceding years, finally became known as the SSDRC in 1959. The purpose of the conference as stated in 1953 was "to facilitate open discussion of recent information and controversial ideas on transistors and related devices in an informal atmosphere."[29] To this end no written abstracts or papers were distributed to attendees, and for many years the press was excluded from the conference. These restrictions were to allow researchers a freedom to communicate across corporate boundaries that they might not have if what they said was to appear prominently on the pages of publications such as *Electronics* or *Electronic News*.[30]

Professional conferences clearly demonstrated that even though semiconductor researchers might work for corporations with vastly different goals, they were part of a single community. In the 1950s and early 1960s the SSDRC, the key annual gathering in the field, was typically held in June or July on a college campus. The SSDRC was a closed conference, with attendance limited to a select group of active researchers nominated by previous attendees. While attendance at the other leading semiconductor conferences might number in the low thousands, attendance at the SSDRC was held roughly constant throughout the 1950s at three hundred. In these conferences researchers from different firms worked on similar problems and techniques. While the fact that a firm worked on a device was no guarantee that it would ever enter production, work that had technically appealing attributes could quickly attract researchers from a variety of companies.[31]

The epitaxial transistor had that kind of attractiveness to the semiconductor community. The origins of the epitaxial transistor date back to 1954, when James Early, one of the leading theoreticians of the bipolar junction transistor at Bell Labs, proposed a new transistor structure in which an additional layer of semiconductor material was added between the base and the collector. In the traditional bipolar junction structure, several important electrical characteristics put conflicting demands on the doping of the semiconductor material used to build the transistor. Because these characteristics could not all be optimized simultaneously, trade-offs had to be made in transistor fabrication. The extra layer gave an additional degree of design freedom so that several characteristics could be optimized simultaneously in a way that had not been possible previously. Although Early could show theoretically that this structure had greatly improved performance, the fabrication techniques available to him then limited his ability to demonstrate this in practice.[32]

Throughout the 1950s Bell Labs and a number of other research labs performed work in epitaxy, or vapor growth. In this process a thin film of single-crystal semiconductor is grown on an existing semiconductor. The advantages this process offered were very close control of the thickness of the semiconductor layer and close control of the doping of the grown layer. However, this work was not directly applied to the production of transistors.[33]

In 1960, Ian Ross, the director of Bell Labs' Semiconductor Device Development Laboratory in Murray Hill, conceived of using epitaxy as a way to permit an additional degree of freedom in fabrication and make practical a junction transistor with the structure previously proposed by Early. He enlisted the support of Henry Theuerer, a veteran chemist in research, to provide epitaxially grown material on a rod of silicon. Workers in the development area then used this material to build actual junction transistors on an existence proof basis. The epitaxial bipolar junction transistors built with production techniques had speeds dramatically faster than those of similar nonepitaxial devices.[34]

Bell Labs immediately knew what to do with this work. In May 1960 Bell Labs engineers were studying the possible use of epitaxial transistors in ESS No. 1, the long-envisioned electronic switching system, an application where the transistor's speed was of critical importance. They requested that the Western Electric facility in Allentown, Pennsylvania, duplicate their epitaxial equipment with a goal that Western Electric be making "significant numbers of miniature transistors using epitaxial techniques" before the end of the summer.[35] And in November of 1960 Bell Labs and Western Electric began a series of joint meetings aimed at speeding the movement of epitaxial transistors into production.[36]

The rest of the industry quickly picked up Bell's work. Bell Labs' disclosure of the epitaxial transistor at the 1960 SSDRC was the high point of the conference, receiving prominent mention in the trade press. The Electron Devices Meeting in October 1960 featured a session devoted to epitaxial devices and techniques.

By March 1961 Fairchild, Sylvania, Texas Instruments, and Motorola had introduced versions of the epitaxial transistor for sale. *Electronic News* titled its article previewing the 1961 International Solid-State Circuits Conference "Parley to Stress Epitaxials."[37]

The epitaxial transistor was the ideal sustaining innovation, continuing technological advance along an established trajectory. A Bell comparison to existing transistors stated: "The epitaxial samples are superior in most respects" and "In no respect are they inferior."[38] Its benefits were along a historically important axis: speed. Given the commitment Bell Labs had made to the bipolar junction transistor, it was important that the epitaxial transistor was a bipolar junction transistor itself. The use of epitaxy was compatible with the existing diffusion and masking techniques and required only incremental changes in the manufacturing process. The epitaxial transistor was an innovation that was very well matched to the state of Bell Labs' development program.

By contrast, although Atalla and Kahng's work came out of the silicon diffused bipolar transistor, their silicon–silicon dioxide surface device was in many ways incompatible with the bipolar transistor development effort. The new device provided inferior performance, particularly a speed roughly one hundred times slower than that of existing bipolar transistors. Bell Labs' strategy had been to develop the silicon diffused bipolar transistor as a technology that would cover a wide variety of uses, particularly with regard to speed and power. Except for a few special applications, Atalla and Kahng's device would be useable only within a subset of the design space covered by the silicon bipolar device. Its main advantage, ease of fabrication, had little relevance to the industry at the time. To call Atalla and Kahng's device an invention was almost a contradiction in terms, for it was inferior by every relevant standard.[39]

The one area in which Kahng and Atalla recognized their device might be advantageous was of no interest to Bell Labs. Kahng mentioned that the device would be suitable for integrated circuits, a new way of making and using transistors that was getting attention through the work of Jack Kilby at Texas Instruments and Robert Noyce at Fairchild. The idea of putting many electronic components on a single piece of semiconductor did not receive universal acclaim, and its harshest opponents were at Bell Labs. Engineers and scientists at Bell Labs made the argument that as one tried to put more and more transistors on a piece of silicon, the yield would go down to a level that would not be financially satisfactory. If the yield for a single transistor were 90 percent, one would expect the yield for producing 10 transistors on a single piece of silicon to be 0.9^{10} or 0.35. Put another way, if one made 1,000 transistors with a yield of 90 percent, there would be 900 good individual transistors. But if one were making integrated circuits with ten transistors each, there would be 100 of these circuits, but only 35 percent would be good. One would get only 350 transistors, a net transistor yield about 40 percent that of individual transistors. Researchers at Bell Labs believed that the best solution was to invent new electronic devices that would be capable of performing complex functions directly, instead of performing these same functions through a combination of conventional electronic components—transistors, resistors, capacitors, and inductors. As late as 1963, Ian Ross and Eugene Reed, another Bell Labs manager, wrote that integrated circuits did not address the fundamental problems facing semiconductor technology and represented only "a palliative rather than a cure."[40]

Perhaps the most striking feature of Atalla and Kahng's device to an industry veteran would have been its sheer perversity. As Jack Morton and his engineers moved forward to establish the transistor within the Bell System and the electronics industry as a whole, they recognized reliability as fundamental. Although the transistor promised near-infinite reliability, reality was something else. Reliability problems had been a major source of embarrassment to Morton, and the semiconductor surface had been the major source of those problems. Over a decade's work had mitigated those problems, without fundamentally solving them. Atalla and Kahng's device, by putting the greatest stresses on the weakest part of semiconductor technology, gave away those hard-won gains and required returning to the point where transistors had been in 1948. Atalla and Kahng's device clearly would not be ready for development until the successful completion of more studies to understand the semiconductor surface, a process that would take years. After that, one would still expect to have a device that was markedly inferior to the existing bipolar transistor in speed.[41]

Jack Morton was a manager who loved to make decisions. And for someone who knew his customers and the strengths and weaknesses of the technology, the decision about Atalla and Kahng's device was implicit. A. E. Anderson and R. M. Ryder, original members of the Bell Labs transistor development effort, wrote that "Morton, who was ever alert to spot a technology loser as well as a winner, was thoroughly convinced of the likelihood of the inherent

TABLE 1.1 CONTRIBUTORS TO EPITAXIAL TRANSISTOR

	Position at Bell Labs (1960)	IRE status (1961)	Year joined BTL semiconductor effort	Contributor, Transistor Technology Vol. I	Vol. II	Vol. III	Participant, BTL diffusion symposium	Participant, Solid-State Device Research Conference 1952	1953	1954	1955	1956	1957	1958	1959
H. Theurer	MTS/R	NM	1949	P	P		P								
H. Loar	SUP/D	M	1955									P			
J. Kleimack	MTS/D	NM	1949	P	P				P		P				
H. Christensen	MTS/D	M	1949						P	P	P		P		
I. Ross	DIR	SM	1952		P				P						
J. Goldey	SUBD	M	1954		P	P			O,P	P	P	P2		P2	P
J. Early	SUBD	F	1951		P		P	P						CH	P

NOTE: BTL = Bell Telephone Laboratories; CH = conference chair; DIR = director, the level of management above subdirector; F = fellow; IRE = Institute of Radio Engineers; M = member; MTS/D = member of technical staff, development; MTS/R = member of technical staff, research; NM = nonmember; O = session organizer; P = paper given; P2 = two papers given; SM = senior member; SUBD = subdirector, a level of management above supervisor; SUP/D = supervisor, first level of technical leadership, development.

SOURCES: The list includes authors and those mentioned in acknowledgments of "Epitaxial Diffused Transistors," *Proceedings of the IRE* 48 (September 1960): 1642–43. For complete source list, see appendix 2.

unreliability of surface devices, as well as being convinced that field-effect devices would be limited to low frequencies."[42]

The acknowledgments in the two papers describing the epitaxial transistor and the silicon–silicon dioxide surface device provide a basis for comparing the researchers at Bell Labs involved in each innovation. Table 1.1 shows the authors and the people acknowledged in a brief article in the *Proceedings of the IRE* describing the epitaxial transistor, while table 1.2 shows Kahng and those acknowledged in his internal technical memo describing the silicon–silicon dioxide surface device. Ross, the chief instigator of the epitaxial work, was a senior manager who had an intimate familiarity with junction transistor theory, fabrication technology, and the needs of the Bell System. The larger circle involved in the epitaxial work were managers and senior members of the technical staff who had been widely recognized both within Bell Labs and in the wider technical community for their work on the junction transistor technology. They had a commitment to the bipolar transistor by virtue of both their positions at Bell Labs and the length of time they had been in the field. They had helped to teach the industry how to make and use transistors by their contributions to the Bell-sponsored *Transistor Technology* and their participation in Bell's transistor symposia. They were regular participants in the SSDRC.

Those involved in Atalla and Kahng's work were relative newcomers, with less standing both within Bell Labs and in the broader technical community. All joined Bell Labs' semiconductor effort after 1955, when the key decision was made to concentrate on the silicon diffused technology. Atalla in particular, coming into the transistor development group as an experienced engineer and leader, may have been unwilling to accept unquestioningly the direction that Bell Labs' transistor work had taken. At the time Atalla arrived, work on the point-contact transistor had largely stopped. Those who had observed the work on the point-contact transistor had learned the difficulties of trying to make a transistor that was totally dependent on the characteristics of the semiconductor surface, a lesson that those involved in Kahng and Atalla's work perhaps had failed to learn.[43]

Kahng's memo, while documenting the birth of the physical structure of the modern MOS transistor, marked the end of work on that device at Bell Labs for a substantial time. Atalla and Kahng were both working on other projects by the time the memo was written. The case number, or technical project number,

TABLE 1.2 CONTRIBUTORS TO SILICON–SILICON DIOXIDE SURFACE DEVICE (MOS TRANSISTOR)

	Position at Bell Labs (1960)	IRE status (1961)	Year joined BTL semiconductor effort	Contributor, Transistor Technology			Participant, BTL diffusion symposium	Participant, Solid-State Device Research Conference							
				Vol. I	Vol. II	Vol. III		1952	1953	1954	1955	1956	1957	1958	1959
D. Kahng	MTS/D	M	1959												
M. Atalla	SUP/D	NM	1956											P	
H. Gummel	MTS/D	NM	1956												P
R. Lindner	MTS/D	NM	1956												
J. D. Goeltz	?	NM	?												
E. E. Labate	TA/D	NM	1956												
E. Povilonis	TA/D	NM	1957												

NOTE: BTL = Bell Telephone Laboratories; IRE = Institute of Radio Engineers; M = member; MTS/D = member of technical staff, development; NM = nonmember; P = paper given; SUP/D = supervisor, first level of technical leadership, development; TA/D = technical assistant (technician), development.

SOURCES: The list includes authors and those mentioned in acknowledgments of "Silicon–Silicon Dioxide Surface Device," Technical Memorandum MM-61-2821-1, 16 January 1961, AT&T Archives, Warren, New Jersey. For complete source list, see appendix 2.

under which Kahng's technical memorandum was filed does not contain another technical memorandum until 1965. And Bell Labs did not have a project aimed at developing MOS transistors for use in Bell System equipment until 1966.[44]

Bell's failure to publish Kahng and Atalla's work offers further hints about its reception at Bell Labs. Bell Labs typically published its most significant conference papers in technical journals, but Atalla claimed to have been denied permission to publish. While Atalla's previous work in using silicon dioxide as a way to stabilize bipolar transistors was recognized as significant and his paper on this subject given at the 1958 SSDRC was published in the *Bell System Technical Journal,* his new semiconductor device was not up to Bell Labs' standards for significant semiconductor research in 1961. It represented no new principles or theory, and its poor performance and poor correlation with theory did not make it something that Bell Labs, which served as a guide and shepherd to the industry, wanted to call attention to.[45]

Atalla and Kahng's writings provide evidence that even they had ambivalence about what they had done. A name is obviously one of the first steps in the serious consideration of any kind of invention, and Atalla and Kahng's failure to name their device implies that they saw it as stillborn. They did not even identify their device as a transistor, suggesting a reluctance to even put their work into the same family line as the work of Bardeen, Brattain, and Shockley.

Atalla and Kahng's paper at the 1960 SSDRC did not establish their device as a promising subject for research or even as something recognized by the semiconductor community at large. The conference chairman made no mention of Atalla and Kahng's work in his brief report on the technical highlights of the conference, although he did mention Bell's epitaxial transistor. No further work on a device like Atalla and Kahng's was presented at either the SSDRC or the Electron Device Conference over the next two years. Two articles reviewing the state of the semiconductor field in 1962 made no mention of Atalla and Kahng's device. Their work seemed to be a dead end.[46]

RCA and the Quest for Radical Technological Change

Bell Labs rejected both the integrated circuit and the MOS transistor. The first two labs to seriously pursue MOS transistor research were two of the earliest sites of work on integrated circuits. While Fairchild's role in the invention of the integrated circuit is well known, RCA's work, predating Fairchild's, is not. At both RCA and Fairchild, researchers saw the MOS transistor's advantages for integrated circuits as sufficient to offset its numerous deficiencies.

RCA's research laboratory had a more radical outlook than that of Bell Labs, based to a large degree on its patron, David Sarnoff. RCA had risen to its position of dominance within the electronics industry in large part due to David Sarnoff's vision of technology. Sarnoff had seen the possibilities of radio as a broadcast medium while it was still primarily conceived of as a method of point-to-point communications and led RCA to the forefront of the new technology. Later he saw the potential of television and gave it his unwavering support through a decades-long development and commercialization effort. In 1942 Sarnoff established RCA Laboratories, a central research laboratory in Princeton. After the close of the war, this laboratory became the center of an enormous effort to develop color television, a project that had cost RCA $65 million by 1953. In 1955 *Fortune* asserted: "Color television represents a monumental technical achievement, embodying more research and engineering at the time of its debut than any other product offered to the public."[47]

For Sarnoff, research was the central engine of corporate growth, where scientists developed new products under the watchful eye of a visionary manager like himself. Sarnoff accepted the need for continual research as essential to the corporation. Sarnoff asserted that RCA did not "fear or resist change" but, as a believer in "research, invention, and pioneering," saw change as a means to progress.[48]

What Sarnoff said mattered, for he had almost total control over RCA, based not on ownership of stock but on his carefully cultivated image both inside and outside the company as "Mr. Radio" and "Mr. Television," which was backed up with a handpicked board of directors. In 1951 RCA held a celebration marking Sarnoff's forty-fifth anniversary in the radio business at its Princeton research laboratory, at which occasion Sarnoff renamed the laboratory the David Sarnoff Research Center, formalizing his close identification with research. But by the late 1950s, with color television not yet proving profitable, there were complaints on Wall Street that Sarnoff was primarily interested in developing new technology, even if it came at the expense of the financial health of the company.[49]

Other factors besides Sarnoff's patronage reinforced the position of research within RCA. RCA Laboratories was responsible to the central corporate office and had only a tenuous relationship with the company's product divisions, giving it the freedom to pursue projects that the product divisions might not find appealing. Furthermore, until the late 1950s licensing revenues from RCA's radio and television patents provided the basis for Research's budget. This independent source of funding, amounting to around $15 million per year in the mid-1950s, encouraged RCA Laboratories to see the entire electronics industry as its constituency.[50]

Not surprisingly, RCA Laboratories was very quick to recognize the significance of a solid-state amplifying device. RCA Research's 1947 annual report, written before Bell Labs' invention of the transistor was made public, asserted that RCA saw solid-state electronics as a new field on the horizon in which it anticipated "doing in solid materials what we have in the past accomplished in radio tubes."[51] The next year RCA claimed to have produced transistors superior to those described by Bell Labs or manufactured by Western Electric. By 1951 the annual report listed semiconductors along with color television as RCA's main areas of research, with about 20 percent of the technical personnel at Princeton involved in semiconductor work.[52]

RCA Laboratories' dependence on patent licensing gave it the incentive to maintain a position of leadership in transistor work. Licensing intellectual property was a major business at RCA, and even though it had not invented the transistor, new patents and applications in the field would show its licensees the benefits of a continuing relationship. In 1952 RCA held an eight-day symposium that was attended by over 450 technical personnel representing over 182 licensees. RCA discussed ways of building transistors and potential circuit applications, such as a cordless microphone, a phonograph, a radio, and a television. *Fortune* reported that some of RCA's competitors were "astonished" with RCA's work in using transistors in electronic equipment, particularly the all-transistor television set. RCA's transistor symposium, claimed C. B. Jolliffe, vice president and technical director of RCA, "appears to have generated greater excitement in the industry than any technical presentation we have ever made, including color TV."[53] In 1956 RCA published *Transistors I*, a collection of articles and papers by RCA engineers and scientists.[54]

By the mid- to late 1950s RCA's sources of research funding were in jeopardy.

Competitors successfully sued RCA, receiving free access to RCA's radio and television patents, which effectively ended RCA Laboratories' independent source of funds. RCA itself could not make up these funds without a great deal of difficulty, for the entire corporation was burdened by Sarnoff's large investment in color television.[55]

To partially make up for this loss, RCA Laboratories turned to the military. RCA had been heavily involved in military research during World War II, and this work had paid dividends in its television work, but prior to the mid-1950s RCA had an ambivalent attitude towards government research. In the late 1940s and early 1950s RCA viewed its military research contracts as a civic duty performed at some cost to itself by diverting research from normal lines. But given the threatened loss of licensing revenues in the mid-1950s, RCA actively sought military funding for work in its main lines of business, with a goal of having government support account for a quarter of its research budget.[56]

By the mid-1950s the semiconductor industry had technologically reached a certain level of maturity, but RCA Laboratories had very little stake in the technology that had developed. RCA had done good research work on junction transistors and, by a number of measures, had established itself technically in the number-two position, behind Bell Labs. But given Bell Labs' body of work in transistors and its much larger research group, it was abundantly clear that RCA would remain a technological bridesmaid in bipolar junction transistors.[57]

All of these factors contributed to a radicalization of RCA Research's semiconductor program beginning in 1957. RCA Research dramatically cut back its work on silicon and germanium bipolar junction transistors of the kind invented by Shockley, where it believed further improvements would be "more a matter of refinement than the introduction of radically new concepts."[58] RCA's move toward "longer range objectives" and "new semiconductor materials and principles" was facilitated by two large military contracts that occupied over half of RCA Laboratories' semiconductor personnel.[59] In a marked change from earlier attitudes, RCA believed that it could use these two contracts to leverage itself into a position of commercial leadership.

The first contract, from the Air Force, was for research in gallium arsenide. The development of a silicon transistor had been important because its wider temperature range had made possible many new military applications where germanium could not be used. However, electrons move more slowly in a silicon crystal than they do in a germanium crystal, so silicon transistors sacrificed speed. Researchers at RCA found that gallium arsenide promised to combine a higher-temperature operation than silicon with a higher speed than germanium.[60]

Although RCA's work on gallium arsenide did not inherently require the abandonment of the junction transistor, that was its effect. In contrast to germanium and silicon, whose properties had been the subject of extensive studies during World War II, gallium arsenide was poorly understood, and a large part of RCA's efforts concentrated on understanding its physical properties. Initial efforts at developing gallium arsenide junction transistors were frustrated by a number of factors, so that by 1958 researchers were looking at "unorthodox transistor types less dependent on impurity content."[61]

The second major contract, from the National Security Agency, offered the potential to give RCA a leading position in computing, a field in which RCA was trying to establish a major presence. Project Lightning, which began in 1957, called for the production of computers one thousand times faster than existing computers. The National Security Agency apparently wanted to use this program to fund work on new and revolutionary kinds of electronic components, which would be the basic building block of this fast computer. The 1957 *Research Report* warned: "Because of the high speed specified, novel and radical approaches will be necessary; many of the approaches tried may prove impractical."[62]

Another program Princeton pursued with government support, although on a much smaller scale, was "integrated semiconductor devices." Although Robert Noyce of Fairchild and Jack Kilby of Texas Instruments (TI) are most commonly associated with the invention of the integrated circuit, an RCA engineer had an idea of the integrated circuit and worked towards its implementation well before either Noyce or Kilby. While Harwick Johnson of RCA had conceived of a specific circuit consisting of resistors, capacitors, and a transistor on a single piece of germanium as early as 1953, sustained work on what RCA would call integrated electronics began in 1957. In that year J. Torkel Wallmark, a Swede who had joined RCA Laboratories in 1953, proposed a "semiconductor logical tree," whereby a number of logic blocks could be incorporated onto a single piece of semiconductor. Wallmark claimed his idea was superior to "separate three-or-more lead units" for six reasons, among them "number of connections and leads reduced," "printed circuit techniques can be used," "small," and "large units can be made in few steps."[63] Wallmark proposed building this device using William Shockley's unipolar junction field-effect transistor, which had yet to be commercialized because of fabrication difficulties. These difficulties seemed to have scared Wallmark off work on this idea, but he and Johnson worked on building an integrated shift register using bipolar-type devices on a piece of germanium.[64]

RCA Laboratories gave Wallmark's work prominent mention in its 1957 annual report, listing it in the front section among the year's highlights. Under a heading labeled "Integrated Semiconductor Devices—A New Concept in Electronic Technology," it discussed an "integrated device which combines in one unit the functions to be performed by a whole assembly of capacitors, resistors, amplifiers and switches." It described the shift register as two units that together replaced 20 transistors, 40 resistors, and 20 capacitors. The report asserted that the replacement of resistors and capacitors by an integrated semiconductor device was a particularly significant achievement. While the report claimed that feasibility had been shown "in principle," it also acknowledged problems of uniformity remaining to be solved.[65]

In March 1958, over a month before Jack Kilby joined TI, RCA made a public announcement of its work, which was carried on the front page of *Electronic News* (figure 1.6). RCA displayed and described a ten-bit shift register, and while RCA claimed some devices had been shipped to the Air Force for evaluation, it admitted that fully functional devices had not yet been made. The Air Force sponsor claimed that the RCA work could eventually lead to computers consisting of a few small elements. RCA Laboratories' vice president of research made the announcement, claiming: "The way is open to the development not only of an integrated shift register transistor, but of an entire class of integrated devices incorporating new standards of miniaturization, economy and reliability."[66]

In an article titled "Integrated Electronic Devices: New Approach to Microminiaturization," published in the *Aviation Age Research and Development Handbook 1958–1959,* Wallmark gave a broad exposition of what could be done with integrated electronics. He defined his basic idea of an integrated device: "Much of modern electronic circuitry is iterated—identical stages follow one another in cascade, 'trees,' or matrices. If one stage can be replaced by an integrated device—a semiconductor, for example—it should be unnecessary to carry on

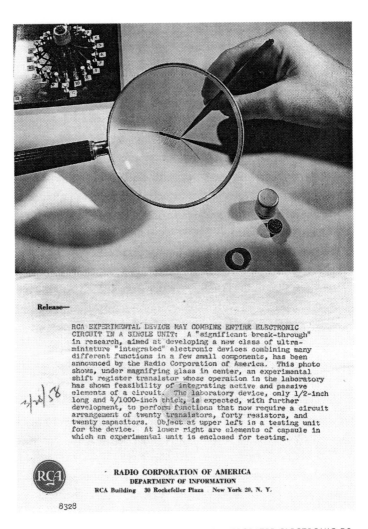

FIGURE 1.6 • RCA'S ANNOUNCEMENT OF ITS INTEGRATED ELECTRONIC DE-
VICE, 28 MARCH 1958. Although the date is handwritten, it can be corroborated
by Jerome P. Frank, "Shift Register Transistor Now under Development," *Electronic
News*, 31 March 1958, 1, 15. Courtesy David Sarnoff Library, Princeton, New Jersey.

the function through an array of separate but identical devices. The entire array may be built into a single, tiny semiconductor."[67] Wallmark described the shift register announced previously and suggested the integrated device concept would be most promising in electronic computing, listing such possible applications as "shift register, matrix switches, transfer trees, adders, and similar widely used circuits."[68]

The techniques used to build the shift register could not be extended broadly to other circuit configurations, and RCA moved its integrated electronics work over to William Shockley's unipolar transistor. The basic attitude of the RCA researchers was to look not for the technology that could be commercialized the most quickly, but the one that would yield the greatest benefits in the long term. Although the bipolar transistor was widely used and its production technology had been well established, RCA researchers believed that it was not well suited for use in integrated circuits. They considered its three-dimensional structure too complex and the simplicity of the unipolar transistor better suited for building and interconnecting a large number of transistors on a single semiconductor slice. While RCA may have been right in the long run—there were better transistors for integrated circuits—ironically, the simpler approach was more difficult in the near term. RCA had to develop simultaneously methods to make unipolar transistors and methods to build integrated circuits. The long-term perspective of RCA engineers also led them not to focus on the most basic implementation of the concept (in a circuit with just a few elements) but on relatively complex digital circuits, where they believed the greatest benefit lay.[69]

In March 1959, in conjunction with the IRE's annual meeting, RCA made another announcement of its work, just days before TI announced Jack Kilby's work on what it would call semiconductor solid circuits. This announcement focused on RCA's work on unipolar transistors. RCA defined its concept of integrated electronics in which "a complete set of circuit functions, including those of both the active and passive elements is built into a single extremely small piece of solid material, such as silicon." RCA further claimed: "The application of a new concept of 'integrated electronics' has enabled scientists to create a logic circuit element—the basic building block of computers—so compact that up to 100,000,000 such elements might be crammed into one cubic foot."[70]

RCA's work on integrated electronics is not well known for several reasons. First, it suffered from grave problems of implementation—several years of work did not yield anything close to a product. RCA appears to have been unable to fabricate a fully working circuit even for its announcements. A key word in both of RCA's announcements was "developing"—it was describing a work in progress, not a completed product. RCA did not try to quickly commercialize the technology but instead was content to work on it in the laboratory for years. RCA's announcements never mentioned when it would have something for sale. Although TI still had problems of implementation to be worked out when it made its announcement, TI was clearly aiming at speedy commercialization. It spoke of a timetable for products and when samples might be available. Even when RCA became aware of TI's work and could evaluate it, RCA still believed its approach was superior in the long run. Many other electronics microminiaturization programs existed at the time, and RCA itself was closely identified with another approach sponsored by the Army, the micromodule, making it difficult for RCA to attract much attention for a project that, no matter how great its possibilities, had practically achieved very little.[71]

While RCA engineers and scientists struggled to implement integrated electronics and make gallium arsenide transistors, William Webster's appointment as head of Princeton's semiconductor research in 1959 marked a significant turning point of RCA's semiconductor research away from the technological frontiers and towards applications that had more near-term prospects. Webster had joined RCA Laboratories in Princeton in 1946 but had later moved into semiconductor development at RCA's Somerville, New Jersey, facility. By the late 1950s the Somerville facility, located just twenty miles from RCA's Princeton research facility, was separated from it by a very wide technological gulf. While Princeton looked for radically new semiconductor devices, Somerville, dedicated to development and production, was more conservative, working on germanium and silicon bipolar junction transistors. Although Somerville received some support from RCA Laboratories, in his time at Somerville Webster took advantage of his friendships at Bell Labs and developed close informal relations with the transistor development group at Murray Hill, which was closer technologically to Somerville than RCA Laboratories was. These connections were facilitated by a feeling on both sides that Bell and RCA were not really competitors, as well as by Somerville's position on the road from Bell Labs' Murray Hill facility to Western Electric's transistor production facility in Allentown.[72]

In the spring of 1960 managers at Bell Labs invited Webster to attend a dry run of their presentations to the SSDRC, including Atalla and Kahng's. Webster, with a desire to move RCA Research back from the frontiers and encourage work in silicon, took a strong interest in Atalla's paper and carried news of it back to Princeton. There appears to have been no response from the senior figures in Princeton's semiconductor work, perhaps because of pressing responsibilities on RCA's large government research projects.[73]

Work on something like Atalla's device was finally energized near the end of 1960 by an intersection of interests between Webster and a recently hired engineer, Karl Zaininger. Zaininger had joined RCA after receiving his bachelor's degree in electrical engineering from the City College of New York in 1959. He was interested in the field of semiconductor surfaces, and after attending the 1960 SSDRC, he immediately began making calculations on an "MOS diode," a device related to Atalla and Kahng's that had been the subject of a paper at the conference by Atalla and a member of his group. In December, Zaininger began an effort to build an "MOS unipolar" transistor, working with Charles Mueller, one of RCA's leading semiconductor researchers, and Mueller's talented technician, Ethel Moonan. In late December 1960, their device displayed voltage amplification.[74]

Webster then moved to secure an Air Force contract for this work, which was picked up by Steve Hofstein and Fred Heiman, two recent hires with undergraduate degrees in electrical engineering. Hofstein recounted his manager's assertion that the well-known difficulties of the semiconductor surface made the project unappealing to the more senior engineers but created a no-lose situation for Hofstein. Failure would be attributed to the inherent intransigence of the semiconductor surface, not his lack of competence; success would redound to his credit. Hofstein and his colleagues, working in the department at Princeton attempting to make integrated circuits, built what they called an integrated logic net, an MOS integrated circuit containing sixteen transistors that could be wired in various configurations to perform logical functions.[75]

While RCA's earlier efforts to develop integrated circuits had used unique methods for fabricating structures, Hofstein and Heiman's device used the same materials and fabrication techniques that Somerville used in producing silicon bipolar junction transistors. The 1961 *Research Report* asserted: "An important factor in the development of this transistor has been close liaison, particularly with regard to fabrication techniques, with the Semiconductor and Materials Division [Somerville]."[76]

Although help from Somerville was essential to realizing the MOS transistor, Princeton's technological environment gave the MOS transistor an ecological niche where it could thrive. With almost all bipolar work eliminated from Princeton, the MOS transistor did not have to face a mature competitor that would have made it seem hopelessly crude. Instead, as Webster looked over RCA Laboratories' gallium arsenide program, he would have seen a program in a primitive state using a new and unknown material that required new fabrication techniques and the invention of new transistor types—a daunting task. Faced with these challenges, the problems of Atalla and Kahng's device would not seem so formidable. It used a well-known material and well-established fabrication techniques. Even though it still required a solution to the problem of semiconductor surfaces, that was a single problem; the gallium arsenide program faced many layers of problems before it would be viable.

The MOS transistor's importance to RCA stemmed from its perceived applicability to computing. In the 1950s RCA launched a major effort in computing, introducing its BIZMAC computer in 1955, and by 1960, computers were one of RCA Laboratories' three main areas of concentration, along with color television and solid-state devices. The work on the MOS transistor was carried out in an interdisciplinary group, which had members from both the Electronic Device Research Laboratory and the Systems Research Laboratory. The group's manager, Thomas Stanley, had previously worked on solid-state circuits but had just returned from a year at Cambridge, where he had worked with computing pioneer Maurice Wilkes. This group envisioned and investigated a number of systems applications for MOS integrated circuits, including a digital frequency synthesizer and an array of MOS logic blocks programmed with light beams.[77]

These were plausible applications, not real products. Research's interest in the MOS transistor was a long-term one. The 1961 *Research Report* stated: "The day will come when hundreds or even thousands of elements will be formed and interconnected in a single processing run. When this occurs, there will be a decrease in cost of two or three orders of magnitude for a given information-handling capability."[78] With the industry struggling to put several transistors on an integrated circuit, "the day" when hundreds or thousands of components could be built on a single integrated circuit would be many years off. In fact, part of the appeal of the MOS transistor to RCA Research was its long-term nature. Its wide implications in systems, circuits, and semiconductor materials could be explored by RCA Laboratories without the immediate pressure to transfer the technology to the Semiconductor Division.

While the MOS transistor was still inferior in many ways to the bipolar transistor, Stanley's analysis of the principles of electrical scaling gave RCA reason to favor the MOS transistor over the long term. Stanley's aim was to determine the effect of shrinking the device's physical dimensions and other design properties by a fixed factor. This work showed that while a bipolar transistor did not scale—that is, one could not improve its performance by shrinking its dimensions by a constant factor—the MOS transistor would yield decreased delays as its dimensions shrunk. This analysis showed (at least in an abstract sense) that a clear development path existed for the MOS transistor, allowing for continued improvement.[79]

A crucial difference between the bipolar transistor and the MOS transistor as RCA researchers conceived it was the orientation of each device. The bipolar transistor, as a three-layer sandwich structure, had its critical dimension in the vertical dimension, a dimension that could be very closely controlled by the diffusion process. The MOS transistor had an essentially flat structure, more like a pizza, and its critical dimension was horizontal, the length of the gate. This dimension, determined by photolithography, could not be made as small or as precise as dimensions controlled by diffusion, but over time improvements in this dimension would come much more rapidly than they would come in the vertical dimension.

In October 1962, Hofstein and Heiman presented their work at the IRE Electron Devices meeting, but it was not noted by the two major trade periodicals, *Electronics* or *Electronic News*. In February 1963 RCA, alerted to the fact that Fairchild would shortly be presenting a paper on a similar device, held a press conference to announce its work on the MOS transistor. RCA highlighted the new device's applicability to integrated circuits, claiming that it was a "new fundamental building block" for integrated microelectronic circuits. RCA claimed that the transistor was "cheaper, simpler, and more reliable than anything yet devised."[80] It stated the device used only a third of the steps required to make bipolar junction transistors and offered yields as high as 99 percent. RCA also asserted that the device promised high circuit densities, claiming that 850 devices could fit on an area the size of a dime and that all the logic for a medium-sized computer could fit on an area the size of forty half-dollars.[81]

Fairchild Semiconductor: The Planar Process, the Problem of Inversion Layers, and an Inventor

Given that the MOS transistor ultimately overthrew William Shockley's bipolar transistor, it is perhaps significant that Fairchild Semiconductor, the other site of early MOS transistor work, was born out of rebellion against William Shockley. In 1955 the ever-restless Shockley, frustrated at both being stuck in middle management and receiving what he considered meager financial rewards for his work, left Bell Labs to found Shockley Semiconductor Laboratory with financing from his longtime friend Arnold Beckman, the founder of Beckman Instruments. Shockley's reputation enabled him to quickly assemble a highly talented team of researchers. After beginning work on producing a silicon double-diffused transistor much like the one Bell Labs had developed, Shockley switched his focus to a four-layer diode, a junction-type device Shockley had invented that had interesting theoretical properties as well as potential for use in telephone switching systems, but no proven market.[82]

While Shockley had phenomenal physical intuition, he had no instincts for managing an entrepreneurial venture. His enormous ego got in the way of running the business. He alienated his workers by subjecting them to an assortment of indignities. Instead of working to quickly get transistors into production, Shockley continued to be preoccupied with the four-layer diode and other diversions. After an abortive attempt to get Arnold Beckman to intervene and install a proven manager, a group of eight arranged funding to start a new company from New York–based Fairchild Camera and Instrument, in return for an option to acquire the company in two years. The new company, Fairchild Semiconductor, with an obvious need to develop a product quickly, concentrated on a double-diffused silicon bipolar junction transistor. The first such device, initially selling for $150 each and targeted at military applications, proved to be highly successful. In 1959 Fairchild Camera exercised its option to acquire the company.[83]

Fairchild's accomplishments in its early years were not based on a fundamentally new semiconductor technology but on taking the basic silicon dif-fused transistor structure of Bell Labs, making significant incremental improve-ments, and implementing it on a production level. Fairchild first used Bell Labs' diffusion and oxidation processes to manufacture the mesa double-diffused silicon bipolar junction transistor. In response to a severe reliability problem in its initial product, Fairchild implemented the planar process, which had been conceived earlier by Jean Hoerni, one of the founders of Fairchild. Hoerni took advantage of silicon dioxide's passivating effect on the silicon surface, which had been demonstrated by Atalla at Bell Labs (but not been made part of Bell's transistor production process), to make transistors that were protected by a layer of silicon dioxide (figure 1.7). In 1959 Robert Noyce built on Hoerni's work with his conception of an integrated circuit, which added a layer of metal to the top of Hoerni's basic structure to connect different components—transistors, capacitors, or resistors—located on the same piece of silicon (figure 1.8). The idea of putting multiple components on a single piece of semiconduc-tor was not new, but the superiority of his concept of an integrated circuit was due to the fact that the planar process provided a very powerful way of imple-menting it. The military was sponsoring a number of high-profile programs aimed at developing miniaturized electronic components during this time, so the market for such a device was clear.[84]

A byproduct of Noyce's integrated circuit was a metal layer over an oxide layer over a semiconductor (which would now be called an MOS structure), so it is tempting to read backwards and assume that Fairchild's MOS transistor work was a logical outgrowth of its work on planar process integrated circuits. This was not the case. Although seven people from Fairchild were registered for the SSDRC in June 1960—including Robert Noyce, Gordon Moore, and Jean Hoerni—and thus would have heard about Atalla and Kahng's device, sustained work at what is now called an MOS transistor did not begin until two years later. Hoerni, obviously inspired by their work, described a device similar to Atalla and Kahng's in his first progress report after the conference. Although he claimed that work was "in progress" to build such a device, which he called a "unipolar transistor," later progress reports show his interest lapsed.[85]

The brevity of Hoerni's work on Kahng and Atalla's device should hardly be surprising. In a 1993 interview, Gordon Moore stated that with the de-velopment of the planar process, Fairchild had gotten into a very rich vein of

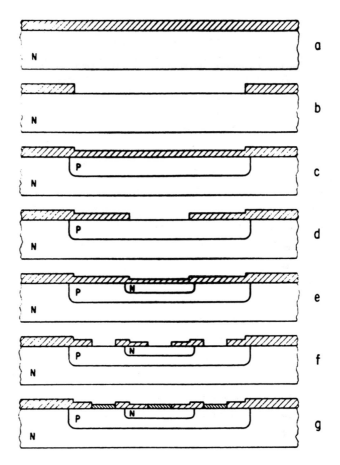

FIGURE 1.7 · THE STEPS INVOLVED IN BUILDING A PLANAR JUNCTION TRANSISTOR USING JEAN HOERNI'S PLANAR PROCESS. The hatched material in (a) is silicon dioxide covering a piece of *n*-type silicon. The oxide is selectively etched off in (b). A *p*-type dopant is then diffused into the open window, and the presence of oxygen in the diffusion furnace causes the surface to be reoxidized (c). Another hole is opened up for a final diffusion (d). During the final diffusion, an *n*-type dopant is diffused through the hole to create an area of *n*-type semiconductor, and the presence of oxygen in the diffusion furnace causes the surface to be reoxidized (e). Holes are opened in the oxide for metal contacts (f), which are formed in (g). The end product is a silicon *n-p-n* transistor (a vertical structure) covered with a layer of silicon dioxide. Notice that the dopants that create the *p* and *n* material diffuse horizontally as well as vertically so that the surface of each *p-n* junction is covered with oxide. Two diffusions are required to make a bipolar transistor with the planar process; only one was required to make an MOS transistor (compare with figure 1.5). From Jean A. Hoerni, "Planar Silicon Transistors and Diodes," Fairchild Semiconductor, Technical Articles and Papers, TP-14, Bruce Deal Papers, Collection M1051, Department of Special Collections, Stanford University Libraries, Stanford, California. Used by permission.

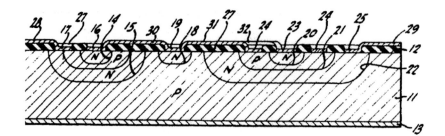

FIGURE 1.8 • ROBERT NOYCE'S INTEGRATED CIRCUIT PATENT USING THE PLANAR PROCESS. The *n-p-n* structure at the right (marked by 20-21-22) is a bipolar transistor. It is connected through the metal (31) on top of the oxide (27) to the rest of the circuit. For example, 19-18 make up a capacitor. Noyce's integrated circuit had a metal layer over an oxide layer over a semiconductor layer (31-27-22, center). In today's terms, this would be called an MOS structure or MOS capacitor. Comparing this figure with Atalla and Kahng's device (figure 1.5), one can see the greater complexity of bipolar transistors and integrated circuits. Atalla and Kahng's device required only one diffusion step, while Noyce's integrated circuit required three. From Robert Noyce, "Semiconductor Device and Lead Structure," U.S. Patent 2,981,877, 25 April 1961, United States Patent and Trademark Office.

technology. Moore was referring to its usefulness in producing individual transistors, bipolar integrated circuits, and, later, MOS integrated circuits. Continuing the analogy, one would obviously begin by mining the material that was easiest to get at and that had a ready market, before going after material that was much less accessible and had an uncertain market.[86]

A number of areas were more attractive to Fairchild research and development. Like much of the rest of the industry, it worked on tunnel diodes, which never became a viable product, and epitaxial transistors, which did. Both of these technologies represented significant advances in speed. Above all, Fairchild was occupied with developing its line of integrated circuits and evolutionary extensions to its line of planar process transistors.

In the spring of 1959 Gordon Moore became head of research and development at Fairchild. Fairchild's original general manager defected to start a new company; Noyce, previously the head of research and development, became the new general manager, while Moore, previously the head of engineering, moved into Noyce's old spot. Moore, the son of a sheriff, had grown up on the San Francisco Peninsula, living primarily in Redwood City. A neighbor's chemistry set began an interest in chemistry, which later led to some fairly sophisticated work making explosives as a teen. He earned a bachelor's degree from the University of California at Berkeley and a Ph.D. from Caltech. He initially worked doing basic research at Johns Hopkins's Applied Physics Laboratory, then in Silver Spring, Maryland, but the desire to do something more practical eventually led him to take a position with Shockley. Although he was technically very talented and played a major role in developing the processes used to make Fairchild's original transistor, his greatest contributions to semiconductor technology were to come as a manager. One sees in Moore's progress reports to his friend and boss Robert Noyce an extraordinarily clear thinker, not easily swept away by enthusiasm for a new technology or technique. Moore had an outstanding ability to assess the strengths and weaknesses of work in semiconductor technology.[87]

Although Fairchild's laboratory was called a research and development laboratory, in practice that meant something different from research and development as practiced at Bell Labs. In contrast to Bell Labs, Fairchild made no clear distinction between research and development. Fairchild's Physics Section, which was the closest to what Bell Labs might call research, worked on problems that cropped up in existing devices, techniques that would lead to evolutionary improvements in devices, and new devices.

The Physics Section's work on the problem of inversion layers in transistors in manufacturing shows both that the section's efforts were occupied in dealing with production problems and that the semiconductor surface remained a most intractable and mystifying problem at Fairchild, which would hardly make Atalla and Kahng's device attractive. In late 1960 Fairchild was forced to stop production of its model 1740 bipolar transistor because of problems with inversion layers. In January 1961 Gordon Moore's progress report to Robert Noyce stated: "The inversion layer problem seems to be completely cured . . . although the mechanism is certainly not understood." The next month Moore reported: "We lost our 'black magic' for removing inversion layers temporarily. This caused considerable more furor than it warranted, since we knew what we changed. The witch's brew is working again. I feel the process is definitely workable, but we are still interested in understanding why." Although Fairchild was able to return the devices to production, they remained plagued by problems. In August 1961 Moore wrote to Noyce: "The inversion layers that appear on pnp mesa transistors are the most elusive things to study we have found. To date, it has been impossible to cut the can off one or to make a hole in the can without the layer disappearing. Another month of this and we will postulate a phlogiston theory to explain them."[88] Atalla and Kahng's device started from the premise that inversion layers could be controlled; Fairchild's experience suggested just the opposite.

Fairchild's problems dealing with semiconductor surfaces were exacerbated by defectors leaving for other companies. In February 1961, Hoerni resigned to head the development lab of a new semiconductor firm, Teledyne Amelco. Between 1959 and 1962 four start-up companies were formed on the San Francisco Peninsula by Fairchild personnel, typically with the intention of duplicating Fairchild's products and processes. This heavy turnover resulted in a substantial loss of capabilities for Fairchild, particularly in areas such as MOS structures and semiconductor surfaces, which relied heavily on art. In June 1962 Moore reported to Noyce: "We have lost the MOS capacitor technology. Hopefully it will return shortly. Evidently there was not an adequate transfer of information before the engineer who originally developed the techniques left."[89]

By 1962 Fairchild Semiconductor had grown from a small semiconductor start-up concerned with quickly developing a product, to a large semiconductor company increasingly concerned with the long term. Fairchild R&D had 275 employees and had moved into a 110,000-square-foot facility in the Stanford Industrial Park. That year it also formed a Basic Physics department, whose first progress report described its charter as "to inaugurate fundamental research in areas deemed to be of long term interest to the company."[90] A list of milestones that Gordon Moore prepared at the beginning of 1962 showed the wide variety of activities Fairchild R&D engaged in, ranging from short-term projects (such as the development of extensions to existing transistor types) to much longer-term projects (such as the exploration of new transistor materials such as gallium arsenide, or the investigation of basic scientific effects such as superconductivity).[91]

In spite of the fact that nothing like the MOS transistor was on Moore's list, early in 1963 Fairchild announced such a device. The development of the MOS transistor at Fairchild was not so much the result of a manager's strategic vision but the result of a creative, individualistic inventor working in an expansive R&D environment where products with less and less immediate application were increasingly allowable. The central figure in developing this device at

Fairchild was Frank Wanlass, who had joined Fairchild in August 1962 after receiving a Ph.D. in physics from the University of Utah. Wanlass, who had been unaware of Atalla and Kahng's work, had his idea of something like an MOS transistor before he got to Fairchild.[92]

When Wanlass joined Fairchild R&D, he was given a great deal of freedom and began to work on his idea. In spite of previous problems with semiconductor surfaces, the planar process made the MOS structure extremely important for Fairchild. Furthermore, Wanlass wanted to make integrated circuits, and Fairchild was very definitely interested in integrated circuits. Although Wanlass's project seems not to have been one that Fairchild managers would have chosen themselves, it fell within the universe of acceptable projects. His first devices were hampered by instability problems, but he continued working and by February 1963 had achieved a substantial degree of success both in building devices and in designing circuits that used those devices.[93]

Wanlass was highly individualistic, and a central feature of his work was that he refused to follow any simple organizational division of labor scheme. Although Wanlass was a physicist and a member of the Physics Section, he had previously worked designing transistor circuits before he got his Ph.D. While much of his work rightly belonged in the Physics Section, at the same time he developed a new class of circuitry (now called CMOS, for complementary MOS) that, by making very effective use of the MOS transistor's characteristics, used almost no power. Wanlass also designed and built flip-flops, the basic storage elements in computer logic (figure 1.9). In an effort to demonstrate applications where his devices could be used to advantage, Wanlass designed and built a current meter that took advantage of the MOS transistor's extremely high input impedance. Wanlass's work encompassed functions normally done by the Physics, Chemistry, Micrologic, and Device Development Sections, and his current meter work was systems work, where Fairchild had just established a formal section. Given that customers were not asking for MOS transistors, Wanlass's broad background in physics, circuits, and systems enabled him to simultaneously develop this new device and applications for it.[94]

Even though Wanlass had a Ph.D. in physics, he was more an inventor than a scientist. He never published in professional technical journals, and his original work on the MOS transistor was made known to the engineering community through conference presentations and an article in the trade journal

FIGURE 1.9 · MOS INTEGRATED CIRCUIT DEVELOPED BY FRANK WANLASS AT FAIRCHILD IN 1963. This circuit consists of six MOS transistors and two resistors. This circuit, measuring 0.03 inches by 0.03 inches, was capable of operating as a flip-flop, a shift register stage, or a basic logic gate. From F. M. Wanlass, "Metal-Oxide-Semiconductor Field-Effect Transistors and Microcircuitry," *Wescon 1963 Technical Papers*, 13.2. © 1963 IEEE.

Electronics. Within Wanlass's progress reports in this period was material that would have provided the basis for many technical papers had he had that inclination, and in fact later researchers wrote a number of papers covering ground that Wanlass had already explored. Wanlass's main interest was in developing products, not publishing papers.[95]

In his January 1963 progress report, Wanlass noted an effect in MOS transistors whereby charge could move from the source of the device through the oxide onto the gate. In spite of its novelty, Wanlass neither published it nor attempted to patent it, apparently because he was focused on the immediate development needs of the MOS transistor. In 1967 two researchers at Bell Labs published an article on this effect, and in the period between 1969 and 1971 Dov Frohman-Bentchkowsky, first at Fairchild and later at Intel, published a series of articles on this effect, with his work culminating in the development of a new semiconductor device, the EPROM (erasable programmable read-only memory).[96]

The MOS transistor was slow but simple. One of the key problems Fairchild had faced as it put multiple bipolar transistors on the same piece of silicon was that extra processing steps were required to isolate the transistors from each other. No such extra processing was required for the MOS transistor, and it required fewer processing steps in the first place. In a January 1963 report, noting that the yield of unpackaged MOS flip-flops was between 80 percent and 90 percent and that there were few process steps involved in making the MOS transistor, Wanlass stated hopefully: "Even though it won't be extremely fast it will be desirable to make because of its low cost."[97]

Although Gordon Moore was not an enthusiast for the MOS transistor in the way Wanlass was, Wanlass's work did capture his attention. In February 1963 Wanlass and his manager, C. T. Sah, presented a paper at the International Solid-State Circuits Conference describing Wanlass's work on CMOS and introducing Fairchild's work on the MOS transistor. In March, Fairchild, aware that RCA was intending to market its MOS transistor, formed a joint program between the Physics Department and the Device Development Department to "push the MOST technology."[98] In June 1963 Moore went on a NATO-sponsored trip to France, Germany, Italy, and the United Kingdom to describe Fairchild's work on MOS devices, claiming they were promising for memory applications. In an article describing the planar process and integrated circuits, Moore wrote that the future of Wanlass's device looked "extremely interesting." In early 1963 Fairchild hired three Ph.D. researchers—Bruce Deal, Andrew Grove, and Ed Snow—to address the problem of MOS stability.[99]

Conclusion

At the transistor level the MOS transistor represented a radically new technology that could not easily be accommodated by the proponents of the bipolar technology. By the second half of the 1950s both the bipolar transistor and the community of semiconductor researchers had reached a level of maturity. Those in the field had a sense of what was promising and what was not. And what would become the MOS transistor was not promising, for it violated a fundamental tenet upon which the bipolar transistor was based: make semiconductor devices from p-n junctions, thus avoiding the effects of the semiconductor surface. Prior to Atalla and Kahng's work, one suspects that what would become the MOS transistor was beyond the pale of what an experienced researcher would even consider. Afterwards, even the briefest consideration would reveal daunting problems, for to embrace the MOS transistor was to move backwards and spurn the progress made with the bipolar transistor. Even if one were to believe that the problems of the semiconductor surface could be solved, the MOS transistor was slower and less reliable and would remain so for the foreseeable future.

The examples of Bell Labs, RCA, and Fairchild (as well as IBM Research, to be discussed in the next chapter) suggest two necessary conditions for MOS transistor research. First, an interest in integrated circuits was an important stimulant to MOS transistor work. Although it was possible to make individual MOS transistors and they had characteristics that made them more desirable than bipolar transistors in some applications, the problems of the semiconductor surface were so daunting that the costs of a research program based on individual MOS transistors were not worth the benefits. Given Bell Labs' skepticism about the integrated circuit, it made perfect sense for Jack Morton to consider Kahng and Atalla's device a technological loser. With the work on integrated circuits at Fairchild and RCA came an application whose simplicity promised benefits so large that they would justify facing the difficulties of the semiconductor surface.

The integrated circuit required a fundamental redefinition of what constituted a good electronic circuit. By most measures, the characteristics of discrete transistors and resistors were superior to those of integrated components, so that circuits made from individual transistors and resistors could operate at higher speeds than their integrated circuit counterparts. Furthermore, discrete components allowed the engineers to design circuits they could control more precisely. Gordon Moore described one customer's response to an early Fairchild bipolar integrated circuit flip-flop (a basic logic element): "This is ridiculous. We need 16 different kinds of flip-flops. We have 16 engineers, each of them a specialist in these flip-flops. There is no way we can use that single design for anything. It's a crummy flip-flop in the first place and it's not specialized for the things we need."[100] But this integrated combination of inferior components led to a product that was much smaller, more reliable, and ultimately less expensive than a circuit made from discrete components. Moore noted that the customer's complaints ended once Fairchild was able to offer the flip-flop for less than the cost of the individual components. The MOS transistor took this redefinition one step further. While it was grossly inferior to the bipolar transistor in most regards, the long-term perspective of MOS advocates at RCA and Fairchild led them to recognize that if the criterion for judging a transistor was its simplicity, the MOS transistor had significant advantages.

The other necessary condition for the MOS transistor was that the organization had to have a long-term horizon, making it more receptive to speculative projects. Experienced research managers knew that this was not a technology that would reach maturity in a few years. RCA Research had this perspective based on the division of labor between it and RCA's semiconductor development division. Given the number of projects Fairchild had promising quick returns, it would have been managerial malpractice to have launched a large MOS program right after Atalla and Kahng's paper. By 1962, with its integrated circuit product successfully introduced and with R&D growing, it could support more speculative projects.

The planar process, the integrated circuit, and the MOS transistor, related technologies built upon a base developed by Bell Labs, but technologies initially rejected by Bell, marked the end of Bell Labs' leadership in transistor technology. Bell's almost total dominance in the transistor's first decade was unsustainable thereafter as more and more firms entered the field. One would have expected innovations to become more widely spread throughout the industry, but something more profound than this happened. Most of the semiconductor industry, with Fairchild in the lead, changed course. Bell Labs did not follow. Bell Labs had been an extraordinary technological leader, generously sharing its knowledge and know-how, but it proved to be a poor follower. There were reasons for it or anyone else to have skepticism about these new technologies. The benefits of the planar process were subtle.[101] Integrated circuits could be seen as a collection of inferior components. But in some ways Morton and the research and development engineers at this company dedicated to improving communications were not good at listening—hearing so as to understand the significance of what others were doing. The unquestionably unique position of the Bell System shielded them from the market and allowed them to argue that what worked elsewhere was not suitable for the phone system. Being a part of the world's greatest industrial research and development organization did not keep one humble. Throughout the 1960s Bell Labs, under

Morton's leadership, explored a number of cul-de-sacs in semiconductor technology. The contradiction between the Bell System, a conservative government-regulated monopoly, thinking and moving on a time scale of decades, and the transistor, a radical technology, where breathtaking developments could be made in a few years' time, became clear in the decade where Moore's Law began its trajectory.[102]

In the spring of 1963, the future of the MOS transistor was by no means secure. The devices that Fairchild and RCA produced were very slow. The problem of the semiconductor surface, which was sure to be of a greater effect than it had been for bipolar junction transistors, had not been definitively addressed. But the fact that two firms were interested in the MOS transistor created a rivalry that spurred both on. RCA and Fairchild had both elucidated clearly what they saw as the advantages of this device, and their very public announcement of their work had opened up a process whereby other companies enlisted to join the research effort on the MOS transistor. It would still be many years before the MOS transistor would become commercially successful, but now it was finally on the agenda.

Back from the Frontier

IBM RESEARCH AND THE FORMATION OF THE LSI PROGRAM,
1951–1965

RCA and Fairchild did not have the MOS transistor to themselves for long; they were soon joined by others, most notably IBM Research. The MOS transistor's attractiveness to IBM Research had much to do with the history of this large but relatively new research lab. Between 1945 and 1963, IBM's research capabilities underwent a profound transformation. What had been a modest group, with few Ph.D. scientists and no pretensions to the kind of work done by Bell Labs or the General Electric Research Laboratory, became one of the country's most important research labs, with hundreds of top-flight scientists working in a new Eero Saarinen–designed building. The beginnings of IBM's MOS research program were related to the growth of IBM Research in an ironic way. Since its formation IBM Research had embarked on several ambitious research programs that had failed. Because of its inability to make a direct contribution to the welfare of the company, by the early 1960s IBM Research was in an increasingly precarious position within the company. The MOS program was a pulling back from the frontiers of science, to an area that seemed to have much more realistic prospects for applications in products.

The technological position of the MOS transistor compared to the bipolar transistor mirrored the organizational position of Research vis-à-vis IBM's development groups. The bipolar technology was the technology of the establishment; the MOS transistor was the technology of the unproven upstart. Because the MOS transistor was so primitive compared to the bipolar transistor and because its characteristics were in so many ways inferior, IBM's development groups were not interested in it. But it offered Research a way to begin work in the important new technologies of integrated circuits and the silicon planar process without competing directly with the development groups.

IBM and Semiconductor Research

By 1963, the year it started its MOS research, IBM was a vastly different company than it had been twenty years earlier. Although revenues from electronic computing did not surpass those from electric accounting machines until 1962, the changes that made the growth of electronic computing possible were put in place in the late 1940s and early 1950s. Both symbol of this transformation and driving force behind it was Thomas Watson Jr., who in 1956 took over as chief executive when his father, Thomas Watson Sr., retired at the age of eighty-two.[1]

IBM's transition from electromechanical accounting machines to electronic computing machines brought substantial changes in IBM's development staff. Prior to World War II the design of new products was done by people Thomas Watson called "inventors," self-educated men with strong mechanical skills working in isolation from the outside world and each other. It was not until 1944 that IBM hired its first Ph.D. scientist, Wallace Eckert, the initial leader of the new Department of Pure Science located at Columbia University.[2]

An internal report written in 1950 describing IBM "Research and Development" gives a picture of IBM's technical staff as it stood on the threshold of this new era. It opened its discussion of IBM's technical personnel with the statement "engineering is the heart of IBM," but the prominence it gave to the fields of dynamics and kinematics was a reminder that IBM was a company based on mechanical technology. It emphasized the experience of the IBM staff, suggesting that experience could substitute for education. Of 619 people in the "category of technically trained members of the organization," 11 held doctorates; 48, master's degrees; and 280, bachelor's degrees. The report noted that while many of the remaining 280 lacked formal engineering education, they had "proved their ability during a long period of experience with the IBM organization as inventors and design engineers." The work of development was done in "Invention Departments." In an era when the Manhattan Project had shown what large numbers of highly trained physicists could achieve, IBM's Physics Laboratory had a staff of 13, "including 8 college graduates in Physics, Electrical or Mechanical Engineering," of whom four had master's degrees.[3]

The pressures IBM's entry into electronic (vacuum tube) computing put on this "Research and Development" organization were heightened by the advent of the transistor. Shortly after Bell Labs announced its invention, IBM began its initial work on the transistor in the Patent Development Department, which was organized at IBM's headquarters in the 1920s to allow inventors to work without the commitments of product development. Men with no formal training in solid-state physics acquired germanium diodes, broke them open, and added a third metal lead to convert the diode into a triode or transistor, work more akin to a *Popular Science* project than to state-of-the-art solid-state physics.[4]

In 1951 IBM hired Lloyd Hunter to head its efforts in solid-state electronics. Hunter, an experimental physicist with a Ph.D. from Carnegie Tech, had previously served as manager of solid-state electronic research at Westinghouse. In July 1951 Hunter's manager admitted that IBM would not be able to compete in "original scientific work" with Bell Labs, and by September, when IBM estimated that Bell Labs had 168 people working on the transistor, IBM was making plans to increase its efforts in this field from 14 to 29 people.[5] Although Bell Labs quickly made the wise strategic decision to focus its efforts on Shockley's bipolar transistor, Hunter concentrated his more limited resources on the point-contact transistor, based on its speed advantage over the junction transistor. One member of Hunter's group later suggested that Hunter's decision reflected IBM's lack of scientific sophistication and vision compared to Bell Labs. The mathematical approaches of the two labs reinforces this judgment: while Shockley developed detailed mathematical theories for his junction transistor, Hunter had a preference for graphical analysis. Although IBM's semiconductor research still lagged Bell Labs' in quality and quantity, Hunter recruited a large number of young Ph.D.'s into his area, including several who would play major roles in IBM's semiconductor work in the 1960s. Before 1956, IBM was a modest force in semiconductor electronics; as measured by papers given at the

Solid-State Device Research Conference, its efforts had yielded three unexceptional papers on point-contact transistors.[6]

Nineteen fifty-six was a watershed year for IBM. Thomas Watson Sr. died in June, shortly after turning over the position of chief executive to his son. Watson Jr., relying heavily on outside consultants, moved to transform IBM from a company built around the person of his father into a managerial one based on the standard organizational structure of a multifunction science- and engineering-based corporation. At the Williamsburg conference later that year, Watson implemented a decentralized organizational structure, with line and staff management. In January, following studies by both outside consultants and IBM executives, IBM formed a research group that was separate and independent from its product divisions. Later that year Watson implemented a further recommendation of the internal study—that research be led by a prominent outside scientist—by hiring former Office of Naval Research chief scientist Emanuel R. Piore to fill that position.[7]

Watson Jr.'s description of his selection of Piore in his memoirs suggests something about both his attitude and that of the scientific elite towards industrial research. He recounts spending a month interviewing candidates for the head of research, finally settling on the dean of an engineering school at a major university. When Watson happened to mention his choice to James Killian, president of the Massachusetts Institute of Technology (MIT), Killian rebuked him. Killian asserted that Watson's candidate would be inappropriate because he was outside the circle of scientists who had been involved in wartime work, making it difficult for him to attract first-rate scientific talent to IBM. Watson, by his own admission, did not know the scientific community and articulated IBM's desires to Killian as "We want a distinguished scientist." Neither Watson nor Killian seems to have been much concerned with whether this distinguished scientist had the ability to translate scientific work into profitable technology.[8]

Research's independence was meant to decouple it from the immediate needs of product development in order to devote more effort to basic research. The laboratory at Poughkeepsie, which had earlier combined research and development in one organization, was divided into separate groups. Beginning in 1957, research operations were moved south to Westchester County, New York, away from the Poughkeepsie development and manufacturing groups and closer to IBM's New York headquarters and the universities in the New York area. IBM commissioned Eero Saarinen to design a home for its new research organization. The facility at Yorktown Heights, christened the Thomas J. Watson Research Center when it opened in 1961, bore a familial resemblance to a laboratory Saarinen had designed for Bell Labs.[9]

IBM's establishment of a research laboratory provides a concrete example of the post–World War II ascendancy of what David Hounshell has called Vannevar Bush's "linear model," that world-class science leads to new technology. In a speech to IBM executives at the Williamsburg conference in November 1956, Piore listed as one of his priorities getting "superior people into our [IBM's] laboratories," asserting that one could have two scientists of broadly similar qualifications, but the one with scientific taste and style would be superior to the other. While Piore complimented IBM on its taste and style in businessmen, he saw his job as bringing the scientific style of Bell Labs to IBM Research.[10] Where previously IBM had no aspirations to do the type of scientific research done by Bell Labs, now Bell provided the model.

Piore's main concern was with long-range research, telling the assembled executives that his first task was "to formulate the research program for five and ten years." In a vision dominated by the revolutionary power of science, technology transfer and the relations between existing technologies remained afterthoughts, with Piore admitting his "only suggestion" at that point was to form a strong liaison group that understood the divisions' problems and that would make sure Research's work was brought to the attention of the divisions.[11]

Shortly after Piore arrived, he shunted Hunter aside from his position as the head of physical research within IBM. In a move that might be seen as indicative of the ascendancy of a new style of scientific work at IBM, G. Robert Gunther-Mohr was made head of semiconductor research. Gunther-Mohr had worked on the Manhattan Project, then earned his Ph.D. in physics at Columbia University working under Charles Townes on a project related to Townes' 1954 invention of the maser, for which Townes would receive a Nobel Prize. Gunther-Mohr had also worked under Harvard's John Van Vleck, one of the country's leading solid-state physicists and a future Nobel Laureate. In 1953 Gunther-Mohr had joined IBM's Watson Labs, which had close ties to Columbia University.[12]

In a 1957 proposal for semiconductor research written by Gunther-Mohr, one might see themes common among post–World War II industrial research organizations. For example, he claimed that one of the functions of basic research was to "provide the corporation's share of the contributions which industry owes to basic science as the ultimate source of profitable technological innovations," which echoed the claim in Bush's manifesto for research, *Science, the Endless Frontier,* that basic research provided scientific capital that paid dividends in practical applications.[13] Gunther-Mohr also noted the strong drive for applications in semiconductor science, but he claimed it was a tendency to be resisted, for it threatened to make a research group lose its contact with the latest scientific work, sapping its scientific vitality.[14]

Gunther-Mohr's 1958 report on IBM's semiconductor research program contained another theme found in *Science, the Endless Frontier*—the importance of the scientist's autonomy. Of his research group, Gunther-Mohr stated: "We, in general, undertake no projects just because they are of interest to IBM, but rather select our work so that it will, at once, be good science and of value to the Company."[15] At a time when the demand for scientists was very high throughout industry and the academy, the promise of autonomy was important to securing top-flight talent. The idea that Bush expressed and Gunther-Mohr reiterated was that leading-edge science led naturally to advances in technology, so that a company enlightened enough to invest in science would automatically benefit.

Acquiring a world-class group of semiconductor physicists was one of the main accomplishments of Gunther-Mohr's tenure. A partial list of those hired by 1960 included J. B. Gunn, who previously worked at the Royal Radar Establishment in England; Robert Keyes, who had a Ph.D. from the University of Chicago and was hired from Westinghouse in 1960; and Peter Sorokin and Marshall Nathan, who were hired after receiving their Ph.D.'s in physics from Harvard. Perhaps IBM's greatest coup came with its 1960 hiring of Sony's Leo Esaki, whose discovery of the tunnel diode had made him a celebrity in the field.[16]

The period between the end of World War II and 1965 was the high tide of industrial research in the United States. "Research has acquired magical connotations since the war," *Fortune* asserted in 1950, while in 1956 *Time* heralded "the Age of Research."[17] There seemed to be no limit to what could be done when the power of science was coupled with lavish funding. While the Manhattan Project was the obvious example, there were others, such as Bell Labs' invention of the transistor and Du Pont's invention of nylon. According to *Fortune*, the U.S. research budget (public and private) had gone from $250 million in 1938 to $1.3 billion in 1950, with about $600 million contributed by industry.[18]

These figures illustrate a new feature of post–World War II research: the expanded role of the federal government. Prior to World War II, the federal government was a minor actor in research, but after the war, it became a major funder of research and development carried on by private industry. Most of this funding was provided by the military. Piore himself was a symbol of this change, for the Office of Naval Research was a postwar creation.[19]

One of the most dramatic changes in industrial research was the institutionalization of the research lab's location in a suburban campus setting removed from the corporation's other facilities. While there were various motives for this physical separation, it symbolized research's autonomy within the corporation as asserted by the linear model. *Fortune* began its 1951 description of the new General Motors Technical Center with the following observations about industrial research in general: "Since the war there has been a phenomenal uprising of research laboratories. Nearly every major corporation, and many a minor one, has launched into building a shining new sanctum of centralized research. Nearly all have followed a pattern of ultramodern decor on parklike acres, far from the pressures and distractions of production departments or corporate offices."[20] Even General Electric, which had established the country's first industrial research lab, now became a part of this trend, moving its research facility from its location at the Schenectady Works, the company's main manufacturing facility, to The Knolls, an estate on the Mohawk River five miles away. At the same time General Electric adopted a new method of funding research that gave research greater autonomy within the corporation.[21]

It is impossible to overestimate what the new setting meant for the consciousness of the researcher. Previously, the daily routines of researchers at General Electric would have brought them in continual contact with the firm's existing technologies. One veteran from research's days at the Works recalled having frequent contact with the other operations at the Works. He remembered being shown "the big shops, especially the turbine shops" by Irving Langmuir, a member of General Electric's research staff who won the 1932

Nobel Prize in chemistry, and said that these interactions gave him "a real education in what was really happening in the company."[22] At General Electric and elsewhere, a new generation of researchers who had come directly from graduate school to a corporate research campus would have been less conscious of the existing technologies' strengths and weaknesses. In an environment where nothing was being manufactured, it would be easier to underestimate the difficulties of overthrowing an existing technology.[23]

Many research programs at the time, with their unlimited confidence in science, did grossly underestimate those difficulties. Du Pont believed that by spending money on basic research, it would be able to come up with "new nylons," new synthetic materials that would be so much cheaper than and superior to existing materials that they would immediately find widespread use. Through the millions of dollars Du Pont lost on unsuccessful programs, it learned how exceptional nylon was.[24]

The atomic-powered airplane was another example of a project based on the premise that science would lead to new technologies that would sweep away the old. On paper the advantages of nuclear fission over conventional fuel were obvious; fission could provide a compact power source with unlimited range and very high speeds. Throughout the 1950s American research labs spent large sums of government money working on an atomic airplane, with *Newsweek* magazine predicting in 1960 that routine commercial atomic aircraft flight was only a few years distant. In 1961, after fifteen years of work, the expenditure of over $1 billion, and precious little progress towards realizing an atomic-powered airplane, President Kennedy cancelled the program.[25]

At IBM, Piore launched two large research programs aimed at overthrowing existing computer technology. IBM was not alone in thinking these projects had potential—they had originated elsewhere and were pursued by others. But both ultimately failed to prove competitive against the established technology. Funding came from the National Security Agency's Project Lightning, which aimed at developing a revolutionary new computer system. The first project, a microwave computer, had originally been conceived by computer pioneer John von Neumann. Von Neumann had worked for IBM as a consultant and assigned his microwave computer patents to IBM. Although the microwave computer had a large number of qualities recommending it, such as the potential for leading-edge science and the possibility of radical improvement in computer performance, a group of roughly forty failed to demonstrate the viability of the technology.[26]

The other project, the cryogenic computer, had its origins in the cryotron, a superconducting logic element made by Dudley Buck at MIT. Staff members at IBM Research believed that a modified cryotron could be the basis for a major advance in computer speed, and in 1956 Research began a program to develop an entire computer using superconducting logic and memory elements. IBM planned to build a follow-on computer to the SAGE air defense computer using this technology, but delays due to technical problems caused the contract to be cancelled. In spite of a hundred-person effort, the program was afflicted with serious problems of performance, reliability, and reproducibility, lingering until 1965, when it was cancelled.[27]

In spite of these failures, IBM had developed an increased ability to perform leading-edge research. This was reflected in a 1963 *Harvard Business Review* article rating the strength of various companies' research efforts by discipline, based on the number of articles published in key technical journals. By this measure IBM had moved from twelfth in physics in 1957 to third in 1962, and from fourteenth in electronics in 1959 to fourth in 1962.[28]

A more substantial measure of IBM's growing prowess in scientific research, particularly semiconductor research, was its work leading to the invention of the semiconductor laser in 1962. Following Charles Townes's 1954 invention of the maser, a device that used differences in molecular energy levels to produce coherent microwave radiation, a number of research labs tried to build an optical maser, or laser. Although Theodore Maiman of Hughes Research built the first one, IBM Research had been in the hunt, producing a laser several months after Maiman. Following this work, a number of researchers considered the possibility of constructing a laser from semiconducting materials. In 1961, Rolf Landauer, recently appointed as the head of Solid State Science Research at IBM, established a study group to investigate the creation of a semiconductor laser. In September 1962 researchers from IBM, independently and almost simultaneously with researchers from General Electric and MIT's Lincoln Laboratory, demonstrated laser action in the semiconductor gallium arsenide.[29]

In the fifteen years since engineers at IBM had first cracked open a germanium diode to convert it into a transistor, IBM had shown its ability to compete with the leading laboratories in the country for scientific advances in solid-state physics. The semiconductor laser came about through IBM's very capable group of research managers, theoreticians, and experimentalists. However, it was still far from clear how this research capability would benefit IBM.

Solid Logic Technology and the Formation of a Semiconductor Organization

IBM concentrated its early semiconductor efforts on research and development rather than manufacturing. In October 1957, IBM's vice president in charge of research and engineering, acting on a request from Thomas Watson Jr., issued an edict that all future IBM machines would use solid-state devices, rather than vacuum tubes. In December IBM signed an agreement with Texas Instruments (TI), making it IBM's primary supplier. IBM's senior management felt that the returns from producing semiconductor components were markedly smaller than from systems and that an arrangement with TI would save IBM capital. The agreement called for IBM and TI to undertake joint development programs and to share "research and development information and manufacturing know-how."[30]

IBM also moved to strengthen its internal capabilities in transistor development. When IBM established a separate research group, it also formed a new transistor development organization and hired as its head B. N. Slade, who had eight years of experience in transistor research and development at RCA. Throughout the 1950s IBM increased its competence in semiconductor development, hiring both recent graduates and experienced engineers, so that by the end of the decade it had established a respectable presence in the semiconductor industry. In 1959 IBM's semiconductor development effort consisted of about eighty people, while IBM estimated Bell Labs' development team at seven hundred people, RCA's at four hundred, Philco's at two hundred, and General Electric's at two hundred. IBM's semiconductor development group had had several significant achievements, such as the invention of current-switching circuits (high-speed circuits that were widely used into the 1980s) and the development of an automated transistor production line, but it was hardly a major player in the field.[31]

With separate research and development organizations came the question of how each would relate to the other. Piore's statement at the Williamsburg

Conference on technology transfer displayed the implication in Vannevar Bush's linear model that the scientists' research would be of such import that merely disseminating information about it would be sufficient to start development work. Although Thomas Watson employed Mervin J. Kelly, the former head of Bell Labs, as a consultant, there were substantial differences in how R&D functioned at the two organizations. In the 1950s Bell Labs included both research and development, which made it easier to coordinate the two. At IBM, research was centralized, with development decentralized so that the two did not share a common executive until the president of the corporation.

IBM Research's early semiconductor work showed some of the difficulties this could create. One early focus, partially funded by the Department of Defense, was epitaxial growth, in which a chemical reaction creates a thin, well-controlled layer of semiconductor on top of the semiconductor wafer. This work generated very little interest from the group responsible for semiconductor development within IBM, and the Research manager then got permission to publish this work in the *IBM Journal of Research & Development*. The articles did not mention application of this method to the actual production of transistors. At Bell Labs work on epitaxial growth in research was quickly transferred to the development group, resulting in the first epitaxial transistor, whose faster speeds led to its adoption throughout the industry (see chapter 1).[32]

In 1960, with the backing of Kelly, IBM formed a separate organization to develop and manufacture electronic components. The proponents of the new components organization, who included semiconductor development engineers and research managers, claimed that IBM needed to produce its own components if its systems were to remain competitive. In July 1961 IBM management upgraded the components organization to division status and by August the Components Division (CD) had begun the development of the semiconductor technology known as Solid Logic Technology (SLT). SLT used a silicon epitaxial, planar transistor structure, which owed much to earlier work at Bell Labs and Fairchild. Individual diodes and transistors were then mounted onto a ceramic substrate, which held discrete resistors (figure 2.1).[33]

One feature of SLT was to prove crucial to IBM's future semiconductor efforts: SLT made use of the silicon planar process as developed at Fairchild. This made SLT at least broadly consistent with the rest of the industry, allowing it to take advantage of aspects of the semiconductor production infrastructure that

FIGURE 2.1 • IBM'S SOLID LOGIC TECHNOLOGY (SLT). While Fairchild's integrated circuit technology put multiple components on a single piece of silicon, IBM's SLT put individual transistors (the two small squares on the right and the one small square on the left) and resistors (the darker shapes) that were connected by a conductive paste (the thin lines) on a ceramic substrate. IBM Corporate Archives. Used by permission.

had been built up elsewhere. A major advocate of the silicon technology at IBM was William Harding, and his experiences and interactions with other companies provided him with most of his exposure to silicon. Harding had worked for seven years on transistors for the Radio Receptor Corporation in New York, the last two on silicon devices for military applications. While there he had attended Bell Labs' symposium on diffusion, during which techniques for developing silicon transistors were discussed. At IBM, Harding made visits to both Fairchild and TI to evaluate their capabilities for supplying silicon transistors to IBM for military projects. IBM's position as a large buyer of transistors enabled Harding to see what others in the industry were doing. While several organizations, including Bell Labs and Fairchild, had cast their future lot firmly with silicon by this time, a contingent within IBM advocated the use of germanium because of the higher speeds it yielded, a decision that would have put IBM on a far different path than the rest of the industry.[34]

It would be difficult to overestimate the importance of SLT, both for the success of System/360, the family of compatible computer systems IBM introduced in 1964, and for the future of IBM's semiconductor work. If System/360 was IBM's $5 billion gamble, the outcome of that gamble depended to a large degree on IBM's ability to make SLT. With SLT and System/360, IBM was committing itself to becoming one of the largest producers of semiconductors in the country. IBM developed SLT under task force–style management, with an overall development effort of over nine hundred people by February 1963. The challenge was to develop the technology and the capability to mass-produce it. To that end IBM devoted substantial resources to the development of specialized automated production equipment. Although not an integrated circuit, its single-transistor approach combined with automated production equipment to enable IBM to manufacture circuits in large volumes extremely cheaply.[35]

SLT gave IBM a semiconductor organization on a scale it had not had before. IBM built a new plant in East Fishkill, New York, for its semiconductor operations. Prior to SLT, IBM had on occasion brought in managers from other firms to fill senior positions in semiconductor development. This almost never happened after 1965. With SLT, IBM developed its own style of semiconductor development and production along with its own pool of semiconductor talent. SLT was the defining experience for a generation of semiconductor engineers within IBM, with key figures in SLT work holding a number of significant management positions in CD and its successors for over twenty years.[36]

SLT also underscored the tensions that IBM faced as an equipment manufacturer that built its own semiconductors. Integrated circuits were a new technology initially limited to relatively slow speeds, and IBM had good reason to be hesitant to stake its future on them. But at the same time IBM engineers made the typical equipment manufacturer's arguments against integrated circuits. They argued that the poor tolerance of resistors in integrated circuits was an insurmountable problem, invariably resulting in slow speeds. As an equipment manufacturer that leased its machines, it was in IBM's interest to have technological progress that was either slow or that made discrete jumps at intervals of five years or more. One of IBM's areas of competitive advantage was its huge volumes, which made possible a highly automated system of production, but this expensive system made sense only for a long-lived technology. Managers in CD launched SLT under the assumption that it would be a viable technology for at least five years. Although integrated circuits were the subject of wide interest within the semiconductor industry, CD argued that they were not the right approach for IBM. By late 1962 the positions of CD managers had hardened to the point where they believed any work on integrated circuits could jeopardize SLT within IBM.[37]

The opposition to even working on integrated circuits is shown by the case of Owen Hill. In 1962, with a Ph.D. in electrical engineering from the University of California at Berkeley, Hill joined Westinghouse, where he worked on integrated circuits. On the basis of presentations of that work made at the 1962 Electron Device Meeting, Hill and two associates were recruited by J. Earl Thomas of CD to establish an integrated circuit development effort at IBM. However, when Hill's group arrived, senior CD management apparently decided that work on integrated circuits could threaten the future of SLT, and Hill was redirected away from integrated circuits.[38]

The Research Large Scale Integration Program

Engineers in CD had conceived SLT without any input from Research, a lack of involvement that came by design. Shortly after the establishment of CD, Research and CD came to an agreement that CD would be responsible for work using the materials germanium and silicon, while Research would be responsible for the newer and lesser-known compound semiconductor materials, such as gallium arsenide. In this division of labor very little overlap existed between the two organizations, and they had little need for cooperation. In an earlier meeting held to discuss the coordination of work between CD and Research, the two groups agreed that Research was responsible for covering the company in the distant future (greater than five years), while CD was responsible for the immediate five-year period. The representatives from CD stressed "their desire to see Research 'leap frog' over present problems by coming up with entirely new solutions not just pursuing a straight evolution from present status."[39] The view that technology would advance rather quickly through discrete revolutionary jumps, such as from silicon to gallium arsenide, was also commonly expressed in the trade press at the time. This division of labor allowed CD a large degree of freedom in its work, for it had jurisdiction over the area of silicon technology and had no need to consider potentially competitive or overlapping Research work.[40]

But by late 1962, after the successful work on the semiconductor laser, several managers in Research, including Rolf Landauer, the director of Solid State Science; Walter Proebster, the director of Experimental Machines; and Proebster's assistant, Donald Rosenheim, began to believe that Research's semiconductor program required a fundamentally new direction, a belief motivated by three concerns. The first was that Research's work up to that point had had little relevance to the company. While some in Research were frustrated at not contributing more to the company, there was also a growing awareness of dissatisfaction on the part of IBM's senior executives with Research's performance—a dissatisfaction that would become all too apparent in the next few years.[41]

The second concern was the staying power of the silicon planar technology. Research's and CD's division of labor assumed a rather orderly and rapid transition from silicon to gallium arsenide and other new semiconductor materials. In early 1963 Landauer wrote that the development of transistor technology placed "severe standards on competing technologies."[42] Later that year he was more specific, stating: "The planar silicon technology has been successful, and sets a high standard for competitive technologies. A close contact with this work is essential, otherwise Research operations may be in danger of competing with obsolete technologies."[43] Up to that time Research had essentially no contact with that work. The entire ethos of postwar research was that science would produce revolutionary new technologies. If silicon transistor technology were to hold new technologies at bay, Research's agreements with CD would doom it to irrelevance.

The third concern was the company's lack of effort in integrated circuits compared to the semiconductor industry's. The commitment of CD to SLT left few resources and almost no incentive to pursue development work in integrated circuits, which could threaten SLT. In an undated memo, apparently written in late 1962, Rolf Landauer expressed concern over the long-term implications of the integrated circuit. While he acknowledged that IBM's SLT was reasonably competitive with current integrated circuits, IBM was "less well protected" in the "highly integrated circuit." Efforts by others in integrated circuits, Landauer claimed, "will be pushed harder, and will be followed by the attempt to integrate whole systems on one sheet of circuits. The paucity of really good engineering solutions will not stop this development, the Department of Defense can pour money into the forced solution of poor ideas and bring about their successful development."[44] By the fall of 1964, Landauer, while not questioning the decision to produce SLT, claimed that commitment to SLT over integrated circuits had become a "religion" that extended even to CD's development activities.[45]

Landauer provided an important bridge between the worlds of practical electrical engineering and world-class solid-state physics. Although a theoretical solid-state physicist with a Ph.D. from Harvard, Landauer also had electrical engineering sensibilities and experience. At Harvard he had taken one of his qualifying exams in the area of vacuum tubes and circuits. He worked summers at Sylvania doing tube work, and during World War II he had studied to be a radar repair technician in the Navy. Landauer joined IBM in 1952, before it had an independent research organization, initially working under Lloyd Hunter. In addition to managing the injection laser effort, he published a number of significant papers in solid-state physics. But in 1963 he was inventing electronic circuits whereby MOS transistors could be used as memory elements.[46]

Landauer's attendance at the 1962 International Electron Device Meeting and the 1963 International Solid-State Circuits Conference was most likely aimed at keeping track of the work of other research laboratories on injection lasers. But it was the work on MOS transistors presented by RCA and Fairchild that galvanized him and the other research managers to action. The MOS transistor was a simple device to fabricate, and extrapolating from the work of RCA and Fairchild suggested that it would be possible to put very large numbers of transistors on a single chip. The program planned by Landauer and other research managers in response, which became known as the Large Scale Integration (LSI) program, was comprehensive, spanning Research departmental boundaries (appendix table A.1), and aimed at exploring the implications of integrating large numbers of circuits onto a single chip or sheet. The work would address questions of new machine organizations required to take advantage of these new chips. New methods of design automation would be required as well as methods for interconnecting the circuits on a chip. The program would be staffed by roughly fifty people, equally divided between professionals and technicians.[47]

The MOS transistor represented the ideal entry vehicle for Research into the silicon planar technology. Because of the large bipolar transistor effort in place for SLT at CD there, East Fishkill had practical jurisdiction over the silicon bipolar technology. On the other hand, for CD, the MOS transistor was not a strategic technology—it was too slow, with applications too far out in the future. CD had done work in MOS transistors, but this work had come about more because of circumstances than because of any firm commitment to the transistor. When Owen Hill was ordered off integrated circuit work, he and his group turned to MOS transistors, which had just been announced by RCA. Hill had a small effort, producing individual MOS transistors, not integrated circuits. Another group in CD made MOS transistors as a way to study effects important in bipolar transistors. Within a short time CD stopped this work and Hill turned to work on extensions to SLT.[48]

But if CD saw MOS transistors as being too far off in the future to merit its attention, for Research MOS transistors were closer to realization than other technologies it had worked on. In many ways the MOS transistor represented an evolutionary advance over the silicon bipolar transistor of SLT and took significant advantage of the technological infrastructure that had built up around the silicon bipolar transistor, which was largely unfamiliar at Research. One of Research's first moves when it started the program was to send an engineer and a technician to East Fishkill to learn the silicon planar process.[49]

In launching the LSI program, Research made a move into CD's territory. Research conceived of its LSI program independently of CD and initiated it out of its own funds, without getting explicit support from either CD or the corporate office. Landauer expected that within two years Research would have made enough progress to convince CD to launch its own program.[50]

But by mid-1963 CD was challenging Research's right to conduct its LSI program. In 1959 IBM had established an R&D board, which consisted of senior technical managers who oversaw and coordinated IBM's technical work. In August 1963, J. Earl Thomas, manager of solid-state development within CD, argued before the R&D board that the silicon planar technology should be centralized within CD, with all such work taken out of Research. Thomas further asserted: "As far as the division between Research and Components is concerned, the general principle should be followed that devices with relatively well understood design theories which are to be fabricated from the familiar materials germanium and silicon should be worked on exclusively in the Components Division. If the design is familiar and the material is not, or if the material is familiar and the design is not, the job is a logical one for Research."[51]

"It is unreasonable," Thomas claimed, "for the Research Laboratory to concern itself with problems which are, essentially, of an engineering nature."[52] He called for the work on the silicon MOS transistor to be placed in CD and suggested that Research concentrate on MOS or other field-effect transistors made from other materials. Landauer, believing engineering work to be crucial to Research, successfully fought Thomas's proposal.[53]

Thomas, who had previously worked at Bell Labs, was suggesting something similar to Bell Labs' R&D model. Bell maintained a very strict differentiation between research and development, with research, made up primarily of scientists, focused on the most basic of work (such as that leading to the invention of the transistor), while development, with a much higher proportion of engineers, focused on converting the work of research into products. But the historical and organizational factors that enabled this division of labor to work at Bell Labs were not present at IBM. Bell Labs' status as the nation's leading industrial research laboratory, with an impressive list of achievements that had improved telephone service, ranging from a workable vacuum tube to the transistor, gave it a position of authority. Any failures in research efforts were diluted by previous successes. Bell Labs was organized to include both research and development, so they could be more effectively coordinated. And even if research programs failed, the organization remained the source of more predictable development work. IBM Research had only the riskier side of R&D. With development in the divisions, were it to forswear engineering work, it could be dooming itself to irrelevance.

Not only did IBM Research reject Thomas's proposal that it halt work on silicon technology, in 1964 it began other programs of an engineering nature, a move that was prompted by the controversy accompanying the announcement of SLT. The 1964 introduction of System/360 provided opportunities for critics of SLT both inside and outside the company to proclaim that it was an obsolete technology compared to integrated circuits, leading IBM's senior management to fear that in its first foray into semiconductor production, it had invested hundreds of millions of dollars in an outdated technology. While some of the criticisms of SLT were motivated by more than the objective facts, what was true was that CD's focus on SLT had resulted in IBM being behind in its efforts to develop newer semiconductor technologies.[54]

Research itself faced severe criticism. If IBM was technologically behind, it stood to reason in the minds of IBM executives that Research bore a major share of the responsibility. In November 1964 Thomas Watson Jr. wrote to his brother, Arthur Kettering Watson, a senior vice president at IBM with responsibility for Research:

> At your convenience I would like to talk with you and Dr. Tucker about the way Research has been organized for the past few years and specifically who has been in charge of the major areas of our future salvation; i.e. circuitry, printing, memory, etc. It seems to me that anyone who has been in charge of these areas over a fairly lengthy period should now be removed and get an opportunity to run their own labs, and that we as a company should get an opportunity to try someone else. Unless I am wrong, we have been able to create nothing of a substantial nature in Research for our investment of dozens of millions of dollars over the past ten years, and I think in addition to redirecting our efforts here we must have some managers who will say that no matter what the level of investigation and research we should have developed some practical thoughts from our investment.[55]

Although the LSI program launched the year earlier was just such an attempt to develop "practical thoughts," it was still in its early stages.

Research managers realized that the autonomy they were granted at Research's formation could be a curse as well as a blessing, for it had no direct connection with what went on in the rest of the company. In August 1964, Gardiner Tucker, the director of Research, wrote to A. K. Watson about the problems Research faced. He identified one as a "discrepancy between the accountability and the authority of Research." "Tom Watson has made it abundantly clear during recent months that he holds Research accountable for the modernity of our product technology," Tucker said, but at the same time Research had "no direct way to influence the decisions of product development regarding technologies."[56]

The MOS transistor promised to deliver a slow-speed technology, and IBM's most pressing needs were for high-speed circuits. A chastened Research made efforts to mix it up with CD and contribute in this area, first by participating in CD's high-speed integrated circuits program. A Research manager justified its involvement in this development program by claiming the criticisms Research had faced for its lack of contributions to SLT required its involvement in applied and technological programs to prove its value to the corporation.[57]

In 1964 Research also started a germanium program aimed at developing an integrated circuit technology with a higher speed than possible with existing silicon transistors. Electrons move faster in germanium than they do in silicon, and germanium had the added advantage of being better understood than such compound semiconductor materials as gallium arsenide, which suggested that a germanium program would yield results sooner than a gallium arsenide program. This work was conducted with the support of CD and eventually was staffed by seventy-five people.[58]

The net effect of Research's involvement in both the CD integrated circuit program and the germanium program was to overshadow the MOS work within Research. IBM's central concern as a manufacturer of large computers was with circuit performances at the highest speeds practically obtainable, and these programs addressed this concern. If they succeeded technically there would be a clear case for using them in products. The MOS transistor was much slower than either of the bipolar technologies, which would make it impractical for use in System/360 follow-ons. Research envisioned MOS as a low-end technology with potential use in terminals and peripherals. In retrospect, had Research's MOS program not been established in 1963, by 1964 concerns of relevance to the company might have dictated against supporting the MOS program, in favor of bipolar technologies.[59]

But Research embraced the MOS technology and undertook an active role in promoting it throughout the corporation. While Research believed the technology could have extremely important applications to IBM, it had no large constituency outside of Research. In 1963 Landauer and a colleague went to IBM's Office Products Division in Lexington, Kentucky, to describe Research's work to a group interested in developing an electronic calculator. On returning from this visit Landauer proposed several new ways to use MOS transistors in small computers or calculators.[60]

As part of Research's efforts to win support for the LSI program, it made an unsolicited proposal to the Air Force in October 1964. The proposal, for "A Research Program in Large Scale Integrated Circuits," was aimed at winning the program greater resources at a time when Research was under budgetary constraints. Some in Research hoped that a government contract would force groups within IBM to take MOS technology more seriously. In 1964 the Air Force responded, requesting proposals from IBM and other semiconductor and computer companies. Ultimately, IBM management decided not to submit a bid to the Air Force, which awarded research contracts for LSI to TI, RCA, and Philco-Ford.[61]

Although by 1965 Research's MOS/LSI program was well established and had made significant progress, it remained marginal to the central needs of the corporation. In October 1965, Dan Doody, who had been the engineering manager for the System/360 Model 75, the highest-performance System/360 yet announced, presented the Corporate Technical Board with his proposed strategy for IBM's semiconductor work after System/360.[62] At this point CD had lost its separate identity and was part of the Systems Development Division; thus, a systems development manager (a user of IBM's semiconductors) had direct responsibility for the semiconductor strategy. He had an unparalleled understanding of IBM's computer systems and their relative importance to IBM. He was primarily concerned with large, high-performance computer systems, recommending that the chief focus for integrated circuit work be a bipolar logic technology having circuit speeds in the 4–6-nanosecond range, which would satisfy "the bulk, if not all, processor requirements." Doody further recommended a "strong, coordinated development effort to provide subnanosecond technology." These higher-speed circuits were crucial to give IBM a foothold in the area of scientific computing, which, although not the most important area in terms of revenues, was the most prestigious area and one in which IBM had visibly lagged in the past.[63]

Even when he turned to lower-performance systems, he did not find MOS attractive. Here Doody claimed that in low-cost products, the semiconductor portion of the cost represented a small total of the whole, so that an extrapolation of the higher-performance semiconductor technology provided the best solution. Since IBM needed the bipolar technology for the high-performance area, it made sense to modify that technology for use in low-cost areas rather than support an entirely new semiconductor technology. Doody acknowledged that MOS offered the prospect of lower cost but estimated the potential savings at only $1 million per year. He recommended that IBM terminate its MOS effort. Considering IBM's business, it was a perfectly rational recommendation.[64]

At its October 1965 meeting, the Corporate Technical Board adopted his recommendation with regard to MOS technology. The minutes of the meeting state: "After considerable discussion, and over Dr. Tucker's [the director of Research's] objections, the Board took the position that Research will reduce their FET [MOS] effort to a minimum holding action. Dr. Fubini or Research may wish to bring this decision back to the Board for review at a later date."[65]

The Corporate Technical Board's action was a test of Research's commitment to MOS. If it did not believe in the potential of MOS, this offered a convenient time to cut its losses and get out. But as Gardiner Tucker's objections show, Research did believe in the MOS technology, from the working-level engineers and scientists, all the way up to its director. Research refused to stop its MOS program. In February of 1966, the secretary of the Corporate Technical Board wrote to Tucker noting that Research had not stopped its MOS work and asked him if he intended to bring the matter up again. The Corporate Technical Board had only advisory powers within IBM, with no enforcement mechanism. Top management at IBM did not intervene to enforce the Corporate Technical Board edict, and the Research program continued unabated. While Research had established the right to have an engineering program that infringed on CD's territory, the question remained as to whether it had the necessary engineering skills to do the job and whether MOS technology would ever prove of any value to IBM.[66]

THREE

Development at Research

IBM Research's managers had argued for conducting an MOS program that blurred the distinction between research and development, which was how the program was carried out at the working level. Research's most important contribution was not its scientific skills but its role as an incubator for the nascent MOS technology. At a time when the immature MOS technology was no match for the bipolar technology then being mass-produced by IBM's Components Division (CD), a group of scientists and engineers at Research became committed to MOS technology and displayed great flexibility in advancing its cause. The MOS transistor suffered from stability problems, which Research addressed not only through scientific research, but also by transferring technology from IBM's production facility and by adding new technological capabilities of an engineering or empirical nature. At the same time Research searched for applications in which MOS technology might prove to be an effective competitor against bipolar technology. Because of its small scale and primary commitment to MOS technology, Research proved to be an efficient organization in searching for applications for MOS technology within IBM. It sought customers and was willing to reorient its program when the chances for success seemed greater in another area. In spite of Research's efforts, it was extremely difficult

TABLE 3.1 KEY MEMBERS OF THE RESEARCH LSI PROGRAM WITH PREVIOUS
EXPERIENCE WITHIN IBM

	Year Joined IBM	Discipline	Degree	Previous Experience
D. Rosenheim	1951	EE	M.S.	701, tunnel diodes
S. Triebwasser	1952	Physics	Ph.D.	Electroluminescent photoconductors
D. Seraphim	1957	Metallurgy	Sc.D.	Cryogenics
G. Cheroff	1955	Physics	B.S.	Electroluminescent photoconductors
A. Fowler	1958	Physics	Ph.D.	Cryogenics, thin-film transistors
F. Fang	1960	Physics	Ph.D.	
M. Smith	1952	EE	M.S.	NORC, microwave computers
P. Balk	1959	Chemistry	Ph.D.	Electroluminescent photoconductors
R. Dennard	1958	EE	Ph.D.	Tunnel diodes
D. Critchlow	1958	EE	Ph.D.	Tunnel diodes
F. Hochberg	1958	Technician	None	Cryogenics

NOTE: EE = electrical engineering; LSI = large-scale integration; NORC = Naval Ordnance Research Calculator.

SOURCES: (biographical information accompanying published articles) *IBM Journal of Research and Development* 8 (September 1964): 478–80 [Triebwasser, Seraphim, Cheroff, Fowler, Fang, Hochberg]; *IEEE Spectrum,* May 1967, 40 [Smith]; *Proceedings of the IEEE* 57 (September 1969): 1647 [Balk]; *IEEE Transactions on Electron Devices* 31 (November 1984): 1674 [Dennard]; *IBM Journal of Research and Development* 25 (September 1981): 835 [Critchlow]; Don Rosenheim, interview by author, 4 January 1996; George Cheroff, interview by author, 21 August 1995.

to establish the new technology within CD, and MOS technology moved into CD through the back door—a joint program that put very few demands on CD. A key factor in the establishment of this program was the advocacy of the corporate staff. The joint program was by no means a commitment to use MOS technology, but it kept the question open, allowing continuing advocacy and technical progress.

At the time IBM's Large Scale Integration (LSI) program started, IBM Research had technical staff members with international reputations in semiconductor physics, but they were not part of the program. These researchers by and large stayed in programs that were considered pure or basic research. Given the difficulties IBM had had in its previous solid-state applied research programs, it is not surprising that none of the LSI program members came into it with established reputations in either the company or the industry at large. The researchers who worked on IBM's LSI program had much in common (table 3.1). Most had joined IBM in the buildup of personnel that had come with the establishment of a separate research organization within IBM in 1956 and had worked on one or more of Research's failed programs. Some had worked on the cryogenic computer, some on the microwave computer, others on the electro-luminescent photoconductor program or tunnel diode program—all programs that had begun with a great deal of promise but that had ultimately failed to prove competitive against existing technologies. Most of the members of the LSI program were in their middle thirties or younger, and one suspects that as they approached the period that would normally be the peak of their careers, there was frustration at not having been involved in a successful program.[1]

Backward Technology Transfer

Because of Research's previous work on revolutionary technologies, very few in the LSI program had experience working with silicon semiconductor technology. Research was able to obviate this lack of experience by taking advantage of work done by CD, fifty miles north of the Yorktown Heights research facility in East Fishkill and Poughkeepsie, New York, where IBM was rapidly establishing one of the largest semiconductor manufacturing facilities in the world. By the time Research started its MOS program, CD already had over nine hundred people working on Solid Logic Technology (SLT), the silicon planar technology used in IBM's System/360 series of computers. In an industry where much was learned by doing, a large production facility was a fount of new knowledge and techniques.[2]

George Cheroff, a physicist in the Research LSI program with responsibilities for processing technology, and Fred Hochberg, a technician, spent time in CD's Poughkeepsie facilities learning the silicon planar process and the techniques of photolithography, oxidation, and diffusion. Hochberg spent roughly four months working in Poughkeepsie and then transferred this technology back to Research. MOS technology, Landauer wrote late in 1963, "is still an area in which Research is receiving information from CD at present, but we hope to reverse the predominant direction of this flow."[3]

Because SLT used a silicon planar transistor that contained an MOS structure, even though CD had little interest in the MOS transistor, the MOS structure was of vital importance to it. CD was the site of work that was crucial to the production of MOS devices. Herb Lehman, a young chemist in the group developing SLT, had already begun making MOS transistors—the first made in

IBM—as a vehicle for studying the causes of leakage currents in bipolar transistors, an important problem in SLT. It was in Lehman's area that Research personnel were trained in the silicon planar technology.[4]

The strength of CD's work on silicon surfaces and MOS structures can be seen by the work IBM presented at the 1964 Solid-State Device Research Conference. This conference, sponsored by the Institute of Electrical and Electronics Engineers (IEEE), was the first in which MOS structures took center stage. Out of seventeen papers presented on MOS topics, five were from IBM, the most of any firm. An RCA engineer's summary of the papers given by competitors stated that IBM's work was "outstanding." Of IBM's five papers, the most significant four were given by CD.[5]

The IBM paper that attracted the most attention (the RCA engineer said that with it, IBM "dropped a bomb") discussed the role small amounts of phosphorus in silicon dioxide had in enhancing the stability of MOS structures.[6] This work had its origins in the leakage problems of SLT—problems that were addressed in production through testing. The SLT group developed an accelerated test, stressing transistors for a short time at elevated temperatures and voltages so as to simulate the transistors' performance over a long period of time at normal voltage and temperature. A sample of each day's production was tested under the accelerated test as a way of ensuring quality. Initially there were no failures, but as production increased, failures grew alarmingly in what was called the "two hundred degree disease," after the test temperature. Engineers finally determined that the failing devices had been reworked after having a misaligned metal layer applied. In the process of etching off the faulty metal layer so that it could be reapplied, a thin layer of oxide was also removed. Later work showed that the etched oxide had a phosphosilicate glass (PSG) layer (that is, phosphorus combined with silicon dioxide instead of pure silicon dioxide), which had been formed incidentally during an earlier diffusion step. This PSG layer proved to be essential in limiting the device's leakage current, which was shown conclusively by an applied research group in CD. The applied research group also showed that the PSG layer greatly improved the stability of MOS structures—it kept their characteristics from drifting over time. The crucial fact was that PSG worked. It was not completely clear how it worked, and the initial article describing the effects of PSG barely addressed this question.

The Research group picked up the work on PSG, and it became one of the cornerstones of Research's work to produce stable MOS transistors.[7]

PSG became particularly important to Research's program because of an early decision to concentrate on *n*-channel MOS transistors. Two kinds of MOS transistors can be made, *n* channel and *p* channel. Based on information available at the time, it would have been reasonable to expect *n*-channel devices to be roughly three times faster than *p*-channel transistors, which is in fact what researchers at IBM found. At the same time IBM researchers found that *n*-channel transistors were harder to make than *p*-channel devices—their characteristics were much more likely to drift over time. Undaunted by the potential problems, and without complete knowledge of what it would take to overcome them, researchers and low-level managers at Research decided to concentrate their work on *n*-channel devices, largely ignoring the simpler *p*-channel devices. This decision, made when most semiconductor companies worked on *p*-channel devices, reflected IBM's preoccupation with speed and its much more distant time horizon—it could do what it thought best in the long run, rather than settle for whatever would get a product out the door the most quickly.[8]

Some of the problems of *n*-channel devices were so severe that they threatened to block all work. For a device to be usable for digital logic, it had to have an on state and an off state. Early *n*-channel devices could not be turned off. Until *n*-channel devices could be turned off, Research could not build functioning logic circuits. Fred Hochberg, a technician in the group, noticed that changing the voltage applied to the substrate of the transistor could change the voltage required to turn the device on. This meant that *n*-channel transistors could be turned off by applying a negative voltage to their substrate. By late 1964 George Cheroff had developed a mathematical theory for how the substrate bias worked. The substrate bias allowed Research to build and test *n*-channel MOS transistors while ongoing work addressed the question of making stable ones.[9]

Although researchers and low-level managers continued to enthusiastically pursue the *n*-channel device, upper-level Research managers questioned whether this was the correct approach. On several occasions, higher levels of Research management pressed George Cheroff, the manager responsible for the technology effort to fabricate MOS transistors, and his manager, Donald Seraphim, to abandon the *n*-channel technology and work on *p*-channel devices.

Cheroff and Seraphim placated research management by promising to work on *p*-channel devices, but they by and large continued the program's work on *n*-channel devices. Research management's concerns about the *n*-channel technology came from the concentration of almost all other MOS programs on the *p*-channel technology, which raised questions about IBM's approach. In such a situation there was a need for Research management to be somewhat skeptical, making sure the *n*-channel decision was well thought out, but part of Research management's hesitancy about pursuing *n*-channel MOS came from deeper considerations. IBM Research's position within the company, with the failure of its previous programs and its unproven value to the company as a whole, made Research management less willing to gamble on a direction contrary to that of the industry. But given the recalcitrance of lower-level Research management, upper-level Research management was also unwilling to take the draconian measures, such as the wholesale replacement of the program's management team, necessary to impose its will on the program, and thus the *n*-channel work continued. The question of whether *p* channel or *n* channel was the proper direction was one that would be raised by IBM management time and time again.[10]

This question was revisited because of the difficulty of making stable *n*-channel devices. Although initial efforts to produce a stable *n*-channel device failed, in late 1964, Fred Hochberg proved that stable *n*-channel MOS devices could be made. Hochberg, a technician, was one of the central figures in Research's early work in fabricating MOS devices. Although he had attended only a local community college and had no in-depth mathematical knowledge of solid-state physics, he was behind several of Research's early accomplishments in MOS. He was a talented experimentalist, and his freewheeling style was an important asset in the early days of the program, when there was much new territory to be explored. Hochberg also provided one of the key links to the silicon work at IBM East Fishkill and transferred important technology from there to Yorktown.[11]

In late 1964 Hochberg, using a solid source diffusion technique developed in East Fishkill, made a wafer of *n*-channel devices whose characteristics did not change with time or temperature. However, the phosphorus diffusion source was prone to absorb water vapor, which made reproduction of the process extremely difficult, and Hochberg was unable to replicate his work. The stable wafer, number 3717, became important as an existence proof that showed upper-level management that stable n-channel MOS devices could be made. Devices from the wafer were kept on stress test for a period of years, with their characteristics remaining constant. But researchers were unable to make stable n-channel devices again for almost two years.[12]

Research continued to produce important scientific work on understanding MOS structures, even though it did not lead directly to stable MOS devices. Highly empirical work in CD had shown that heat treatments of MOS devices could enhance the surface conduction and thus the speed of MOS transistors. This work was extended by Pieter Balk, a chemist at IBM Research, who provided a convincing theoretical explanation for how this annealing process worked, isolating the critical role of hydrogen and aluminum. Scientists from both Research and CD collaborated on a study that showed the ⟨100⟩ silicon crystal orientation, not commonly used in the industry, had characteristics that made it advantageous for MOS devices.[13]

The Search for a Product

Research began its work on MOS transistors under the name "LSI program," and although the name remained the same, the LSI program gradually became an MOS program. This change can be seen in two memos written seventeen months apart by Don Rosenheim, the manager in charge of the overall program. In June 1964 Rosenheim wrote to Gardiner Tucker, the director of Research:

> Although many people in the company have tended to think of the Large Scale Integration program as a field effect transistor [MOS transistor] effort, this is hardly the case. The Large Scale Integration concept is fundamentally based upon the following:
> 1. A computer controlled versatile interconnection scheme for devices or circuits on a semiconductor wafer.
> 2. Less than 100% yield of these devices or circuits.
> 3. The possibility of cost reduction through avoiding handling and packaging individual devices or individual low level integrated circuits.
> 4. An advanced design automation system which will allow for computer control of device interconnection, (i.e., fabrication automation).[14]

In this memo Rosenheim claimed that the LSI program was not fundamentally based on MOS transistors—it was about putting large numbers of any kind of transistor on an integrated circuit. The key parts of the program were the design automation infrastructure.

In November 1965, however, after IBM's Corporate Technical Board ordered Research to halt work on MOS transistors, in a memo explaining the importance of MOS technology to IBM, Rosenheim wrote to Tucker: "Silicon integrated circuits using FET's [field effect transistors] (also called MOS transistors) will allow for a significant packaged cost advantage over monolithics which utilize bipolar transistors. . . . All indications are that the monolithic FET technology will be the forerunner LSI technology and it is therefore of prime importance to the IBM Corporation."[15] At this point the primary commitment was not to large-scale integration generically but to the MOS transistor as the vehicle that would make large-scale integration possible. The driving force behind this change was the enthusiasm and commitment that the working scientists and engineers had for MOS technology. They believed in MOS technology and were able to convince their management of its significance. As the Corporate Technical Board considered halting work on MOS transistors, Research managers claimed that in their organization MOS transistors formed a "morale focal point."[16] The organization of the group as reflected in a 1966 organization chart (appendix figure A.1) recognized the MOS transistor as the fundamental unit of importance, and it contained under a single management structure the wide range of skills needed to make the MOS technology into a viable product— from physicists, to chemists, to electrical engineers. All were committed to the MOS technology. While many items were open to discussion and modification, with some questions passionately debated, no one in the group questioned the merit of the MOS transistor itself—that had been established. The MOS transistor held the group together and gave it its identity.

The engineering group's job was to devise ways of using MOS devices in circuits and systems, and specifically, implementing the programmable interconnect pattern (PIP) wiring system. Research had begun the LSI program when integrated circuits were just finding general acceptance, and so managers had made assumptions about the direction of future semiconductor technology. In particular they assumed that as one put more and more circuits on a chip, one would reach a point where the chip would be so large that few chips would have all their transistors (and thus all their circuits) working. The PIP approach attempted to improve the yield by testing all the circuits and then wiring the good circuits together. The freedom this gave from the constraint that the chip had to be 100 percent good to be usable was assumed to offer substantial advantages. Many in the industry saw something such as PIP as essential to progress beyond very small chip sizes.[17]

While many of the working scientists and engineers were relatively new to IBM, having joined since the founding of Research in 1956, the higher levels of management were more experienced. Three of these managers, Don Rosenheim, Merlin Smith, and George Cheroff, did not have Ph.D.'s. Both Smith's and Rosenheim's initial work at IBM had been on projects that were purely engineering. Rosenheim was involved in the development of the IBM 701 Defense Calculator, IBM's initial entry into the field of electronic computing. Smith had worked on the NORC (Naval Ordnance Research Calculator), a one-of-a-kind scientific computer built for the Navy. These managers had a keen appreciation of the importance of engineering, and engineering considerations played a large role in shaping the program.[18]

As the engineering group worked to implement the PIP system, it also made studies to ensure that its work would be competitive with other technologies. A key figure behind this approach was Merlin Smith, the manager of the engineering group. After NORC, Smith worked on the microwave computer, eventually having responsibility for the overall program and finally killing the program. The microwave computer program had proceeded without detailed analyses of factors (such as cost, architecture, reliability, and power usage) that were essential to proving the program's viability to potential users. As Smith came from the microwave program to the LSI program, he put a priority on demonstrating that MOS technology worked in a realistic environment.[19]

A crucial matter for Smith was the position of MOS technology vis-à-vis its competition, the bipolar transistor. Digital logic circuits made from MOS transistors were significantly slower than almost all circuits made from bipolar transistors, and in Smith's mind Research had to do everything possible to make MOS attractive to the rest of the company. The two to three times speed advantage of n channel over p channel was compelling in a company where speed was a primary consideration in all its digital logic. Even though n channel did not result in MOS circuits faster than bipolar, it at least made them more competitive with the lower-performance bipolars. Meeting the perceived needs of IBM was seen as more important than following the industry, which had concentrated on p channel. Smith supported the decision to focus on n-channel devices and clarified why it was necessary.[20]

Smith and the engineers also evaluated the PIP approach against a fixed interconnect pattern (FIP) approach, where the fixed wiring required the entire chip to be good in order for it to be usable. While the benefits of the PIP approach might seem intuitively clear, just as the microwave computer's had, its implementation added complexity that nullified many of its advantages. For example, the PIP approach required two levels of wiring, while the FIP approach used only one. A comparison of the circuit density of the two approaches showed that FIP promised two to three times the density of PIP. The PIP approach resulted in longer wires between circuits and therefore much slower speeds.[21]

The PIP approach, with its heavy dependence on computerization and techniques of design automation, resonated with IBM Research. In the early years of Research, key managers spoke of the need for automation research, which later came to be known as design automation. The effort to build Stretch, a large scientific computer completed in the early 1960s, had included a design automation group. Design automation promised multiple benefits to IBM. In a period of general optimism about what computers could do, design automation offered the prospect of greatly speeding up the process of designing computers. Success would show the value of computerization, winning additional customers for IBM as this work spread. IBM's cryogenic computer program had a design automation group attached to it, and when the program was killed, the design automation group ended up in the LSI program. The PIP program appears to have been launched based on the compatibility of the program and the skills of the people available. In fact, the design automation group brought parts of its previous work into the LSI program. However, design automation was to become less central as the program developed.[22]

As the engineering group undertook studies of PIP, any work comparing costs with other approaches was based to a large degree on paper studies because IBM had no experience manufacturing integrated circuits. Engineers from Research worked closely with people from CD in getting cost data on SLT transistors. Research also received data from CD on the distribution of good circuits within a wafer from integrated circuits being made on a pilot line. These wafer maps showed that it was possible to have large areas within a wafer where all the circuits were good, which contradicted a basic assumption of the PIP approach—that with a random distribution of defects, as the chip area increased one quickly reached a point of near-zero yield. This suggested that the PIP approach might not be necessary.[23]

Questions about the PIP approach were reinforced with the fabrication of the first PIP chips in the summer of 1965. These wafers, built on a crude pilot line with a process that was still being developed, showed chip yields as high as 26 percent without using the PIP approach. Furthermore, the PIP approach assumed that wiring together all the good circuits would be a step with nearly 100 percent yield. But the wiring yield proved to be lower than the yield of building the rest of the chip, which vitiated any advantages of the PIP approach. By November 1965, after consultation with CD, a consensus was reached to drop the PIP approach.[24]

Research's ability to reorient the LSI program quickly and easily came from its autonomy—it had made no commitments to anyone else to carry the PIP approach to completion, and it could discontinue the program without the approval of higher levels of management or outsiders. Research had almost given up this autonomy, however, by making a proposal in 1964 to the Air Force requesting military support of its PIP program. The Air Force responded very favorably to the IBM proposal and put out a generalized request for proposal (RFP) modeled after IBM's. IBM Research prepared a formal bid to the RFP, but in the summer of 1965, higher-level management at IBM made the decision not to bid on the program, based on a belief that the Air Force had little money available and on a desire to avoid sharing IBM's LSI work with the industry at large, as would have been required under the government contract. The very process of writing the proposals had raised further doubts among the engineers about the wisdom of the PIP approach. Had IBM competed for and won the government contract, the momentum it would have given to the PIP approach would have made it much more difficult to reorient the program. Texas Instruments, one of the contract winners, spent three to four years working on its version of PIP, called discretionary wiring, an approach that never proved commercially viable.[25]

With the demise of the PIP approach, Research then focused its attention on developing an MOS logic chip using fixed wiring (FIP). This reorientation had the most effect on the design automation group, whose previous work had been most specifically directed by the PIP approach. The FIP program had much in common with the PIP work, and it too would have been inconceivable without the design automation group. "An efficient design automation system," claimed the group's manager, "is in a sense the heart of LSI development."[26]

The continuing importance of design automation resulted from basic economic problems that were perceived in any implementation of large numbers of logic circuits on a single chip. When each integrated circuit contained only one or two logic circuits, it was possible to build up a large digital system from these primitives using only a handful of different chip types. But when each chip contained upwards of one hundred logic circuits, as envisioned in LSI, each chip became less generally applicable and more confined to one specific application in a specific system. Since one hundred logic circuits were still a small fraction of the circuits in the overall system, under LSI the system would be made up of many chips, but also many different chip part numbers. Across different systems, there would be very little commonality, so that part numbers could not be shared. This explosion of part numbers, where the semiconductor plant would have to make small quantities of numerous chip designs, threatened to undermine any economic advantage of moving to higher levels of integration. This was compounded by a turnaround-time problem. In small-scale integration, engineering changes were by and large made not at the chip level but at the printed circuit board level. A design was changed by adding a new integrated circuit to (or by taking one away from) the printed circuit board. But with LSI, any logic changes resulted in an entirely new chip design that might require months to go through the manufacturing process and could delay the development of a new system by many months or years.[27]

To address these issues, Research developed a master slice. All designs would be exactly the same through the final level of metallization, where personalized wiring would create different logic blocks and interconnect them according to the design of the specific part. For most of the production process commonality would be maintained. Turnaround time would be reduced, as changes required redesigning only a single layer. To further speed changes, Research developed an automatic mask generator, which under computer control generated the mask for the final metal layer. The design automation group developed computer programs that would place and wire circuits on a chip, partition the logic design of a digital system so that it could be efficiently built using LSI chips, and control the operation of the mask generator. Research designed and built a fifty-five-circuit chip using these techniques.[28]

As Research worked to implement FIP, it also sought customers for MOS technology. The interest of systems development groups was a necessary if not sufficient condition for the establishment of MOS technology within IBM. Research held several LSI users meetings, describing its work for systems designers who might be interested. By 1966, Research was working with IBM development groups in Raleigh and San Jose to provide MOS circuits for them. Research had also identified sixteen different groups within IBM that had contacted Research and "expressed interest" in MOS technology. The systems represented varied from a biomedical analyzer to an electronic cash register, a terminal controller, and a large computer system. The most common reason listed for interest in MOS technology was low cost, but there was no common circuit application in these systems, which required functions such as shift registers, control memory, logic, and analog-to-digital converters.[29]

Even with this interest, the fate of MOS in IBM rested ultimately with CD, which had the authority to decide whether to produce MOS. In 1966 CD in East Fishkill established a group responsible for making LSI into a product that could be used in IBM's low-cost, low-performance systems. CD had not decided whether its first entry into LSI should use bipolar or MOS technology.[30]

IBM's semiconductor business operated according to a different calculus than the rest of the industry. A merchant semiconductor producer with a flexible manufacturing system could take a chance and introduce an MOS integrated circuit on a small scale, increasing production later if demand warranted. In contrast, IBM's semiconductor operations were oriented to the high-volume production of a small number of basic types of semiconductors. IBM's orientation to mass production made the development of a new technology extremely costly. CD was therefore reluctant to produce a new technology unless it had assurances that it would be used in large volumes. On the other hand, although systems development groups had expressed an interest in MOS, no responsible systems development manager would commit to using MOS unless CD had committed to producing MOS on a schedule. And unless systems development groups had experimental or preproduction MOS parts to evaluate in their systems, they would be unable to determine if MOS technology had any benefits for their systems. This system, which tended to minimize risk, made the injection of new, inherently risky technologies difficult.[31]

Furthermore, IBM's systems organizations were by and large captive users of IBM's semiconductor technology. If CD offered only bipolar technologies, that is what IBM's systems would use. CD would base its selection of LSI technology on what was satisfactory, not on what was optimum. There were no important systems in development that absolutely needed MOS technology in order to be viable. For managers in East Fishkill, going to LSI was itself a risk; they had no desire to compound that risk by using an unproven transistor technology.

By June 1966 CD had decided that its initial work on LSI would be confined to bipolar technology, causing some in Research to believe that their work on MOS would be ignored, leaving Research with another failed program. In July 1966 the manager of the Research LSI program, Sol Triebwasser, wrote to his manager that he was apprehensive "that the content and perhaps the identity of the LSI program" would be "lost very quickly upon transfer of the program to CD." Triebwasser suspected that CD wanted to extend an existing bipolar program and call that LSI. He closed by claiming: "It is going to take some doing to protect the value of the work we have done."[32]

Friends in High Places: MOS and the Corporate Staff

In August 1966 Gardiner Tucker, the director of Research, called for an audit of the LSI program. Although this audit was not primarily intended to advance the MOS work, it ultimately had that effect. Tucker had established a policy that all Research programs be audited before they were transferred out of Research, so that a formal evaluation of the worth of Research's work could be made by representatives of both Research and the division to which the program was being transferred. Such an audit would provide Research protection from claims that its work had been of no value—even if the work never made it into production.[33]

The audit committee consisted of personnel from Research (although not from the LSI program), CD, divisions that would be users of LSI, and the corporate staff. The staff representative was Americo DiPietro, who had responsibility for semiconductors. The corporate staff was a watchdog, ensuring IBM's competitive position while it also worked to educate IBM's top management, most of whom did not have technical backgrounds. The staff, based at IBM's corporate headquarters in Armonk, New York, held power to the extent that it won the confidence of IBM's top executives; otherwise, it risked being an annoying irrelevance.[34]

DiPietro came to his corporate staff position in January 1966 with a background that perfectly suited him to the oppositional role that the corporate staff might be expected to have. DiPietro had a Ph.D. in physical chemistry from Cornell and had worked at General Electric's (GE's) semiconductor operations prior to joining IBM. At GE, DiPietro got an indoctrination into how the merchant semiconductor industry worked, which was by following closely on the heels of the firm that had established itself as the industry leader. In the early 1960s this was Fairchild, and DiPietro reverse-engineered Fairchild parts so that GE could produce copycat versions.[35]

In 1963 DiPietro was hired by IBM's Federal Systems Division (FSD) in Owego in order to establish an integrated circuit capability there. Although IBM had decided not to use integrated circuits in what would become its System/360 family of computers, the military's enthusiasm for integrated circuits forced FSD to develop integrated circuits to compete for military contracts. As DiPietro interviewed for the position, he learned that IBM's commercial division was not planning to use integrated circuits, a move he believed to be a serious error. In joining IBM Owego, DiPietro began what would be more than a decade-long crusade to reorient IBM's semiconductor effort to make it more consistent with the direction taken by the rest of the industry.[36]

Although Owego was in a separate division from IBM's SLT effort, DiPietro's work on integrated circuits put him in the center of controversy, for it posed a threat to CD, which had no control over his work. Managers at CD managed to have DiPietro's work put under their chain of command, even as he remained in Owego. With this takeover, DiPietro was unable to pursue his own direction and became embroiled in a series of battles with his managers.[37]

In spite of DiPietro's battles with CD—or perhaps because of them—DiPietro increasingly came to the attention of IBM senior management. DiPietro became one of IBM's key non-CD consultants on questions related to SLT. In 1964 he made a presentation to a task force considering whether integrated circuits represented a competitive threat to SLT. In 1965, when CD began having trouble manufacturing SLT transistors at reasonable yields, DiPietro was called to serve on a task force. DiPietro, partially capitalizing on his work at GE, played a key role in identifying and correcting the source of SLT's yield problems.[38]

DiPietro was a voice calling for IBM to move closer to the rest of the semiconductor industry by following industry-standard practices. IBM had set off on a course different from that of the rest of the semiconductor industry, choosing to implement a hybrid circuit technology at a time the industry was moving to monolithic integrated circuits. When System/360 was announced, the negative reception given to SLT by the outside technical community set off a firestorm of controversy within IBM. Several competitors announced computers using integrated circuits, and it appeared that IBM had bet on the wrong technology. CD and Research were put on the defensive for their lack of work on integrated circuits. The president of CD was removed from his position, and CD itself was abolished. IBM created for use by its sales force a definition of the term *integrated circuit* that was broad enough that it encompassed SLT. In *Time,* IBM ran advertisements for System/360 describing the company's latest research accomplishments, even though none of them were part of System/360. The advertisements then went on to call SLT an "advanced integrated circuit" and assured the potential customers among *Time* readers that IBM had the framework to solve their problems, "no matter how we send electrons through circuits."[39] Rolf Landauer, the head of Solid State Science at IBM Research, analyzed IBM's underestimation of integrated circuits in the following way: "The semiconductor industry as a whole, unified in its direction by the availability of government money, constituted a development force which far exceeded in its total ability that of any one of its constituents and caused progress which exceeded the expectations that almost anyone in IBM ever expressed."[40]

Landauer's analysis was trenchant, but the long-term lesson IBM took from SLT was not that it was dangerous to move in a direction different from that of the rest of the semiconductor industry, but just the opposite—the semiconductor industry could be ignored with impunity. IBM's System/360 was an enormous success. The RCA computers using integrated circuits proved not to be serious competitors. Ultimately, IBM's semiconductor technology had little to do with whether a customer chose IBM equipment. More prosaic factors such as IBM's sales and service organizations, and IBM's abilities to fit its systems to a particular task, were much more important. In October 1965 Thomas Watson concluded that the decision to use SLT had been correct, and in March 1966 CD was reestablished, with John Gibson restored to his division presidency.[41]

Although there was a brief period when it might have been possible to implement the reforms DiPietro wanted, many factors militated against such change. CD's course had substantial momentum based on its infrastructure and the hundreds of people who had developed SLT. But fundamentally, IBM's position as a vertically integrated corporation producing semiconductors for internal use gave it a freedom merchant producers did not have. Merchant semiconductor firms' very survival depended on an awareness of what others were doing and an ability to quickly copy competitors' successful products. Moreover, merchant firms had to be able to compete on price, matching cuts with competitors during price wars meant to drive less efficient firms from the market. IBM's semiconductor operations were controlled by the visible hand of IBM management and almost completely isolated from the invisible hand of market forces in the semiconductor industry.[42]

DiPietro's beliefs and outspokenness circumscribed his opportunities at IBM. Argumentative by nature, in less than three years he had already made himself anathema to many within CD. CD offered him a job, but the offer seemed primarily designed to co-opt him. Given his fundamental opposition to CD's direction, any position there would have frustrated him. In early 1966, based on DiPietro's high-profile semiconductor work, J. A. Haddad, the director of technology and engineering at IBM, offered DiPietro a position on his staff. Haddad, one of IBM's most senior engineers, had joined IBM in 1945, led the development of IBM's first entry into electronic computing, and then held several division presidencies. The staff position was ideal for DiPietro, for while he had no chance of changing CD from within, he had a position of potential leverage if he could win Haddad and IBM's other top executives over to his views. Within six months of joining the corporate staff, DiPietro wrote a proposal to the president of CD suggesting that IBM develop an integrated circuit technology that would be largely compatible with industry-standard approaches and would lend itself to evolutionary improvements. The proposal received an icy response from CD and no serious consideration.[43]

In addition to his views on industry-standard approaches to integrated circuits, DiPietro came to the corporate staff with settled views on the MOS transistor. Part of DiPietro's work at Owego had involved visiting and evaluating other semiconductor manufacturers seeking to sell products to FSD. In 1964 DiPietro had visited General Microelectronics, which had developed a 20-bit MOS shift register it was seeking to sell to IBM. There DiPietro met Frank Wanlass, the originator of MOS work at Fairchild. DiPietro was well aware of the stability problems in MOS devices and asked Wanlass for proof that these problems had been solved. Wanlass was able to convince DiPietro that these problems had been addressed, and DiPietro became a strong advocate of MOS technology.[44]

As the LSI audit committee met, DiPietro was alone among his colleagues pushing for MOS transistors. The committee endorsed CD's decision to concentrate on bipolar technology, stating: "The process for fabricating IGFET's [MOS transistors] is still undergoing iteration, consequently, reliability, uniformity, reproducibility, and yield are not established."[45] The committee expressed concern over Research's inability to make stable MOS transistors. The committee added: "Industry is leading IBM in making LSI devices" in MOS technology.[46]

While Research's decision to concentrate on n channel rather than p channel had slowed it down, resulting in a technology that could be dismissed as not yet ready for development, there was also no demonstrated need for MOS circuits within IBM. DiPietro argued for the inherent importance of MOS technology, ignoring the details of the Research program. He presented data from General Microelectronics showing that MOS transistors could be made to work reliably. Emerson Pugh, the chairman of the committee, recalled that initially DiPietro was the only person pushing MOS technology. Both DiPietro and Pugh remembered DiPietro as being very strident in his fervor for the MOS transistor. DiPietro was finally able to win assent for a statement in the audit report that asserted that the importance of the MOS technology to IBM had not been established and recommended that "Research and Components be charged with making a joint evaluation of the importance of IGFET technology across the whole of IBM's product line by the end of March 1967."[47] While DiPietro had succeeded in keeping the MOS question open, the audit committee's recommendation was a painless one for CD; it was required only to study MOS, not to do anything. It did not threaten East Fishkill's basic decision to implement bipolar LSI.

The Search for Stability

In spite of the fact that the three inventors of the transistor were among the first Americans trained in quantum physics, semiconductor processing in the 1950s and 1960s could not be fully contained by theoretical science. At its center was a large degree of empiricism and a heavy reliance on technique. By 1966 IBM Research had published a number of papers on MOS physics and made at least as many significant discoveries related to MOS devices as any other laboratory in the world. But IBM Research still was unable to produce stable *n*-channel devices. While their work was important, the scientists who published in the leading technical journals could take the program only so far. Ultimately, engineers and technicians had a big role to play. That Research managers recognized this can be seen through their personnel decisions.

In 1965 efforts at processing MOS devices in IBM Research entered a second phase aimed less at finding new scientific phenomena than at implementing MOS technology in such a way that it would appear credible to CD. The foremost requirement was a reproducible process that would yield stable devices. A sign of this new phase was Fred Hochberg's removal from the program. Hochberg had been the prime fabricator of devices in the early stage of the program and had been responsible for several key breakthroughs. His freewheeling experimental style, in which he made frequent undocumented changes to the fabrication process, was replaced by a more methodical approach. At this time several people joined the program who did not have typical Research profiles. They lacked Ph.D.'s and were not expected to perform leading-edge research or publish papers. They were hired for their hands-on experience in processing silicon. Ernie Wurst, who had a bachelor's degree in physics, joined the LSI program after having worked at Philco and Texas Instruments, while Joe Shepard, a chemist, came to Research after working for Philco and Singer Kearfott.[48]

IBM Research's route to stable MOS transistors shows the importance of both scientific research and technique. In the second half of 1966, Research made several key advances, none as significant scientifically as some of its previous work, and most not of the sort to warrant publication in a scientific journal. But it was through this work that stable devices came to IBM Research. The central aspects of this work included the development of a thermodynamic understanding of the principles of PSG, the work of a technician in making stable devices without PSG, and the acquisition of new process equipment.

In 1966 Jerome Eldridge joined the IBM Research group after receiving his Ph.D. in metallurgy and materials science from New York University. Shortly after he arrived, Eldridge was assigned to work on PSG, the film that had produced the only stable n-channel devices seen at IBM to date. Earlier that year Ed Snow and Bruce Deal of Fairchild had cast doubt on the efficacy of PSG layers in producing stable MOS devices. They showed that while PSG layers were effective in trapping sodium ions, which they had previously reported as being a key contributor to instability, at the same time PSG itself introduced polarization effects. These effects, they claimed, "lead to MOS devices less stable than those which can be made using pure uncontaminated SiO_2."[49]

Snow and Deal could afford to be skeptical about the efficacy of PSG because Fairchild, working on p-channel MOS, could make stable devices (in the laboratory) without it. IBM's only success in making stable n-channel MOS devices had come through the use of PSG. It became Eldridge's job to try to balance the good effects of PSG (its trapping of sodium) against its deleterious effects (its polarizability, which led to instabilities). One of the first things Eldridge did was to substitute a $POCl_3$ diffusion process for the solid source P_2O_5 diffusion. The $POCl_3$ diffusion system, commonly used elsewhere, was much more reproducible than the P_2O_5 diffusion source. Eldridge then made a thermodynamic study of the P_2O_5-SiO_2 system, in which he was able to show that layers of PSG could be made that had controlled thicknesses and concentrations, making it possible to minimize the polarization effects reported by Deal and Snow, for the polarization was proportional to the thickness of the PSG layer.[50]

Another breakthrough was the ability to grow clean oxides that led to stable structures without the use of PSG. While IBM researchers knew the importance of minimizing sodium, accomplishing this was far from straightforward. Sodium's omnipresence added to the difficulty. Specific diffusion furnaces could be contaminated, as could specific diffusion tubes. If a furnace was contaminated with sodium, the high temperatures used in processing could cause the sodium to diffuse through the tube and into the wafer being processed. Researchers from IBM visited Bell Labs, which had claimed to have a furnace that yielded clean oxides, and they attempted to copy the Bell apparatus. In its attempt to produce stable oxides, IBM Research also used a technique of etching the wafers it processed in a solution of HCl (the idea being that the Na would react with the Cl). In late 1966, a very talented and observant technician in the group, David Dong, was able to make stable n-channel devices using the HCl technique without the use of PSG. Dong's work in producing stable MOS devices was initially confined to one particular furnace; while the HCl technique played a role in producing stable devices, exactly what Dong had done was unclear to his colleagues. Dong's coworkers claimed he had gone on a low-salt diet.[51]

The difficulty in determining what Dong had done was that the work on MOS stability was a very slow process that consisted of many elements that added incrementally to the device's stability. One big element that consisted of many little elements over a period of several years was the establishment of an infrastructure to support the fabrication of stable MOS devices. When Research began its MOS work it had only the most rudimentary facilities available for fabricating integrated circuits. Even before it had been conclusively proven, researchers believed that a clean environment and pure chemicals were even more important for MOS than for bipolar transistors. Research installed laminar flow hoods and air filters to ensure the purity of the room air. Research also developed its own methods for supplying pure water, oxygen, and nitrogen. This work was led by George Cheroff, a second-level manager responsible for MOS physics and processing who in this instance more than made up for his lack of a Ph.D. with his practical experience in plumbing. Cheroff worked closely with the plumbers installing the gas and water lines, directing them in the modifications necessary to ensure the required purity. The development of diffusion processes, particularly the $POCl_3$ process, was another crucial element in device stability. In July 1966 an electron beam system for evaporating aluminum was installed; it was said to be "necessary to obtain stable Si-SiO_2-metal systems."[52] The impetus for the electron beam system was the knowledge that Fairchild was using it with good success, and a comparative study by Research showed its superiority.[53]

In many ways the achievement of stable MOS devices was a triumph of technique and empirical studies over theoretical science. Some aspects of wafer processing simply required empirical study, with theoretical science adding little. In general, the split or divided run, where one or more parameters were varied in the hope of discerning these parameters' effects on stability, was the key tool in the drive to make stable devices. This was true of wafer preparation techniques, rinses used to clean the wafers, metallization techniques, and chemicals used in the fabrication process.[54]

Based on these cumulative developments, by late 1966 Research had become more aggressive in its advocacy of MOS technology. In December, Sol Triebwasser, the manager of the LSI program, wrote a memo to file stating that since June there had been a change in attitude "away from a cautious, limited useful objective attitude to one of asking for more bold approaches." Identifying one key factor contributing to this bolder approach, Triebwasser wrote: "Progress in P. Balk's group on achieving very stable oxides has been such that we can demonstrate that this area of uncertainty has been removed. . . . Basically our own internally derived confidence level that the FET [MOS] technology is viable has risen sharply during the past six months." Triebwasser closed by recommending that the LSI audit committee "be reconvened for the express purpose of auditing the FET vs. Bipolar decision."[55]

The Switch to Memory

When Research began work on MOS transistors, Research management recognized that MOS transistors could be used to build a semiconductor memory, but Research focused its MOS efforts on digital logic. The managers responsible for conceiving the program had no responsibility for memory within Research. In the early 1960s computer memory was largely magnetic, and Research's memory projects were largely staffed by "professional magneticians who had spent their lives working in magnetics."[56]

In 1965 two engineers in Research outside the formal boundaries of the LSI program began working on MOS transistors for use in memories. Lewis Terman joined IBM in 1961 after completing his Ph.D. in electrical engineering at Stanford. Terman, the son of Stanford engineering dean Frederick Emmons Terman, wrote his dissertation on the MOS capacitor, which became the fundamental tool in assessing the stability of MOS transistors. Terman initially worked at IBM Research on computer systems architecture but then became part of a large group working on magnetic memories. Terman was part of a team that designed a magnetic memory with an access time of nineteen nanoseconds—much faster than magnetic core memories. Terman and a colleague, Peter Pleshko, then began looking for ways of reducing the memory delays even further and quickly turned to semiconductor memories, particularly MOS memories.[57]

Terman and Pleshko tried a number of ways to implement a memory cell (the circuit configuration that holds one bit) as well as a number of different architectures for the memory. In the middle of their search they heard of a memory cell proposed by Jack Schmidt of Fairchild Research and Development. Although their search continued, the cell they finally settled on was Schmidt's cell. Pleshko and Terman chose a memory design that would yield an MOS memory that was as fast as possible. MOS transistors suffered in speed particularly when they were forced to drive a signal across any significant distance of wire (which meant they were driving a significant capacitance), because MOS transistors generated much less current than bipolar devices. Terman and Pleshko designed a hybrid system, where an array of individual cells was composed of MOS transistors but all the peripheral circuits were on separate bipolar chips. The MOS array took advantage of the simpler fabrication process of the MOS transistor to provide density, while the bipolar peripheral circuits gave increased speed. With this configuration and the use of n-channel MOS devices, Terman and Pleshko were able to report an extremely fast memory with an access time of 12 nanoseconds and a cycle time as low as 35 nanoseconds. (The MOS memory designed at Fairchild had used p-channel devices and reported a cycle time of 2–3 microseconds.) Terman and Pleshko's focus on speed was typical of IBM.[58]

By the late summer of 1965, Terman and Pleshko had informally joined forces with the Research LSI program. In the fall, members of both groups participated in a study of integrated circuits for use in computer memories, which included both integrated circuits used as peripheral circuits in magnetic memories and memories made up entirely of integrated circuits. The report suggested that given the difficulties of competing with the much larger resources of CD devoted to bipolar transistors and Research's position as the site of MOS work within the company, it made sense for Research to concentrate on MOS memories.[59]

Both the LSI MOS program and Terman and Pleshko actively sought customers for MOS memory within IBM. This involved not just disseminating information about MOS memory but modifying the program in response to the needs of a potential customer. Over time the Research group established a relationship with a system development group in Poughkeepsie. One of the innovative aspects of System/360 was its use of control stores (or microprogramming) to facilitate the development of a compatible family of computers. But in its implementation of System/360, IBM had not found a totally satisfactory technology for the control store. Control stores were relatively small and had to provide speeds that were significantly faster than the main memory, which made magnetic core memories impractical. System/360 Models 30, 40, and 50 each used completely different control store technologies.[60]

The systems development group found MOS memory an attractive way to implement control stores, and its engineers worked with the Research engineers. By August 1966 the systems group in Poughkeepsie offered Research a contract to develop an MOS memory. While this contract showed Research's relevance to IBM's main line of business, it also posed a challenge. The Poughkeepsie group wanted firm commitments as to when hardware would be available, commitments that the Research group had never had to make before, as it had not worked so closely with development groups. Furthermore, the contract would have bound Research to one particular application of MOS memory and taken away its freedom to pursue new approaches it thought were promising. In September 1966 the LSI group turned down the contract while at the same time promising to work "toward satisfying memory needs with or without a contract" and offering to accept a contract "that was worded in a way consistent with the exploratory program of a kind" it was in fact conducting.[61]

By the summer of 1966 managers in the Research LSI program increasingly found interest in MOS memory within IBM and also realized that CD had no intention of picking up its work on MOS logic. In July 1966 Merlin Smith, the manager of the engineering group within the LSI program, wrote a memo describing the reorientation of the program towards memories. Smith's memo suggested that he was not reorienting his program completely voluntarily but that the press of circumstances left him little choice in his effort to establish the MOS technology within IBM. By the end of the year Research had designed and built a forty-bit MOS memory chip.[62]

The reorientation of the Research MOS program into semiconductor memory brought it into an area where IBM already had major innovative activity under way. While CD's existing semiconductor memory program was in some ways in competition with Research's MOS memory work, the two programs were more complementary than competitive. In 1964 Bob Henle, one of the first-generation transistor circuit designers in IBM, was appointed an IBM Fellow, giving him the freedom to work in an area of his own choosing. Henle, who worked in CD, chose semiconductor memories, which had been discussed in the industry before but had yet to be reduced to practice. In January 1965 Henle collaborated on a report, "Potential for Monolithic Megabit Memories," that proposed a three-phase bipolar semiconductor memory program. That same month, CD began implementing a slightly scaled-back version of Henle's program, first producing a sixteen-bit chip for use in the storage protect circuit for the high-end System/360. The storage protect element was a perfect vehicle for semiconductor memory—a very small memory that would be impractical to implement in core, requiring speeds that could be attained by only semiconductors. Phase Two of the program was a sixty-four-bit memory chip qualified in 1967 that was used to implement a cache memory, an architectural innovation whereby a small, fast memory allowed for enhanced performance of the overall computer. Two keys to the program's success were that it scaled up gradually from small memories and quantities to larger ones, and that semiconductor memories were coupled with architectural innovations, which practically required innovations in the memory technology.[63]

As CD worked to develop bipolar semiconductor memory, it met continuing pressure to consider the potential of MOS for both memory and logic. This came not just from the LSI audit committee's request, but directly from the corporate staff as well. One of the corporate staff's leverage points was its sign-off authority on the plans and strategies of IBM's divisions and groups. Every year each division and group produced a strategy or plan detailing areas of future work. Each major constituency within IBM, such as the other divisions, Research, and the corporate staff, had to concur with a division's strategy. Non-concurrences had to be hammered out by the parties involved or be taken to IBM's top management for resolution, a result likely to discredit both sides. In the fall of 1966 the corporate staff had refused to concur with the operating plan of the Data Processing Group because of its lack of effort in MOS circuitry. In response, CD, the division responsible for the particular item, repeated that it had promised to undertake a study of MOS circuits for memory and that shortly after the completion of this study a decision would be made on whether MOS circuits would be needed for memory. (See appendix figure A.2 for an organization chart of the company.)[64]

In early 1967 the corporate staff was joined by the Corporate Technical Committee (CTC) in its advocacy of MOS technology. The CTC, although in a different form, had a function similar to that of the corporate staff. The CTC was headed by IBM's chief scientist, Emanuel R. Piore, and its membership was taken from the senior members of IBM's technical community. It had oversight and coordinating responsibility for all of IBM's technical work. In January 1967 the CTC requested that CD provide an answer within 30 days to the question of whether MOS technology would be needed for either logic or memory. Although both the CTC and the technical staff had merely requested that CD study MOS, the implications were clear—both believed that MOS was important and if the study were to return a negative answer, continued agitation might be expected.[65]

This was the background under which Robert Meade, the chief strategist in the memory group of CD, undertook to study the potential of MOS memory. Meade's original work at IBM had been in systems design, and then in 1964, he moved to CD, where he helped plan the move into semiconductor memories. Meade's primary role had been to bridge work in the systems area and the semiconductor area. After evaluating Research's work, Meade concluded that the potential of MOS for memories warranted a program. He accepted Research figures that MOS memories could cost from 20 percent to 60 percent of what bipolar memories cost. Meade's endorsement of Research's work enabled it to get a hearing in CD that it could not have earned itself.[66]

By this time a rivalry, perhaps a hostility, had developed between East Fishkill and Research, where those in East Fishkill often considered Research impractical, and those in Research often considered East Fishkill too conservative. But beneath the level of name calling were vast differences in the way each did business. The central questions for East Fishkill were ones of production, cost, and reliability. While East Fishkill had a production line with a capacity of up to ten thousand wafer starts a day, making it one of the largest semiconductor production facilities in the world, Yorktown had what it called a pilot line with several wafer starts a day, a difference in scale so great that some in East Fishkill would question the validity of anything done on such a line. Even though Research had produced stable MOS transistors, it was unclear if that work would be directly transferable into large-scale production. Although Research had gone beyond the bounds of its normal practice in considering questions of cost, it would never be able to satisfy East Fishkill with its cost data, which were based on paper studies or work done in its pilot line.[67]

The tensions between Research and East Fishkill were clearly evident in a meeting held in East Fishkill in February 1967 to consider the response CD should give to the CTC on its plans for MOS technology. The fundamental question under consideration was whether work done by Research had any relevance to CD. Sol Triebwasser's summary of the meeting noted the frustration of Pete Fagg, the CD director of development, with Research's work: "He [Fagg] also asserted that he was dissatisfied with the point to which Research had brought the FET technology, namely that we [Research] had not proved feasibility. He wasn't sure we [Research] had the capability to prove feasibility, but believed that Research should have this responsibility." Other CD managers chimed in on specifics. One commented that "reliability was not a Research function and asserted that Research people are novices in the area of reliability." Another asserted that "Research would never be able to demonstrate yield to CD's satisfaction." For CD management the risks of a technology that had unproven yield (and therefore cost) and reliability were too great to justify a large MOS program at that point.[68]

Participants at the meeting also wondered where CD would get the resources for an MOS program. CD claimed that crucial bipolar development programs would be jeopardized if it assigned any resources to the MOS memory work. Fagg authorized an MOS program under the restrictions that no new fabrication facility would be set up for MOS technology and "no work would be initiated until he understood what would suffer in order to initiate FET [MOS] work." Fagg proposed a joint program with Research that would not only take advantage of Research's work in MOS, but also its MOS budget. Research would continue its spending at a level of $1 million a year, while CD would spend $250,000 on the program.[69]

The program that emerged was a joint program aimed at MOS memories involving four different IBM locations. The plan was for an IBM laboratory in Boeblingen, West Germany, to take ultimate responsibility for MOS technology development, and a group from that laboratory was sent to Yorktown Heights,

New York, to work with Research personnel. Personnel from IBM's Systems Development Division in Poughkeepsie would assist in memory design. These groups were chosen so as to avoid taking personnel from IBM's main semiconductor effort in East Fishkill—a sign that, understandably, the MOS work took a back seat to the existing bipolar programs there. Research would continue its involvement in the program until the feasibility of the MOS technology had been established to CD's satisfaction. Although joint programs of this sort would become standard at IBM to smooth technology transfer, the shape of this one owed much to the relative unattractiveness of MOS technology.[70]

As the joint program began in May 1967, the future of MOS technology within IBM was far from clear. Although Research had done much important work on MOS transistors and had succeeded in producing stable transistors, it was uncertain how much of that work would transfer into a high-volume manufacturing environment. Although Research had shown its concern about making the MOS program practical and had proven very flexible in steering the program in directions likely to attract customers, in a strict sense the feasibility of the MOS technology had not been established. CD was unwilling to commit a large program to MOS technology, but in response to pressure from others in the company, it was willing to supply noncritical resources to keep the program going. What had been decided was that MOS technology work would continue to move forward, and there would be an arena for the continued advocacy of the technology within IBM.

FOUR

MOS in a Bipolar Company

FAIRCHILD AND THE MOS TRANSISTOR, 1963–1968

From 1963 to 1968 Fairchild succeeded in developing a scientific understanding of MOS structures but utterly failed in developing an MOS business. Fairchild's MOS program faced difficulties common throughout the industry, such as the challenge of finding a role for the new technology and the difficulties of making stable MOS transistors. But the crucial problem in Fairchild's MOS program was organizational. Initially, Fairchild R&D's MOS program suffered from being decentralized, spread out across departments where bipolar technology represented a satisfactory (and simpler) solution for a given application. As R&D centralized its MOS work, a competing site for MOS work arose at the company's Mountain View plant. The Mountain View work was not only not under the management control of R&D, it was under no management control whatsoever. The R&D and Mountain View groups had very different conceptions of MOS technology, particularly how it should be made and what it was best suited to do. The two sides battled to a stalemate over a period of years, to the detriment of Fairchild. In the industry at large, despite the publicity given to it, the MOS technology remained in an inchoate state, with no firm able to establish an unequivocally successful MOS product. Thus, despite the frustration both sides at Fairchild felt, the situation was not clearly unsatisfactory and persisted until the calamitous summer of 1968. Ultimately, Fairchild had enough MOS talent for two successful companies, but too much for one.

Studying the Silicon–Silicon Dioxide System

In early 1963, after Frank Wanlass's initial MOS transistor work, Gordon Moore, the director of Fairchild R&D, began putting together a team to understand the MOS structure and the silicon–silicon dioxide system in a systematic way. Moore's main reason for this study was to produce better bipolar transistors—Fairchild's main area of business. But this work would also be expected to address the problems of stability of MOS transistors. Up to this time the problems of MOS stability were so great—an MOS transistor's characteristics might vary by over one hundred volts with changes in time or temperature—that they made MOS transistors useless as a product. If these problems were solved, MOS transistors would be technically viable. In 1963 Moore and the manager of the Solid State Physics Department, C. T. Sah, hired three people who would play key roles in developing a systematic understanding of MOS structures.[1]

The first member of the group was Bruce Deal, who joined Fairchild in March 1963. Deal had earned his Ph.D. in chemistry from Iowa State University in 1955, working on the oxidation of uranium and thorium. He then worked for Kaiser Aluminum on the oxidation of aluminum, and in 1959 he joined Rheem Semiconductor, the first of the start-up companies formed from Fairchild. In 1962 Rheem was acquired by Raytheon, a Massachusetts-based electronics firm, which sent a group of its Massachusetts semiconductor team to California to direct Rheem. In the ensuing turmoil, large numbers of pre-Raytheon Rheem employees left, including Deal. Deal came to Fairchild with four years' experience in the semiconductor industry and many more years' experience with the processes of oxidation, a key part of the MOS structure. Deal was an extremely talented and methodical experimentalist.[2]

Andrew Grove joined the group later that spring after earning his Ph.D. in chemical engineering from Berkeley. Grove, a Hungarian, had emigrated to the United States following the Soviet suppression of the 1956 uprising. He earned a bachelor's degree in engineering from the City College of New York and then went to Berkeley, where he received his Ph.D. in chemical engineering, concentrating on fluid mechanics. Grove was to be the chief theoretician of the group.[3]

The final member of the group, Ed Snow, joined later that year, after receiving a Ph.D. in solid-state physics from the University of Utah. Although he had no explicit background in semiconductors as such, his dissertation had been on the migration of ions in quartz. Snow had initially accepted a job with one of the groups at IBM working on MOS structures, but then Frank Wanlass, who had known Snow in Utah, brought him to the attention of Fairchild.[4]

Although all three had been hired to work on the same problem and their complementary abilities must have been considered by Moore and Sah when they were hired, they were left to find each other. After they had been at Fairchild several weeks they discovered through casual meetings in the hall that the other two were working in the same general area. They agreed to cooperate and publish their work jointly. Grove and Sah (who was still involved in technical work) developed methods of comparing theoretical and experimental CV (capacitance-voltage) curves, with Grove writing computer programs to facilitate such comparisons. Deal and his engineer, Maija Sklar, worked predominantly on oxidation and surface states, while Snow worked on transient instabilities. Sah, the nominal manager of the group, left shortly after the group was assembled to take a teaching assignment at the University of Illinois, and thereafter Gordon Moore directed weekly meetings of the MOS group, which included personnel from the Solid State Physics, Device Development, and Materials and Processes Departments.[5]

As Deal, Grove, and Snow began their work, Frank Wanlass, nominally in the Solid State Physics Department, continued what was essentially his own personal MOS program. Wanlass had little supervision, although he informally reported to Moore. Wanlass had many ideas and a strong inclination to pursue them, with his progress reports showing him frequently working on different projects from month to month. He designed a current meter using MOS transistors, a tester for MOS transistors, and several circuits using MOS transistors, in addition to reporting on phenomena involving the physics of MOS transistors. Wanlass worked in the lab experimenting with ways to make and use MOS transistors. By late 1963 Wanlass, an impatient young man, came to believe that Fairchild was more interested in studying MOS structures than in selling MOS transistors, and he left Fairchild to join the start-up General Micro-electronics, which had hired him to help it move into the MOS business.[6]

The primary tool used by Grove, Deal, and Snow was the CV technique to make measurements of MOS capacitors. The MOS capacitor was first described in 1959 by John Moll, then at Stanford, who initially proposed it as a circuit element. Then one of his graduate students, Lewis Terman, used the MOS capacitor as a way of studying surface states. While the MOS transistor was slow to catch on with the technical community, the MOS capacitor was quickly explored by other research labs because it offered attractive properties as a circuit element and a method of studying the semiconductor surface. Between 1959 and 1963 the MOS capacitor was the subject of a number of papers presented at technical meetings and published in technical journals.[7]

The MOS capacitor was a very simple, yet powerful structure. It consisted of a piece of silicon, on top of which a layer of SiO_2 was grown and on top of which a thin metal layer was evaporated. As one varied the voltage across the capacitor, the capacitance was measured as the dependent variable. One could calculate the CV relation and thus measure deviation between theoretical and experimental values.[8]

Much of Deal's early work involved developing or evaluating techniques that related to fabricating MOS structures. He concentrated on determining what worked best; theory played very little part. For example, in August 1963 Deal provided a laundry list of procedures that would lower the surface charge on an MOS structure. Deal also performed studies to determine the best method for packaging MOS transistors. In December 1963 Deal reported on the development of standardized conditions for applying a metal layer over an oxide.[9]

In May 1963 Ed Snow began a research program entitled "Metal Ion Migration in Oxides," suggesting a connection with his dissertation work. Snow's premise was that the metal making up the plate of the oxide capacitor migrated into the oxide, causing instability. He began by making MOS capacitors with a silver plate, and his early experiments gave evidence suggesting that the metal was indeed migrating into the oxide. Snow then began looking for metals that would not migrate into the oxide. In October 1963 Snow was reporting "Platinum looks especially promising," with a platinum capacitor showing no drift after ten hours in conditions under which an aluminum-plated capacitor would have drifted five to ten volts in ten minutes.[10]

In October 1963 researchers at Fairchild made a discovery that was central in ordering its work on MOS stability. In general researchers evaporated metals onto the oxide using a tungsten filament evaporator (much like a light bulb filament), but because of the extremely high melting point of platinum and tantalum, they had been evaporated using an electron beam evaporator (the metal was in a crucible that was bombarded with an electron beam). Snow had found that these metals yielded much more stable MOS devices. He then evaporated aluminum onto oxide using an electron beam and found that with this method, aluminum too yielded stable MOS structures—the key was not the metal used, but the method of evaporation. Snow hypothesized that the "differences might be explained by dissolved gasses or other impurities in the aluminum" rather than the aluminum itself, as he had originally suspected. Although the discovery of electron beam–evaporated aluminum had been significant, it had not solved the problem of stability completely. Snow reported: "MOS structures are quite stable against drift at 125°C, but still drift badly at 200°C."[11]

Snow then ran a split run comparing filament-evaporated aluminum and electron beam–evaporated aluminum, showing that the electron beam led to a more stable device. He then fabricated devices where the oxides had been purposely contaminated with sodium, lithium, magnesium, or calcium. The devices contaminated with sodium drifted more than fifty-six volts in one minute at two hundred degrees centigrade, while the other contaminants led to smaller but still significant drifts. This experiment identified sodium as a major cause of MOS instabilities.[12]

Snow's work led to a search for sodium in any of the materials used in making MOS transistors. Researchers found that the tungsten filament wire used previously for evaporating aluminum had been extruded through a die lubricated with sodium, and therefore the tungsten filament was evaporating sodium along with aluminum. Fairchild began using the highest-purity aluminum available to minimize sodium contamination. By October 1964 the Materials and Processes Department had a special section of its progress report devoted to its work on "Low Alkali Materials" and was spending a great deal of time analyzing the sodium content of materials used in semiconductor fabrication. By April 1965, the group had developed a comprehensive list of all the materials it had analyzed for sodium content.[13]

In addition to his specific studies of optimal techniques, Deal conducted several more general studies. One, undertaken with Grove, was a study of the kinetics of the oxidation of silicon. In some of the more mathematical work the group did, Grove and Deal developed a general relationship for oxidation as a function of time, showing that it followed a linear-parabolic model.[14]

Another major focus of the group was surface states. These were understood at the time as forbidden energy levels between the allowed energy bands. The surface states, which occurred as a result of dangling bonds at the surface, had been predicted theoretically by Tamm in 1932 and Shockley in 1939 and then given a fuller exposition by Bardeen in 1946, in the light of the failure of Shockley's field-effect experiments. Previous experimental studies of surface states had included caveats that reproducibility had been a problem and that the results were valid for that specific case only. Although it was clear that surface states would affect the performance of MOS transistors, they defied generalization. Deal's work had led to techniques to eliminate surface states, and Grove, Deal, and Snow published an article claiming that in fact the surface state density was highly reproducible, although the article did not detail the techniques they had used to achieve reproducibility. Later work by Deal showed the various dependencies of surface state charge—specifically, the role of dry oxygen heat treatments in reducing the surface state density. The work of Deal, Grove, and Snow thus identified two types of surface charge as important in MOS transistors: the fixed charge due to surface states, which they called Q_{ss}, and the mobile charge due to ionic contaminants such as sodium, which they called Q_o. Each of these charges had to be minimized to produce optimal MOS transistors.[15]

Between 1964 and 1967 Deal, Grove, and Snow authored or coauthored twenty-five papers related to the silicon–silicon dioxide system, which to all appearances put the MOS structure firmly on scientific footing. Where previously MOS transistors had been susceptible to huge random drifts, Snow had shown the important role sodium contamination played in these drifts. Grove had developed mathematical models covering many aspects of the silicon–silicon dioxide system and detailed the failure modes of MOS transistors. Deal had provided extensive studies on the process of oxidation and the properties of oxides. These three had established themselves within the industry as the leading authorities on all matters relating to the physics and chemistry of the silicon–silicon dioxide system.[16]

The achievements of Grove, Deal, and Snow were most clearly shown in a semiconductor physics course that they and other Fairchild researchers put together. The centerpiece of the course was their work on the silicon–silicon dioxide system. The course, more thorough and advanced than anything offered in a university, was conducted like a graduate-level course, with participants receiving grades. Grove published the course notes as a book, which became the authoritative text in the field of semiconductor physics for a generation. At R&D, the course served as a rite of initiation for new scientists and engineers.[17]

However, it would be wrong to overestimate the extent to which Deal, Grove, and Snow had put the actual fabrication of MOS devices on a scientific basis. Although a very broad foundation of scientific work lay somewhere below, the actual fabrication of MOS devices rested on a knife edge of technique that was liable to sudden and inexplicable upset. While the published articles suggest a monotonic increase in knowledge, progress reports show that the actual fabrication process periodically experienced setbacks. For example, in June 1965, when the group had been together for two years and completed much of the work that would be published on MOS stability, Bruce Deal reported:

> Experiments have been started to establish a better cleaning procedure prior to final sealing of MOS devices. For over a year the canning procedure used by our pilot line was satisfactory for producing devices reasonably free from drift effects associated with canning processes. This has not been true for several months and we are trying to determine where the difficulty lies. A series of various washes and bake-out treatments have not produced any clear-cut results. . . . Considerable effort may be required to return to the era of reproducible canning procedures.[18]

Three months later the department reported little progress: "The cleaning process prior to final seal still presents problems involving device stability. Results of special tests designed to understand and improve the situation continue to be inconclusive."[19] The problem of MOS stability would often recur at Fairchild.

QUALIFICATIONS OF NEW DRIFTS

- HAS $\mathcal{BED}$ OBSERVED IT ? · · · · · · · · *YES*   5 Points

- HAVE OTHER FAIRCHILD
     DEPARTMENTS OBSERVED IT ? · · · · · · *YES*   5 Points

- HAS COMPANY R OBSERVED IT ? · · · · · · *YES*   −3 Points

- IS IT REPRODUCIBLE ? · · · · · · · · · · · · · *YES*   5 Points

- IS IT UNIQUE COMPARED TO
     OTHER DRIFTS ? · · · · · · · · · · · · · · · *YES*   5 Points

- WAS IT DISCOVERED BY A
     WOMAN SCIENTIST ? · · · · · · · · · · · · *YES*   3 Points

IF TOTAL IS 20 POINTS OR MORE, ...
......THEN NEW DRIFT CAN BE OFFICIALLY RECOGNIZED

FIGURE 4.1 • BRUCE DEAL'S METHODOLOGY FOR QUALIFYING NEW MOS DRIFTS CIRCA 1968. This chart was presented at the Electrochemical Society Spring Meeting in 1974 but presented within Fairchild earlier. "BED" is Bruce E. Deal. "Company R" is RCA, Fairchild's archrival in efforts to explain the causes of MOS instability. While the fact that extra points are awarded for a drift being discovered by a "woman scientist" suggests something about the attitudes prevalent at the time, Deal had a woman engineer working for him (a rarity in the industry), and he personally intervened with Fairchild management so that she could come with him from Rheem to Fairchild. (There had been no previous women engineers at Fairchild.) (Bruce Deal, interview with Henry Lowood, 12 July 1988, Stanford Oral History Project, Department of Special Collections, Stanford University Libraries, Stanford, California.) From Bruce E. Deal, "The XIX Drifts of Silicon Oxide Technology" (in author's possession). Courtesy Bruce Deal.

Another example of the incompleteness and tentativeness of the understanding of MOS structures springing from its empirical base was Bruce Deal's effort to maintain a systematic catalog of the causes of MOS transistor drifts. Each drift was identified with a Roman numeral and by 1968 Deal was up to Drift XII. While Deal's drifts included some that all could agree on, such as the role of sodium, even within Fairchild disagreements existed over how many of these drifts were real. Around 1968 Bruce Deal presented his methodology for qualifying new drifts (figure 4.1), which was funny precisely because it acknowledged the process was not as magisterial as the roman numerals might imply.[20]

TO THE DIGITAL AGE

Developing an MOS Transistor Product

At Fairchild, the development of an MOS product went on in parallel with the research that established a basic understanding of the MOS structure. The most obvious initial MOS transistor product was an individual MOS transistor. It had electrical current-voltage characteristics much closer to the vacuum tube than to the bipolar transistor, making it desirable for some applications. While the MOS transistor's greatest potential clearly lay in integrated circuits, the individual transistor would allow prospective customers to get acquainted with the device and design circuits that might be integrated onto a single chip.

Although the individual MOS transistor was not without its problems, it transferred rather smoothly between the Solid State Physics Department (research), the Device Development Department, and production. As part of this process, people at Fairchild gradually came to realize that while the MOS transistor had similarities to the bipolar device, it had peculiarities that had to be recognized. The key challenge of MOS transistors was not simply making ones that worked but making ones that were reliable over time. In August 1963, before the importance of electron beam metallization had been discovered, the Device Development Department reported that little difference existed between the process used to make MOS transistors and the process used to make diodes, but also noted continuing reliability problems with the MOS transistors made this way.[21]

As R&D continued its MOS work, it developed processes for making MOS transistors that diverged substantially from the processes previously used to make bipolar devices. The Solid State Physics Department developed many techniques unique to the MOS transistor that may or may not have had theoretical foundations. The 1965 product manual for the MOS transistor, which was prepared by the Device Development Department and contained detailed fabrication procedures, noted four main differences between MOS processing and traditional bipolar processing. This included surface preparation, where special rinses were used; oxidation steps, which required extra cleanliness; heat treatments, which implemented Deal's work on eliminating surface states; and metal evaporation, which implemented Snow's work on electron beam evaporation of aluminum.[22]

The oxidation steps give an example of how some of the rituals surrounding

MOS processing developed. A key step in making integrated circuits was heating the wafers in a diffusion furnace to diffuse impurities into the wafer or to grow an oxide layer on the wafer. While the very high temperatures used in diffusion allowed the dopants to penetrate the silicon wafer, they also made other impurities more mobile and likely to contaminate the wafer. In June 1963 Bruce Deal had noticed that the most reproducible MOS surfaces had been made "in a special furnace used only for these experiments." By October, Device Development had special furnaces reserved for MOS work. Although having dedicated furnaces was not specified in the product manual, it appears to have been the common practice.[23]

The CV plot, not typically used in bipolar transistor fabrication, became the final arbiter of the quality of any technique for making MOS transistors. CV plots were used to fix the surface cleaning procedures, verify the cleanliness of oxidation tubes, and monitor the process of making MOS transistors generally. The equipment and the techniques for making CV measurements were transferred twice within Fairchild, from the Solid State Physics Department to the Device Development Department in August through October 1964, and then to the production group in Mountain View by January 1965.[24]

Transferring the MOS transistor was far from trouble-free, however. In October 1964, Device Development reported that it was "learning every day something about these devices." Faulty MOS devices discovered after assembly in Mountain View were traced back to overvoltages inadvertently applied to the transistor's gate, which had not been an issue in bipolar technology. Fairchild had to develop special handling procedures to minimize this problem. Furthermore, as some process steps were optimized for MOS, problems that had been solved for bipolar transistors occasionally reappeared.[25]

In October 1964, Fairchild formally introduced its MOS transistor called the FI-100. The announcement was timed to coincide with a paper Fairchild presented at the IEEE Electron Device Conference on the reliability of MOS transistors, which included data on 500 units under a total of 150,000 hours of tests to show there were no reliability or stability problems with MOS transistors. Fairchild released application notes showing how the MOS transistor could be used to advantage in designing both AM and FM radios. Along with its MOS transistor, Fairchild also introduced the Planar II process, an improved version of the original planar process, which controlled surface effects. While the Planar II process was absolutely essential for the production of MOS devices, it also resulted in bipolar transistors with superior characteristics, and Fairchild produced bipolar devices using Planar II. Although Fairchild had not been the first to introduce an MOS transistor as a product—Wanlass and General Microelectronics had that distinction—Fairchild certainly was among the leaders in MOS, and there was reason to believe that its device was superior to General Microelectronics's.[26]

Fairchild's initial announcement of the MOS transistor was made based on the production capabilities of the Device Development Department, not on the capabilities of the manufacturing line in Mountain View. This was a process similar to what Fairchild had used in its initial introduction of integrated circuits. By January 1965 Device Development reported: "Mt. View is now processing wafers routinely and obtaining results equivalent to ours." The engineer in charge noted: "I think this device is about ready to be decreed transferred."[27]

After Fairchild introduced its individual MOS transistor, it still faced the challenge of conceiving an integrated circuit product. Although one could imagine dozens of applications, no one knew which one would be successful. The trick was to find ones that could be implemented in the near future, that were economically attractive, and that performed some function that could not be achieved using bipolar devices. Fairchild R&D attempted to develop an assortment of MOS products in a decentralized fashion. Given the many problems of MOS technology and the countless ways it could be used, there was a certain logic to the decentralization of MOS work at Fairchild. One could find MOS work in almost every department at R&D, including Solid State Physics, Chemistry, High Speed Memory Engineering, Device Development, and Digital Integrated Electronics. Fairchild R&D worked on MOS memory, MOS analog circuits, MOS digital logic, and individual transistors. But in each department MOS work was carried out alongside bipolar work, and in no department was the department's success vitally connected to the success of an MOS product. In most cases the adjacent bipolar projects represented satisfactory solutions, and the MOS work languished.

Fairchild's MOS program was undertaken within the context of an R&D lab that was increasingly concerned with computer systems. After Fairchild completed its initial development of the integrated circuit, computer systems offered Fairchild R&D new areas to explore. And the integrated circuit itself, which blurred the line between components and systems, made computer systems a logical field. While previously computer manufacturers had responsibility for all aspects of the design of the computer beyond the component level, with the integrated circuit, semiconductor manufacturers had encroached on the territory of systems designers. They took over the design and manufacture of the elementary circuit blocks (e.g., AND gates) from systems firms. Although the change was hardly revolutionary in terms of its immediate implications for systems firms, they had met it with significant resistance.[28]

Beginning in 1962 Fairchild R&D established departments that worked on systems, not just components. First was the High Speed Memory Engineering Department, formed to design an entire memory system. The next year marked the creation of the Digital Systems Research Department, under Rex Rice, who had been hired from IBM Research. Robert Noyce, at the time general manager of Fairchild Semiconductor, explained the new department by asserting: "It is becoming necessary in designing computer components to know more about and participate in the over-all computer design."[29]

Increasingly, Fairchild hired engineers with backgrounds in computer system design to staff its older departments as well as its new ones. These new people represented not just Fairchild management's nibbling at the edges of computer technology as it recognized computers' relevance to the company with the growing convergence of computers and components, but also the realities of the local labor supply. Throughout the early and mid-1960s Fairchild R&D grew at a rapid pace, and at times it had problems finding qualified personnel. While the company had been started by 8 people in 1957, by October 1962, it had 275 at R&D, a number that grew to 541 by August 1968. One source of engineering talent was local facilities of computer companies that had fallen on hard times. In July 1962 Gordon Moore reported that a difficulty in recruiting personnel had been partially eased by General Electric's decision to cut back at its nearby Sunnyvale research laboratory. One of the former General Electric computer engineers Fairchild hired was John Schmidt, who became a senior member of Fairchild's effort to build thin-film magnetic memory systems.[30]

In 1964, in parallel with Deal, Grove, and Snow's work on MOS stability, Schmidt began working on the design of an integrated MOST memory (IMM). This was to be a semiconductor memory, in contrast to the magnetic memories

Schmidt had designed at General Electric and had been hired by Fairchild to design. The idea of semiconductor memory had first been proposed at Fairchild by Robert Norman in the late 1950s. At the time, Gordon Moore decided not to pursue the work or seek a patent on it, believing the memory would be too costly. By 1963, with the success of the bipolar integrated circuit, which showed that multiple transistors could be built economically on a single piece of silicon, and the promise of the MOS transistor, which offered simpler fabrication techniques, Moore had changed his mind. At a NATO-sponsored conference, Moore stated: "The possibility of using active [semiconductor] memories for micropower systems sometime in the future looks especially inviting. I do feel it is not necessary to consider flip-flops made of active devices inherently expensive."[31]

Schmidt's chip contained 64 bits of memory made up of 447 MOS transistors on a chip 115 mils by 145 mils. (A mil is one-thousandth of an inch.) At this time, production bipolar integrated circuits contained roughly sixteen circuit elements (transistors or resistors) in one-tenth the area. Schmidt's IMM took advantage of one of the key differences between MOS transistors and bipolar devices. In most circuits using bipolar transistors, resistors were also needed to design a complete circuit, and they affected speed, power, and the size of the chip. In general, the lower the value of resistor used, the faster the circuit would operate and the more power it would consume. But high-value resistors took up more space on the integrated circuit, making the chip larger and more difficult to build. Schmidt was able to use MOS transistors to function as resistors, which was particularly important here because larger resistors were needed to reduce the total power consumed. The entire chip contained only four conventional resistors.[32]

In September 1964 Schmidt described his pioneering work at an IEEE-sponsored Computer Memory Workshop. At the time almost all computer memories were magnetic, and most attendees were from magnetic memory areas. Several other papers advocated using bipolar semiconductor technology for computer memories, but they were paper studies only—nothing had been constructed. Furthermore, while the studies using bipolar technology promised faster speeds than Schmidt's, they also anticipated greater costs. An IBM proposal for a small memory made from Solid Logic Technology modules predicted costs of two to three dollars per bit, while Schmidt expected ultimate costs of ten cents per bit. Although all cost predictions were speculative, Schmidt showed that if it were possible to develop a viable MOS process and also to make good chips of large size—two very big ifs—semiconductor memories using MOS transistors were possible. Because of the chip's size, Schmidt would venture to predict only that the ultimate yield of the chip could be as high as 15 percent, although some participants at the conference were skeptical that one could make a perfect unit at all with so many transistors, and at the time of the conference Schmidt was unable to report the fabrication of a fully working memory chip.[33]

The IMM required great strides forward in many areas and ultimately proved to be beyond Fairchild's capabilities at the time. Aligning the masks for this very large chip posed problems. Fairchild managers knew that building the IMM would be difficult because of the problems of fabricating reliable MOS transistors over a large area. Further work showed more processing problems that arose from the complexity and size of the chip. By November 1964, the High Speed Memory Engineering Department, which consisted primarily of circuit designers without experience in semiconductor processing, turned the project over to the Device Development Department, which continued to work fighting process problems for about eight months longer, but by the middle of 1965, the IMM was mothballed without ever having produced a fully working chip. In the meantime the High Speed Memory Engineering Department moved to building a semiconductor memory from bipolar transistors.[34]

In late 1964, Leslie Vadasz began a new program within the Device Development Department to develop MOS integrated circuits. Like Grove, Vadasz was a Hungarian who had fled his homeland in 1956. He then went to McGill University, where he received an engineering degree, and thereafter worked for Transitron, a small Massachusetts semiconductor firm. At Transitron he had made studies of the MOS transistor and had become interested in its possibilities. In 1965 Vadasz moved to Fairchild because of its position as the leading semiconductor R&D lab in the industry. He initially worked on bipolar circuits but moved back into the MOS area at the first opportunity. Although he was not designing computer systems, the approach Fairchild R&D chose in MOS bound it tightly to the computer.[35]

As Fairchild considered bigger and more complex integrated circuits, several economic considerations loomed large in the minds of those responsible for determining Fairchild's technical direction, just as they had among those who had planned IBM's Large Scale Integration (LSI) program. In October 1964, Robert Seeds, the head of the Digital Integrated Electronics Section of the Device Development Department who had come from IBM just two years earlier, wrote that two questions needed to be answered, "assuming that technology would allow fabrication of circuits two orders of magnitude more complex than at present." They were:

1. What standard functional building blocks are necessary and sufficient? (probably none!)
2. If the number of different functional blocks becomes very large (almost a sure bet) how can tooling costs, turn around time, and inventory be minimized such that such an approach is practical.[36]

Seeds's first question and its probable answer meant that Fairchild's previous way of selling integrated circuits, where a small number of parts could be used to build up an entire system, would no longer be viable, making an answer to the second question urgent. Vadasz provided an answer to the second question after several months of feasibility studies, which included the study of different design alternatives, analyses of the performance of MOS circuits as a function of processing parameters, and fabrication of test structures. Under Vadasz's proposal, each chip would consist of twenty cells. Depending on how it was wired, each cell could be made to serve as four three-input NAND gates, two six-input NOR gates, two latches, or two flip-flops. Two levels of metal (all previous integrated circuits had used only one) would be used to make the intracell connections to define the cell's function and to provide the intercell connections. The entire chip was to be forty mils by forty mils in size, about equivalent in area to that of production bipolar integrated circuits.[37]

Although initial runs had suggested that the two-level metal process would pose no major technological problems, this optimism proved highly misleading. The two-level metal process involved substantial problems in the insulator separating the two layers of metal, and a great many different types of dielectrics were tried. Two-level metal wafers were first processed in October 1965, but throughout 1966 engineers battled one problem after another—primarily open connections between the two metal levels. By the end of 1966, in a tactical retreat, Fairchild started work on an MOS array with only one level of metal. Work continued on the two-level process for bipolar arrays, which reached the market in 1968 after a Herculean effort.[38]

Both the bipolar and the MOS logic array projects required significant computer skills to make their implementation possible. In 1966, Fairchild R&D hired a number of engineers into its Digital Integrated Electronics Department who had experience designing or programming computers. Rob Walker was hired from Ford Aerospace; Bob Schreiner, Harold Vitale, and Jim Downey from General Electric; Bob Ulrickson from Lockheed; and Hugh Mays and Jim Koford from IBM. Mays led the Computer-Aided-Design Section, which worked on developing an infrastructure to make logic arrays feasible. This included the development of programs to simulate logic, to generate test patterns, to lay out circuits on a chip, to route wires on a chip, and to analyze a circuit's performance. Another group developed a tester that would test logic arrays. An IBM System/360 Model 40 computing system arrived for the use of the Computer-Aided-Design Section in November 1966, a time when the department had seven people developing computer programs to support logic arrays.[39]

Lee Boysel, MOS Maverick

In 1966 Lee Boysel, a true believer in MOS technology and a disciple of Frank Wanlass, joined Fairchild Semiconductor. But where all MOS development work had been previously concentrated at R&D in Palo Alto, Boysel was located in Fairchild's production facility in Mountain View, where he was to form a circuit design group. Boysel brought Wanlass's individualistic style back into Fairchild's MOS program. Although Boysel had one of the clearest visions of the possibilities of MOS technology as well as the abilities to make it a reality, and although he saw the future of MOS technology as closely linked to computers, he had a dramatically different vision from R&D's. Boysel's arrival at Mountain View made Fairchild's MOS program bipolar, leading to conflict between the MOS work at Palo Alto and Boysel and his allies. Boysel stayed true to his own vision of MOS technology and refused to compromise or even communicate with others so that his work might be more easily adopted at Fairchild.

In 1966 Boysel was one of the most experienced MOS circuit designers in the industry, but throughout his career he had been on the periphery. It was fitting that when Boysel joined Fairchild, the industry's leader in MOS research, instead of working with the large and renowned MOS group at R&D, he worked by himself in Mountain View. Just as it had earlier, Boysel's position at the margins gave him the opportunity to pursue his own agenda—a freedom he would not have had if he had been at the center.

Boysel had joined Douglas Aircraft in Santa Monica, California, in 1963 after receiving his bachelor's and master's degrees from the University of Michigan. He had come to college an electronics enthusiast, having built his own oscilloscope and hi-fi equipment. At Michigan, Boysel was an indifferent student, finding the coursework too theoretical for his taste, and he continued tinkering on his own. As one of the few electrical engineers at Douglas, Boysel got a hands-on education in the Pacific preparing a missile system for testing. He then moved to writing proposals for government contracts. At this time he also began studying inertial navigation using digital electronics for potential application in guidance systems.[40]

Boysel encountered Frank Wanlass, who was at General Microelectronics (GME), during a visit to Douglas to promote MOS technology. However Douglas management responded to the new technology, for an impressionable young engineer it was a life-changing experience. Wanlass had a brief conversation with Boysel, electronics enthusiast to electronics enthusiast, and showed him a twenty-bit shift register chip he had designed in MOS. At that time, Boysel had been designing a shift register from discrete circuit elements; it had taken him four circuit boards to accomplish what Wanlass did in one chip. The possibilities of MOS became very clear to Boysel.

He then began a self-education program in MOS. He had no ongoing relationship with Wanlass, and he had no other contact with engineers working in the field, so he collected and read all the published articles he could find. At the same time Boysel began to design MOS integrated circuits on paper. Among the proposals he was preparing were several in which MOS integrated circuits could prove advantageous. Boysel then worked out an agreement with GME under which it would lay out and fabricate an MOS circuit Boysel had designed. Boysel complemented his paper design with lab work measuring the characteristics of individual MOS transistors acquired from Fairchild and GME.[41]

After Boysel had finished his initial circuit designs, he had come to a dead end at Douglas. Although Boysel had been able to use Douglas to further his interest in MOS, it lacked MOS fabrication capability, and its limited mandate to work in digital electronics did not match Boysel's expansive interests. He began looking elsewhere for work as an MOS circuit designer and applied for a position advertised by IBM in Huntsville, Alabama. When Boysel was offered the job, he accepted.

While the presence of Wernher von Braun and his team made Huntsville the center of rocketry work in the United States, in semiconductor electronics it was a backwater. IBM's Huntsville location had been established in 1964 to manufacture the Instrument Unit, a critical guidance system in the Apollo program. IBM wanted to explore the use of MOS in telemetry systems, although the MOS circuitry was not required for the operation of the Apollo guidance system. Boysel's manager directed him to work on a multiplexer chip nearly identical to the one he had designed at Douglas. But Boysel had much grander plans than his assignment allowed for, and he began work on a circuit capable of performing three functions: (1) multiplexing (as requested), (2) converting analog signals to digital signals, and (3) converting digital signals to analog signals.[42]

Boysel had work habits befitting an electronics enthusiast. He worked very long days, typically seven days a week. After putting his time in at the IBM facility, he frequently worked at home in a laboratory he had built himself. At Douglas he had acquired government surplus electronics equipment for pennies on the dollar, and when he moved to Huntsville, he converted a bedroom in his house to a lab. Here, after completing the design of his circuit, he built a breadboard of it using discrete components as well as a few integrated circuits. Boysel's home lab enabled him to work outside normal bureaucratic channels and to assume total control over his work without being dependent on technicians. After demonstrating the operation of his breadboarded circuit to management, who had been unaware of exactly what Boysel had been doing, he received authorization to continue and to make arrangements to have his circuit fabricated.

To get his chip built, Boysel turned to Frank Wanlass, who was by this time with General Instrument (GI) on Long Island, one of the few firms in the industry producing MOS circuits for sale. Even though IBM Research had an MOS program under way, Boysel and Research were unpromising partners. For Research, Boysel's program was not connected to anything in IBM's main line of business. On the other hand, Research had no proven track record in MOS, and its program was still in a state of flux, making any commitments suspect. IBM Huntsville reached an agreement with GI whereby GI would fabricate the circuit for a greatly reduced price in exchange for GI's receiving the rights to sell the circuit to other customers. As an added benefit, GI, a relative novice to semiconductor work, became publicly associated with IBM.[43]

Whatever the benefits of this arrangement to IBM, it offered Boysel great personal benefits. In spite of his enthusiasm for MOS technology and his self-education, the craft nature of MOS meant that there was much that could not be learned from articles. He now began an apprenticeship under Wanlass, who at this point had more experience in all phases of MOS work than anyone else in the industry and could connect Boysel with the wider community. Boysel's circuit design had been made under the assumption that the same design techniques that worked in bipolar transistors held true in MOS. When Boysel took his work to Wanlass, Wanlass showed him how to redesign it in the optimal way for MOS technology. Wanlass laid out Boysel's circuit on the dining room table of his home, instructing Boysel in the skills of the craft. Wanlass took Boysel's circuit and, working without any sophisticated design methodology or computerized tools, produced a layout that had only one mistake.

The chip was fabricated at GI's Hicksville, New York, plant by Wanlass, with Boysel assisting. The two did most of the work on weekends, when they were typically the only people in the plant. Here Boysel began to learn how to process MOS circuits by watching a master at work. After the initial chip was designed and built, Boysel, having absorbed many of Wanlass's tricks, then designed another MOS circuit under Wanlass's tutelage, this time doing both the circuit design and the layout work himself.[44]

Wanlass and Boysel shared a technical style that included a preference for hands-on work over analysis, an eagerness to circumvent formal bureaucratic channels, and a willingness to work extremely long and odd hours. While both had abilities in a wide range of areas, Wanlass's main interests were semiconductor processing and circuit design. Boysel's were circuit design and systems. At IBM Boysel continued his study of digital electronics, particularly digital differential analyzers, while Wanlass fed him information on MOS work at other firms.

Boysel also began to study IBM's System/360 family of computers and became impressed with the possibilities of computers organized in parallel rather than in series. Initially, those interested in implementing computers in MOS technology had focused on a serial architecture, such as in a digital differential analyzer, because its low input/output pin count and the fewer circuits it required made it possible to implement a processor on a chip. At a time when hardware was very costly, a serial organization reduced costs by using the same hardware to process different bits of a word. The tradeoff was that a serial computer required much more complex control circuitry than a parallel organization. As Boysel considered how to implement the arithmetic parts of a computer, he began to believe that a parallel implementation was better suited to MOS than a serial one. IBM System/360's use of a read-only control store also interested Boysel, particularly after he discovered that it could be built using MOS technology.[45]

Much as had occurred at Douglas, at IBM Boysel was rapidly approaching a dead end. His interests went much beyond anything IBM Huntsville had a charter to do. And while IBM had numerous positions designing systems or circuits, moving to the center of IBM's electronics work in East Fishkill or Poughkeepsie, New York, would have meant accepting a much more regimented approach with much less responsibility and freedom than he had at Huntsville, in the marchland of IBM.

In 1966 Boysel began looking for another job. Largely on the strength of the circuits he had designed at Huntsville, he was offered and accepted a position with the applications group at Fairchild's Mountain View site. His mandate was to begin an MOS circuit design effort. All previous MOS circuit work had been done at Fairchild's R&D lab in Palo Alto. (Mountain View was roughly four miles from the Palo Alto lab and was the site of most Fairchild Semiconductor operations.) R&D's computer-aided-design approach was a long-term project that would take years to complete. By 1966 Fairchild had only two MOS integrated circuit products in its catalog. Boysel was hired to add to that. Fairchild management expected that Boysel would continue the work he had been doing at IBM, designing similar circuits for them.[46]

Boysel's position within the Fairchild organization was analogous to the one he held within IBM: he was to develop a new capability that was not seen as critical to the overall corporation. Fairchild was doing very well selling bipolar circuits; a successful MOS product line would be nice, but it was not necessary.

Boysel's assignment had some ambiguity in it, and no one above him in the organization had enough knowledge of MOS technology to adequately supervise him.

Now that Boysel finally was at a firm that was in the MOS integrated circuit business, his own vision of the possibilities of MOS again expanded beyond his mandate. Boysel set out to design computer subsystems in MOS. Boysel had several factors working in his favor as he pursued his vision. As a new hire, Boysel was given a period to get acclimated, but he knew precisely what he wanted to do and started working quickly to design a 256-bit MOS read-only memory (ROM). Boysel's other advantage came from his apprenticeship with Wanlass. His knowledge of how to perform most of the steps in the design process allowed him to do most of the work himself (up to the device fabrication), without having to rely on other groups, which would have required formal management authorization. For example, he cut the masks himself, circumventing the group with that responsibility. After the circuit had been designed and laid out, Boysel presented it to Fairchild management as a fait accompli, making it much harder to reject than had it been just a proposal. Boysel followed this with the design of an eight-bit MOS arithmetic unit. Again Boysel kept his design secret from management until he had the circuit completed.[47]

In September 1967 Boysel wrote a manifesto for MOS computer design, which explained his work at Fairchild to date and his intentions. Boysel's manifesto was a single page long, two-thirds of it consisting of a schematic diagram of a computer (appendix figure A.3), with text on the remainder of the page. The entire computer described by Boysel was implemented in not bipolar technology (which would have been well beyond the state of the art) but MOS technology. Six different MOS chips, ROM, random access memory (RAM), a "basic CPU [central processing unit] element," and three different types of register chips would make up the bulk of the computer. Boysel described the CPU element as a "4 Bit Wide Slice With All Op Code Decoding and Branch Instruction Control Built In." The ROM chip would have 4–8 kilobits per chip, while the RAM chip would have from 512 to 1,024 bits per chip. Boysel claimed his computer represented "the general IBM 360 system philosophy" but noted: "Using new LSI-MOS circuit techniques and proper machine organization, the relatively slow speed MOS devices can produce an instruction execution speed of 5 to 10 μs as compared to a 30 to 50 μs rate for most current machines made from bipolar circuits, e.g. IBM 360/20, 30, 1401, 1132, etc."[48] Boysel's claim was extraordinary. At a time when MOS was considered to be low performance, best suited for calculators, Boysel argued that a computer organized around MOS technology and its unique characteristics could result in higher performance than small mainframes using bipolar transistors.

Boysel included in his manifesto a business analysis purporting to show why his approach made sense for Fairchild. Taking advantage of the possibilities of MOS would "invariably lead to low cost, standardized subsystem blocks" that could "be produced in Fairchild volumes." Furthermore, Boysel claimed: "It is the opinion of many that the significant aspect of the fourth generation machine world will be a drastic reduction in machine electronics costs (by a factor of 5 to 20). This will introduce the small and medium performance machines into the volume usage areas of education, industrial and numeric control, consumer goods, etc. This is substantiated by studies made by IBM which indicate that about 75% of the dollar volume of the machine market will be in small to medium size machines, i.e. model 360/40 and down."[49]

But for all the strategic analysis presented and claims about "Fairchild volumes," Boysel's manifesto ultimately represented a very personal vision, not closely tied in to considerations of Fairchild strategy. Boysel's vision was one that ultimately would not be satisfying to either Fairchild or its customers. Fairchild was then supplying large numbers of small-scale integration chips, each consisting of a few logic gates, for use in computers. Hundreds if not thousands of these chips would be required for each system. Boysel was suggesting that all these bipolar chips could be replaced by a handful of MOS chips and still yield a computer system of the same performance. Such a move would require careful management, for it would destroy Fairchild's bipolar business.

The far bigger problem was that Boysel's vision would have been unattractive to computer systems designers. The use of small-scale integration allowed computer makers to design and implement a computer in the way they believed optimal. Boysel was proposing to reparse the task so that his group took over the responsibilities of computer system designers. The question was more than just a jurisdictional one, for Boysel's plan threatened to take away one of the computer makers' main sources of added value, their design of the computer. If system designers had been unhappy when Fairchild took over the business of designing flip-flops, they would be apoplectic with a plan that would give Fairchild the design of almost the entire system. In fact, Boysel's manifesto had little to do with what made business sense for Fairchild, for he was planning to start his own computer company and was using his work at Fairchild to fund its preliminary development.[50]

In its short life Fairchild had had several brilliant engineers who followed their own instincts, whether or not it was exactly what their managers wanted them to do, and Boysel stood in that line. Wanlass was another example, but the best known and most important to Fairchild's success was Bob Widlar. Widlar joined Fairchild in 1963. With his partner, Dave Talbert, Widlar created Fairchild's highly profitable linear product line. Widlar had a genius for designing analog integrated circuits, which were much more complicated than the more common digital circuits. Widlar's exceptional abilities gave him a great deal of leverage within the company. He did what he wanted to and worked when he wanted to. Fairchild management had a deep respect for talent and gave it wide latitude.[51]

Although John Hulme, who managed both Widlar and Boysel, had been able to work with Widlar, tolerating his behavior and at the same time channeling his creativity into areas that were profitable for Fairchild, he was unable to do the same with Boysel. Hulme, a quiet Mormon, and Widlar frequently had long discussions, sometimes arguments, but there was no such intense communication with Boysel. Boysel worked by himself and rarely talked to Hulme, with Hulme considering Boysel virtually in a different company, because it was so hard to find out what he was doing.[52]

Boysel did talk to Wanlass, who occasionally visited him surreptitiously at Fairchild. Through his communications with Wanlass, he learned of two important innovations in MOS technology—four-phase logic and dynamic memory. Wanlass, the first MOS missionary, acted as a clearinghouse for innovations and new ideas. At GME he had shown one of his chips to Bob Booher, an engineer from Autonetics, an avionics firm. Booher then went off on his own and over a period of several years designed both his own chip and a new class of circuits to perform digital logic. Booher then went back to Wanlass, by now at GI, to see about getting his circuit built. Booher's new type of logic circuitry, later called four-phase logic, took advantage of the unique characteristics of

MOS transistors to produce circuits that operated much faster and with much less power than conventionally designed MOS circuits. Wanlass and other GI engineers then actively promoted four-phase logic throughout the industry.[53]

Wanlass was much more cryptic in telling Boysel the possibilities of what is now called dynamic memory, perhaps because he had not completely worked out the implications of it himself. As circuit designers had considered the possibilities of creating memories that could be randomly accessed (as opposed to shift registers, which were serially accessed) using MOS transistors, the first type of memory designed was what is now called static RAM. In static RAM the basic memory element (the cell) held information indefinitely as long as power was applied to the chip. The static RAM cell required six transistors to implement. What is now known as dynamic memory is one in which the information stored on a cell gradually leaks away and requires periodic restoration. The advantage of dynamic memory is that the basic memory storage element can be implemented with fewer transistors. Wanlass either conceived of the idea of a dynamic RAM or had heard of work by others.[54] Wanlass passed on some broad hints to Boysel that it was possible to implement a memory using only three transistors instead of six per cell. Before dynamic memory would be workable, many details had to be resolved, such as how the memory would be restored within the cell and how to keep this restoration process from affecting the performance of the overall computer system. Both four-phase logic and dynamic memory became central aspects of Boysel's work at Fairchild.

Boysel had one characteristic that made him different from Wanlass: he published. While Wanlass practically never published his work, Boysel was a relatively prolific author, writing four articles during his two-year tenure at Fairchild. Although those at R&D published their work in professional or technical journals, such as the *Journal of the Electrochemical Society* or the *IEEE Transactions on Electron Devices,* Boysel published in trade journals, such as *Electronics* or *EDN.* Boysel also published for vastly different reasons than did the people at R&D. Publication assured Boysel that he could continue to use his own work, even if he moved to another company. Even from his days at Douglas Aircraft, Boysel believed he would be an itinerant, and he made a conscious effort to ensure that his patents, which were assigned to his employer, would not limit his freedom of action later on. Boysel did not cooperate with patent attorneys. After Boysel left Douglas, Douglas patent attorneys attempted to get Boysel to file a patent on his MOS work. Boysel replied evasively, stating "I am somewhat puzzled about what you are trying to patent," and claimed that although he would be glad to have a patent, if it was merited, the work he had done had been widely known beforehand.[55] Fairchild did not vigorously pursue patents in circuits, so his efforts to avoid patenting his work caused no direct difficulties. Publishing formed the second part of Boysel's patent avoidance strategy. Although Boysel could keep Fairchild from patenting any of his work through noncooperation, he could not be sure it would not be patented by others. By publishing, Boysel put his work in the public domain and guaranteed his rights to use his own work no matter where he was employed.

Boysel might have cooperated with Fairchild managers if they had been unequivocally enthusiastic about his plans to design MOS computer subsystems, but Boysel was better suited temperamentally to the role of insurgent. As his work developed, he was able to retain full control of his project with no need to compromise or communicate with others in Fairchild, both of which would have been necessary had Fairchild been serious about Boysel's work. Boysel had the best of both worlds, with enough management support to continue on but without enough management interest to interfere in his work.

Boysel's work did serve a purpose for Fairchild. Boysel's designs showed customers that Fairchild was serious about developing products in the new MOS technology. In 1967, Fairchild's marketing director, Jerry Sanders, launched a program where Fairchild introduced a new integrated circuit product every week for a year. Several of Boysel's designs were launched under this program, although they might not have survived the scrutiny of a formal review. The failure of Boysel's parts in the market pleased him, for it meant that he did not have to turn away from designing chips to the more mundane task of supporting customers.[56]

Conflict: Mountain View versus Palo Alto

As Boysel showed his ability as an MOS designer and as the MOS technology became more popular in the industry as a whole, Fairchild management authorized Boysel to add people and form a group. Boysel looked for those who had experience in computer system design and who shared his hands-on "build it and see if it works" approach to engineering. He hired two technicians who functioned as engineers and two engineers with college degrees. One of the engineers had served in the Navy as an electronics technician, then worked at General Dynamics designing telemetry systems while earning his college degree from San Diego State University at night. In Boysel's area, having an advanced degree was just as likely to be a handicap as an advantage. One engineer who had a master's degree did not work out in the group because he was more interested in the analytical approaches commonly used in R&D, which Boysel considered a waste of time. Boysel believed that the key to designing computer subsystems using MOS circuits was an understanding of computer hardware; whatever semiconductor knowledge was required could be picked up on the job.

Both Boysel's manner of working and his position within Fairchild created a situation where Boysel's new hires formed ties to him rather than to the corporation as a whole. Boysel saw himself as an outsider fighting against the entire company, and he passed this attitude on to his engineers. They were MOS engineers in a company whose profits came primarily from bipolar technology, and they were not given the same treatment as bipolar engineers. At a time when Fairchild was growing rapidly, many of Boysel's people were not assigned offices and were forced to sit in desks in the halls. In general, Boysel had fewer resources to work with than did managers involved in bipolar work, but he compensated for this with a fierce loyalty to his people.

The one group that could have been an important ally for Boysel's team, the R&D MOS group, instead became its major antagonist. In their educational backgrounds, their approaches to problems, and their conception of MOS technology, Boysel's engineers were nearly the perfect antithesis of the engineers and scientists working on MOS at R&D, and conflict was almost inevitable. At R&D most of the engineers had advanced degrees. More important than the degrees themselves was what they represented, which was that both the individuals involved and the R&D lab as a whole put an extremely high value on analysis. Nothing was done unless it was first analyzed mathematically. This could best be seen in the years of study, resulting in numerous papers, that Bruce Deal, Andrew Grove, and Ed Snow had devoted to the silicon–silicon dioxide system. A significant portion of this work was devoted to developing mathematical models for oxidation. The same impulse towards analysis guided Fairchild's work on circuit design, as researchers devoted a great deal of time and effort to using computers to analyze circuits. Fairchild bought IBM's ECAP (Electronic Circuit Analysis Program), but the Digital Integrated Electronics R&D group also developed its own programs for circuit analysis, and this project occupied most of one engineer's time for over a year.[57]

Boysel and his group eschewed analysis. The only thing that mattered was whether the chip worked when it was finally built, a point Boysel wanted to reach as soon as possible. Not only was Boysel temperamentally and philosophically opposed to detailed mathematical analysis, it was a luxury an insurgent could ill afford. Fairchild R&D had dozens of engineers and could easily devote a few to analysis, or even developing analytical tools. Its work was supported by a dedicated IBM System/360 Model 40 computer costing over $5,000 a month to lease, as well as several other computer systems. Boysel had fewer than half a dozen engineers, and analysis would have taken them away from the pressing business of designing chips. Boysel designed his parts to have wide operating margins, so that detailed analysis was not necessary to prove their operation. He limited his analytical work to simple mathematical approximations that could almost be worked out in one's head.[58]

The computer was at the heart of what both groups were doing, but the differences in the groups are shown by how they appropriated the computer. Boysel's group was ostensibly building standard MOS parts, but in reality it was building a computer. R&D had many people with backgrounds in computers who were writing programs that would be used in designing chips. Palo Alto had a very intense division of labor, with the whole project held together by management and structured communication between groups. No one at R&D could do anything working alone. Although Boysel's group was planning on building a computer, those in his group did not use a computer in their work, which gave them a freedom that no one in Palo Alto had. Boysel's approach was much more holistic, with all his engineers trained to do all the work in designing a chip by themselves. While Boysel and his group needed cooperation from others, they needed much less than those at R&D did.

This difference in approaches with regard to analysis can be seen in the two groups' responses to four-phase logic, the circuit innovation that Wanlass had passed on to Boysel. Although it offered much faster speeds, its drawback was that its four separate clock signals made for an extremely complex system. While it was challenging to build, proving analytically that four-phase logic would work was even more difficult. One could spend years analyzing four-phase circuits—which was what R&D did. In February 1967 an R&D engineer began work on four-phase logic, which remained his primary assignment throughout the year. By December he reported: "A test vehicle to test some of the concepts of dynamic [four-phase] logic is being considered." By the spring of 1968 work had begun on the design of a four-phase circuit, and the March technical summary announced: "Technical Report #330 entitled 'The Design of MOS Dynamic Logic Circuits' is now available in the R+D Library. This report describes in detail the generation and application of design equations suitable for quasi-four phase MOS dynamic circuits."[59]

For Boysel, four-phase logic represented a way to greatly improve the performance of MOS circuits using circuit techniques, one of his strengths, without requiring major changes in processing techniques, an area where he was much weaker. Boysel embraced it fully. In his MOS computer design manifesto he noted that several parts would use four-phase logic, and by the latter part of 1968 his group had designed five different four-phase parts, without having spent time doing detailed mathematical analysis or building test structures.[60]

The differences between Boysel's group and Fairchild R&D are further revealed in an incident involving the Fairchild R&D semiconductor physics course. As the work of Boysel and his group became known at R&D, they were required to take the R&D semiconductor physics course, under the assumption that they could not design MOS circuits without a knowledge of the work of Deal, Grove, and Snow on MOS structures. Not only were Boysel and his group not theoretically inclined, but they had very little formal background in physics and chemistry. Boysel's group members did not see the course as a way to prove themselves but as an affront, and they did not take the course as seriously as R&D engineers and scientists did. A number of them did poorly in the course, which gave them a grievance to put beside their differences in style with R&D.

Mountain View itself had two groups involved in MOS technology, Boysel's and one responsible for MOS manufacturing that focused on MOS processes. Ideally, this group was to receive MOS technologies developed in R&D and implement them on a production level as received. In the early days this is what happened, but Boysel's arrival marked a turning point. Boysel, talented and charismatic, gave the manufacturing group an alternative to the technologically passive role expected in its relationship with R&D. Boysel, as a follower of Wanlass, provided another way of proceeding, for he concentrated on designing MOS circuits that could tolerate instabilities. At R&D, the work of Deal, Snow, and Grove was central and resulted in a goal of producing stable MOS devices through an ultra-clean process minimizing Q_{ss} and Q_o and thus eliminating the source of instabilities. Boysel's approach provided an opening for someone to design a compatible process for fabricating MOS circuits that was less pristine than R&D's.

The MOS manufacturing group grew greatly in its technical competency over time. In late 1966 Jim Kelley came from R&D, where he had led the development of individual MOS transistors. He later brought over another R&D engineer, Marshall Wilder. In 1967 the group hired GME's Bob Cole, a materials scientist who had learned MOS technology from Frank Wanlass. By early 1967 the Mountain View group had developed its own MOS process. The cornerstone of the process was the use of vapox, a chemically deposited silicon dioxide. Previously all oxides in MOS transistors had been thermal oxides, formed by heating a silicon wafer in a diffusion furnace. Vapox, originally investigated by R&D, could be deposited more quickly in thick layers than thermal oxide, whose growth was a set function of time. The use of thick oxide on some layers of the chip (but not in the actual MOS device) had been pioneered by Frank Wanlass at GI, and it offered faster circuits and a way to avoid an important parasitic effect. However, vapox had many more impurities in it than a thermally grown oxide, making it a potential source of instability.[61]

In the spring of 1967 MOS work at R&D had reorganized and centralized under the Digital Integrated Electronics Department. But by this time Digital Integrated Electronics was concentrating not on its MOS work but on the work at Mountain View, reporting: "The most important project becomes working with Mt. View to establish a reliable and controlled manufacturing process. During May we began testing Mt. View MOS devices and material with equipment we have but they don't. We have begun running the Mt. View process in our area in order to evaluate process differences."[62] R&D had great skepticism about the Mountain View process based on the impurities in the vapox, suspicions that were confirmed by analysis. "Work on this process has been discontinued," reported the Digital Integrated Electronics Department in July, "because its weaknesses have now been accepted without further proof."[63]

Mountain View was not willing to give up the vapox process, however. Boysel had designed his circuits to work even with a wide range of drift. R&D's alternative to vapox, a technique called channel stops, increased the area of Boysel's already large chips to the point where he considered them unbuildable. Mountain View pressed on with the vapox process, receiving very close scrutiny from R&D and periodic orders to shut down the line when Q_{ss} and Q_o went beyond what R&D considered acceptable values.[64]

But R&D had its own MOS stability problems. Although in February 1967 Digital Integrated Electronics spoke of its success in making stable devices, by July things had taken a turn for the worse: "We are still learning a lot about what not to do when trying to make stable MOS circuits on a semi-production basis. Although our best furnace and evaporator monitors progressively improve, we still make bad ones. Our confidence that an MOS circuit oxidation will be stable, given that the last monitor was good is only 50%. This is an impressive improvement, but we are still going uphill."[65] Three months later these problems were solved by installing a ceramic liner into the diffusion furnace and adding filters to the gas line into the furnace. The liner separated the diffusion tube from the rest of the furnace and prevented impurities in the furnace from reaching the tube. Although R&D could solve its stability problems, it could not do so easily or quickly.[66]

In early 1968, Gordon Moore, frustrated at the problems involved in transferring the MOS process to Mountain View, received authorization for R&D to build its own facility to make production quantities of MOS circuits. While R&D carried on alone, thinking it would be better off without the recalcitrant Mountain View group, its MOS program suffered from a major flaw. R&D's computer-aided-design approach to customizing MOS circuits for each user required significant organizational and technical innovation for success. Previously, two approaches to integrated circuits dominated. In the most common one, the semiconductor maker designed a part, wrote a specification for it, and offered it for sale. Customers examined the specification and used the part if it met their requirements. In the other case, customers presented a set of specifications to a semiconductor manufacturer and had it design a part for them. In this case the customer-manufacturer relationship would be closer, but in either case the boundaries between the two were very clear. The computer-aided-design approach blurred these boundaries. A preliminary users manual stated:

"There exists at Fairchild a very strong conviction that the successful development of complex arrays will require a very close (and somewhat unfamiliar) working relationship between the user and device manufacturer."[67]

Almost all parts of the design process involved both parties, and the lines of responsibility were not always clear. For example, once a chip had been built, the customers had to provide Fairchild with the patterns of the electrical signals applied to the chip inputs for testing. Fairchild warned customers that they would have to pay the cost of having overly tight test specifications applied to a chip. To understand what constituted overly tight specifications, the customer had to be intimately familiar with Fairchild's process for making integrated circuits. Previously, testing had been all the manufacturer's responsibility; now users had to participate and accept some of the risk involved. Furthermore, if a part did not work the first time it was built, there were likely to be recriminations back and forth over who was at fault. As time went on, Gordon Moore became increasingly skeptical about the viability of this approach. In March 1968 at the IEEE Convention, Moore described Fairchild's bipolar and MOS computer-aided-design approach but ended with the following caution: "Customized LSI for low unit volume applications is possible at reasonable cost with reasonable turn around time. The question is, Will it be an economically and emotionally satisfying interface between customer and supplier?"[68] In spite of the mounting doubts, Fairchild R&D continued on.

For his part, by the middle of 1968 Boysel had designed a wide range of MOS products, which filled a catalog for Fairchild. Some of the parts were very innovative, most notably a 256-bit dynamic RAM and a 4K-bit ROM. But there was no one (Boysel included) effectively advocating his work either inside Fairchild or to potential customers. Inside the company there was a need for someone to take what was good in Boysel's work (and there was much that was good), structure it so that it would attract customers, and then marshal the necessary forces within the company to make the project successful. Boysel's parts, particularly the ROM and RAM, not only were new but had no close counterparts in the industry yet, so Fairchild could not simply expect to announce a product, pass out data sheets, and have customers buy the product in droves. For Boysel's work to succeed, close contact with customers was essential, both to find out what they wanted and to inform them of the possibilities of his designs.[69]

Here the position of MOS vis-à-vis bipolar technology at Fairchild and bipolar's status as an established technology were most apparent. A relationship such as existed between Boysel and R&D would have been intolerable in bipolar technology, where Fairchild depended on a steady stream of new products that customers were willing to buy to maintain its leading position. Management would have taken decisive action to force Boysel to work with R&D. But in MOS technology, where total industry sales were still small and no clear product direction had been established anywhere, the uncertainties involved and lack of current importance to Fairchild allowed Boysel the freedom to do what he wanted.[70] Even if they did not result in sales for Fairchild, the products Boysel added to Fairchild's line made it seem like he was making a credible effort for the company. On the other hand, R&D continued to muddle along with both the computer-aided-design approach and an MOS memory project.[71] Fairchild had at least three different approaches to MOS, and while it had the resources to keep all three going, management was unwilling or unable to decide which was the most promising. No crisis forced drastic management action.

No group or individual wanted MOS technology to fail at Fairchild. Gordon Moore and his R&D labs certainly did not, for Moore had committed substantial resources to the technology over a period of five years. (Moore noted in retrospect, however, that had he known how important MOS would become, he would have devoted more resources and attention to it.)[72] The Mountain View managers did not either. Although they may have given less resources to MOS than bipolar, MOS certainly had enough technical resources to succeed. The problem was that no organization had a strong enough interest in MOS succeeding. Such were the conditions until July 1968, when a crisis of a different sort hit Fairchild.

It Takes an Industry

THE MOS COMMUNITY

Although the fate of the MOS transistor ultimately hinged on its success or failure at specific companies, its development was a cooperative effort, sometimes intentionally, sometimes not, between companies who were nominally competitors. This industrywide effort had the potential to benefit all who were working on MOS, both through transfers of information and through the creation of an atmosphere that supported the new technology. Transfers of information occurred through conferences, interfirm meetings, confidential exchange agreements, acquisition of artifacts, and movement of personnel. Major research labs received information from start-ups, and vice versa. In conferences and professional journals, researchers (generally Ph.D. scientists) described their latest work using the conventions of academic science. But because of the craft nature of semiconductor production, much information of an operational or engineering nature was transmitted informally. The planar process used to make both MOS and bipolar transistors provided a common foundation that allowed for cooperation. Several firms went out in directions contrary to that of the rest of the industry, but they did so to their own detriment, as they were unable to take advantage of work done elsewhere.

Conferences and Papers

Particularly in the early days of MOS, professional conferences and journal articles were important vehicles for sharing information. They provided a means for the industry's leading researchers—mostly Ph.D.'s—to come together, report their latest findings, and discuss technical issues candidly. At a time when there was uncertainty about the causes of MOS instability, conferences allowed rapid dissemination of information and helped researchers achieve consensus. Conferences, modeled on the conventions of academic science, simultaneously served individual ambitions and corporate purposes.

Two conferences that were particularly important at this stage were the IEEE Solid-State Device Research Conference (SSDRC) and the Electrochemical Society meetings. Contrary to its name, the SSDRC typically consisted of work at a level more abstract than actual devices, detailing the results of experiments on simple test structures. The physicists, chemists, and electrical engineers in attendance were interested in phenomena affecting the fabrication of devices, but they were not, in general, the ones responsible for fabricating devices.

The papers presented at these conferences and published in professional journals were concessions by industrial research labs to the aspirations of their researchers. Most had Ph.D.'s and had come from academic scientific environments that had strongly rewarded publication. Researchers wanted to publish so that they could be seen as advancing knowledge generally, not just within their company, and gain wider recognition. Bruce Deal, a chemist at Fairchild, came to the semiconductor industry from Kaiser Aluminum. At Kaiser, Deal conducted research on the oxidation of aluminum but was not allowed to publish most of his work. Deal's frustration with Kaiser's secrecy was one factor that led him to the semiconductor industry. A group of researchers at Shockley Semiconductor specifically asked Shockley for permission to publish their work as a way to improve morale. Individual researchers would push to publish their work quickly, for if they did not, someone else might well publish before them and receive the recognition.[1]

At the same time, the industrial research labs greatly benefitted from being able to appropriate this openly exchanged information. A typical practice among labs was to publish internal reports on conferences detailing the key papers given by other labs. Even though the SSDRC went to great lengths to simulate an open academic environment, the proceedings of this conference were of more than academic interest. Researchers from RCA (and one imagines other companies) organized to cover all the sessions, taking copious notes on each presentation, so that details could be spread throughout the corporation. The key figures organizing these conferences were senior researchers from major industrial research labs, and not surprisingly, these conferences reflected the agenda of these labs.[2]

The 1964 IEEE SSDRC, a conference typically attended by the top semiconductor researchers in the country, showed the growing interest in MOS structures. In a review of the conference, an RCA engineer asserted: "The conference is usually dominated by one major topic which seems to be the hottest thing in that particular year." And in 1964 "it was the MOS structure which was 'it.'"[3] Seventeen papers were devoted to MOS, representing work from such major industrial research sites as IBM, RCA, Fairchild Semiconductor, Bell Labs, and Clevite (formerly Shockley Semiconductor). The MOS structure, as an integral part of the planar process, the dominant method used to fabricate bipolar transistors, was important even to organizations that were not working on MOS transistors, such as Bell Labs or IBM's Components Division.

Prominent features of the conference were informal "rump" sessions devoted to the main topic of the conference and even more informal discussions between the participants outside of conference sessions. The RCA reporter noted that "the MOS rump session was honored by the presence of W. Shockley." The conference's invitation-only attendance policy and its attendance limit of around 500 made for a situation where many of the attendees came year after year and knew each other fairly well. The RCA reporter observed that a standard practice at the conference was "'pumping' acquaintances that work[ed] for other laboratories."[4] Donald Young of IBM, who presented one of the most important papers at the conference, recalled being "absolutely besieged" by researchers afterwards. A researcher who had a number of contacts and worked them diligently could come away from the conference not only with a good idea of what other companies were doing but with a sense of what other researchers suspected might be promising.[5]

At the 1965 SSDRC, held at Princeton University, three out of ten sessions were devoted to MOS structures. The laboratories that had presented work the previous year described their more recent work and were joined by General

Electric, Sprague, and Texas Instruments. The conference included twenty-six papers and "a very spirited rump session" on MOS topics.[6]

While the MOS structure had dominated the SSDRC in 1964 and 1965, the conference was designed to focus on the most interesting new subject in the field each year, and it could not be expected to maintain its MOS emphasis indefinitely. In 1965 the IEEE Electron Devices Group sponsored a new conference devoted to the physics and chemistry of MOS structures, the Silicon Interface Specialists Conference. The specialist designation meant that it was intended to appeal to a much narrower group than the SSDRC. The initial leadership for the conference came from seven different companies, most of whom had substantial MOS programs. The conference's Las Vegas location (other major conferences were held in Philadelphia or Washington, DC-or on college campuses elsewhere in the nation) was a sign of the spirited younger leadership of the conference and the growing importance of the West in the industry. The first conference was limited to roughly seventy invitees, including the three key figures in MOS work at Fairchild, Deal, Grove, and Snow; the managers of the physics and chemistry groups in IBM Research's MOS program, Alan Fowler and Pieter Balk; and the three leading figures in RCA Research's MOS work, Hofstein, Heiman, and Zaininger. The Silicon Interface Specialists Conference became the key conference for MOS physics work.[7]

While it is impossible to determine the cash value of information exchanged at conferences, the effectiveness of the conference/journal exchange network might be seen by a specific case—the work of Snow, Deal, and Grove at Fairchild on sodium migration in silicon dioxide. The experimental work was done at Fairchild between October 1963 and February 1964, first presented to the larger technical community at the October 1964 meeting of the Electrochemical Society, and published in May and December of 1965. The work was published in two forms. First came a more academic description of the work in the *Journal of Applied Physics* and then a more popular description in the trade journal *Electrotechnology*. This later work, directed at customers, not fellow researchers, came out after Fairchild had introduced its MOS transistor and attempted to demonstrate that Fairchild had solved the problems of MOS stability, so that Fairchild MOS transistors could be bought with confidence.[8]

Although researchers at other laboratories had proposed other mechanisms for instabilities in MOS structures, the idea that sodium was a major contributor to instability was in the air at the time, and one can find anecdotal evidence of researchers who suspected that sodium was a factor prior to the Fairchild work. But it was one thing to have a hunch about something or to suggest it to colleagues at one's company, where many ideas might be brought forward and quickly discarded. It was quite another to do the formal experimental work that would persuade others in the field. By 1964 Fairchild was rapidly becoming known as the leading semiconductor industrial research lab, and its work gave strength and shape to the partially formed hypotheses of others. For example, while electrical engineers at RCA's David Sarnoff Research Center had been working on a number of possible mechanisms for MOS instabilities, chemists at RCA's semiconductor manufacturing and development facility in Somerville, New Jersey, suspected the role of sodium even before the announcement of Fairchild's work. The RCA chemists used the Fairchild work to win management support for a program to investigate the role of sodium and minimize sodium contamination in wafer processing. The Fairchild work gave the RCA chemists leverage in breaking free from the influence of RCA Research, and by April 1966 they had developed a process that yielded stable oxides with minimum levels of sodium.[9]

Although conferences and publications promoted an atmosphere of collegiality, they also fostered a spirit of competition. While most of the companies presenting papers at these conferences were competitors, the competition at the corporate level was abstract, with the success of a company's program dependent on a number of factors outside the control of a researcher, such as marketing, manufacturing, and management. At the SSDRC, the competition was personal: the conference provided an annual opportunity for researchers to see how their work compared with that of other labs. The best-known rivalry during this period was between researchers from RCA's David Sarnoff Research Center and Fairchild Research and Development (figure 5.1). Each group rushed to get its results reported before the other did, and a standard feature of most conferences was a heated debate between the two groups on the causes of instability in MOS structures. This competition proved a healthy boost to MOS work.[10]

The *IEEE Transactions on Electron Devices* published special issues devoted to MOS topics in 1965, 1966, and 1967. The articles, which were primarily expanded versions of papers given at SSDRCs, offer a view of the progress of

**Proposed Mechanisms for
MOS Instabilities**

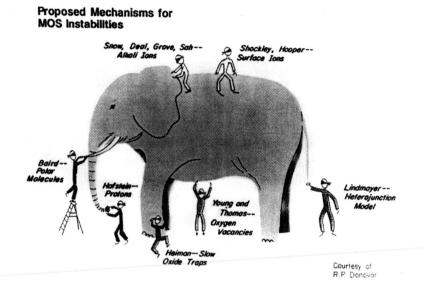

FIGURE 5.1 · BLIND MEN AND THE MOS TRANSISTOR: DRAWING PRESENTED BY ROBERT P. DONOVAN AT THE 1966 PHYSICS OF FAILURE IN ELECTRONICS CONFERENCE IN COLUMBUS, OHIO. This drawing, based on the poem "The Blind Man and the Elephant" by John Saxe, was meant to convey that each of the various conflicting theories of the causes of MOS instabilities was getting at a part of the truth. The presence of Fairchild's Ed Snow, Bruce Deal, Andrew Grove, and C. T. Sah on the head of the elephant was meant to suggest the general consensus developing in the industry that they had identified the key cause of MOS instabilities. Stephen Hofstein and Fred Heiman were the RCA researchers who were the chief rivals of the Fairchild group. Donald Young and J. Earl Thomas were from IBM. Steve Baird was from Texas Instruments. J. Lindmayer was from Sprague Electric. William Shockley and W. W. Hooper were with Shockley Research Laboratory. I thank Robert Donovan and Bruce Deal for helping me to identify the groups represented in the drawing. Drawing courtesy Robert Donovan and Bruce Deal.

the industry as a whole. The special issues were edited by George Warfield, a professor of electrical engineering at Princeton and a consultant at the David Sarnoff Research Center. In the 1965 issue, Warfield concluded that a "plateau" had been reached in the understanding of MOS structures, but he acknowledged: "Many people feel that the stability problem is still not completely understood." Warfield admitted that the journal contained papers that approached "the SiO_2 problem from different, sometimes conflicting points of view." The next year Warfield asserted: "It is the guess of many people working in this field that the goal of obtaining stable, high reliability, silicon devices is imminent." By 1967 Warfield made this claim: "This new field has progressed from its black magic phase, in which various and sundry mysterious potions coupled with assorted witchcraft were used to achieve 'good' devices, and has reached technological and scientific maturity."[11]

In view of the scientific maturity of the field and the thoroughness with which the silicon–silicon dioxide system had been studied, Warfield predicted that future work would be based on MIS structures (metal insulator semiconductor, a more general case of metal oxide semiconductor), where the insulator was something other than silicon dioxide. Although Warfield's claim that the field had moved beyond its black magic phase proved overly optimistic, the issue's bibliography of over 550 MOS papers gave credence to Warfield's assertions about the field's development.[12]

This bibliography, which was prepared for distribution at the 1967 Silicon Interface Specialists Conference by E. S. Schlegel of Philco, allows for an analysis of the field of MOS studies as it was perceived by those inside it. While the transistor might be in its adolescence, the bibliography showed that the MOS structure was much younger. Only thirty-five of the papers were published prior to 1962 (figure 5.2), and these thirty-five were clearly ones that were

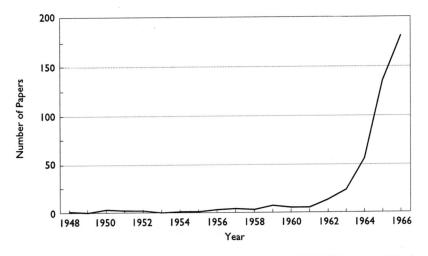

FIGURE 5.2 · GRAPH OF MOS PAPERS BY YEAR. From E. S. Schlegel, "A Bibliography of Metal-Insulator-Semiconductor Studies," *IEEE Transactions on Electron Devices* 14 (November 1967): 728–49.

retrospectively found to be relevant. Their authors were by and large not members of the community of MOS researchers in the mid-1960s: of the forty-eight authors of pre-1962 papers, only four contributed papers to the field in the post-1962 period. And only two of the pre-1962 authors were attendees at the 1965 Silicon Interface Specialists Conference. Perhaps the best example of this discontinuity is John Atalla, the inventor of the modern MOS transistor. After his initial work on the MOS transistor had received a cool reception at Bell Labs, he moved on to other areas. By the time the MOS transistor became a hot topic, he had moved from Bell Labs to hp associates, the semiconductor affiliate of Hewlett-Packard, and he never published another MOS paper (although he did attend the 1965 conference).

The Broader Community and the Benevolent Despotism of the Planar Process

The explosive growth in MOS papers after 1962 can be attributed to the development of the planar process, the rise of the integrated circuit, and the accompanying interest in the MOS transistor. In the early 1960s the planar process became the dominant means of producing semiconductors, and the MOS structure was an integral part of it. And so the MOS community was not limited to an elite few who attended the Silicon Interface Specialists Conference or published papers on MOS topics. Because the materials and processes used in making MOS transistors were essentially the same as those used in making the silicon planar bipolar transistor being manufactured by the millions in the 1960s, the thousands of people making bipolar transistors contributed to the cause of the MOS transistor, even if they had never heard of it before. The industrywide commitment to the planar process and the continual learning that came with large-scale manufacture resulted in a kind of despotism before which no competing technology could stand.

The planar process had been developed at Fairchild Semiconductor in 1959, and its chief advantage initially had been that it resulted in transistors whose characteristics were more stable over time than the transistors that had been available previously. While Fairchild vigorously promoted the advantages of the planar process, the planar process acceded to its position of dominance with the industry's acceptance of integrated circuits, for the planar process was an extremely effective way to make them. Furthermore, silicon transistors made with the planar process could be used in most applications where the previously ascendant germanium transistor had been used. By 1963 thirty firms were using the planar process to make silicon transistors or integrated circuits. Most, if not all, of these firms were infringing on Fairchild patents, and as the importance of the planar process became clear, Fairchild sought to force makers of planar devices to buy a license to use it.[13]

Prices of silicon transistors and integrated circuits fell sharply throughout the 1960s. The transistor that Fairchild had sold for $150 in 1958 sold for less than 10 cents in the mid-1960s. In 1964 Fairchild lowered the price of an integrated circuit flip-flop (which had been $120 in 1961) to $6.35. Two factors made these price cuts sustainable. First, the market proved extremely elastic, and as prices fell, users (aided by semiconductor producers) found many new ways to use transistors and integrated circuits. Second, costs decreased dramatically as manufacturers oriented their operations to larger-scale production, the skills of those manufacturing the devices increased, and everyone learned more about silicon devices.[14]

For the sake of contrast one might look at work on thin-film transistors (TFTs) and planar germanium transistors, two technologies that had great potential to some but, because of their weak links to existing semiconductor technologies and their inability to attract industrywide support, ultimately failed to live up to the promise that they had shown for digital integrated circuits. The TFT, invented by Paul Weimer of RCA in 1961, was very similar in the physics of its operation to the MOS transistor. But instead of being made from silicon, it was made from polycrystalline cadmium sulfide, or other II–VI compounds,[15] evaporated onto a glass plate. Although these materials were used as photoconductors in display applications, they were not commonly used in the semiconductor industry to make transistorlike devices. While the TFT offered many interesting and exciting technological possibilities, the semiconductor industry did not have the commitment to cadmium sulfide, or the other II–VI compounds, that it did to silicon. Fairchild R&D dabbled briefly in TFTs but kept its main focus on single-crystal silicon and never launched a sustained program in the area. IBM Research had a TFT program around 1962 but halted it two years later, considering it uncompetitive with either the MOS transistor or the silicon bipolar transistor. Although the decision to stop a research program is often difficult and traumatic for research managers, IBM, with no stake in the materials or techniques used in the TFT, cancelled the program quickly and painlessly. In 1979 a senior manager at IBM Research's semiconductor area stated: "I don't believe that those of us who had positions of serious responsibility in this program took the TFT too seriously."[16]

Peter Brody, a key figure in later TFT work, asserted that eight industrial research labs were working on TFTs in the early 1960s but that "by the mid-1960's the battle was over: MOS was the victor and TFT the vanquished, with only RCA and Westinghouse continuing any significant research beyond this date."[17] RCA's continuing commitment to the TFT owed much to its work in television and its hope that TFTs could be used for displays, an area outside the scope of most semiconductor companies. Brody, calling the MOS transistor and the TFT "dissimilar twins," offers the following analysis of why the MOS transistor flourished and the TFT did not:

> This outcome was not surprising: not only were the resources devoted to the respective efforts greatly different in size, but the MOS also had the advantage of relying on familiar and well-understood materials and processes, in comparison with the poorly characterized polycrystalline II–VI compound films and evaporated insulators used in the TFT. By 1966, the stability and fabrication problems of MOS were more or less solved through a massive, industry-wide effort, while the TFT's were still poorly reproducible, unstable, and effort on them diminished day by day. This 'negative feedback' effect rapidly became fatal for any prospect of TFTs being used for mainstream logic circuits.[18]

In the mid-1960s, some in IBM Research believed germanium held promise. Silicon's rise to dominance in the early 1960s came about due to its superior high-temperature characteristics and its native oxide, which allowed for oxide masking to produce transistors using the planar process. The native oxide of germanium was water soluble and thus not useable in transistor production. (The wafers were washed during processing, removing the germanium oxide.) But electrons move faster in germanium than they do in silicon, giving rise to the hope that if some way were found to grow a non-water-soluble oxide on germanium, allowing the use of planar fabrication techniques, transistors superior to those possible in silicon could be made. IBM Research formed a program based on that promise in 1963. After a failed effort to persuade Texas Instruments to collaborate, Research proceeded alone in a seventy-five-person effort, a large program by IBM Research standards. In 1968 the program was cancelled after both Research and the Components Division determined that germanium offered no competitive advantages over silicon technology. Although there had been a great deal of work on germanium done in the 1950s, IBM Research, as the only one in the field in the 1960s, was forced to solve all the technological problems of making planar germanium integrated circuits on its own—a daunting task even for IBM.[19]

Furthermore, the dearth of effort in germanium outside of IBM made it difficult for Research to convince others in IBM of germanium's value. The lack of any mention of research in germanium in the trade press would hardly create an atmosphere conducive to germanium research within IBM. If other companies had passed on it, top IBM management would surely wonder why IBM should pursue it. A competitor's plan to use germanium integrated circuits in a new computer system, for instance, would have motivated serious consideration of germanium by IBM, and the absence of such a challenge made it easier to ignore germanium.[20]

This is not to ignore the inherent technological advantages of silicon. Silicon dioxide was easily grown and had many outstanding properties, such as its passivating effect on the surface and its ability to mask out dopants. But these advantages were amplified by the much greater resources applied to silicon. The seventy-five people at IBM Research had no chance of competing with the thousands of people, in R&D and manufacturing, working on silicon both within IBM and elsewhere.

The Talented Individual: Frank Wanlass, MOS Missionary

Although thousands of people—operators, technicians, engineers, and scientists—worked on the silicon planar technology, the number working on the MOS transistor was small enough that one talented individual could have an enormous impact on the field. Frank Wanlass was such a person. Although not a prolific author, he proved to be one of the most important figures in MOS technology in the 1960s. Those who presented papers at the SSDRC were by and large physicists and chemists. And although they knew much about the physics and chemistry of silicon and silicon dioxide, they did not necessarily know the optimum way to manufacture MOS transistors, and they surely did not know how to design MOS integrated circuits. Wanlass, who considered himself more an inventor than a scientist, did, and he played an important role in spreading this information, which was less likely to be the subject of papers at conferences, throughout the industry. Wanlass, completely committed to the MOS transistor but not an entrepreneur, was more interested in the technology than business success and spent most of his career working just beyond the edge of the possible.

Wanlass provided the driving force behind the two firms that made the first major push to sell MOS large-scale integrated circuits, General Microelectronics (GME) and General Instrument. In December 1963, Wanlass, who had been at Fairchild a year and a half and had become frustrated at what he considered Fairchild's emphasis on studying MOS physics as opposed to developing and selling MOS products, left to join GME. GME, formed in the summer of 1963 by technical and marketing people from Fairchild, hired Wanlass to start its MOS business.[21]

At GME, Wanlass continued his highly impatient, highly individualistic style of work. Wanlass, extremely creative but uncomfortable in large organizations, followed his own lead rather than attempting to conform to the dictates of a corporate organization or strategy. He avoided detailed theoretical studies and concentrated on working in the lab to make useful MOS transistor circuits. While a standard procedure in industrial research labs was to have technicians doing the hands-on work for Ph.D.'s like Wanlass, Wanlass did his own experimental work and device fabrication. Wanlass was unwilling to be dependent on a technician, for if he did things himself, he could be sure they were done his way. Wanlass attempted to attack simultaneously the problem of making stable MOS transistors and the problem of finding a profitable MOS product. The centerpiece of Wanlass's efforts to produce stable MOS transistors was the use of electron beam–evaporated aluminum, the significance of which had been discovered just prior to his departure from Fairchild. At GME Wanlass appears to have come to the realization that sodium was an important component in the drift of MOS devices. While the Fairchild Solid State Physics Department used the reductionistic MOS capacitor and the capacitance-voltage plot to measure stability numerically, Wanlass generally used actual packaged transistors. His typical technique was to put the transistor in a curve tracer, heat the transistor with a soldering gun or a cigarette lighter, and then make a quick assessment of the device's stability based upon how much the heat caused its characteristics to jump.[22]

Wanlass's main focus was MOS products. GME had introduced an individual MOS transistor by May 1964, several months ahead of a comparable product from Fairchild. Wanlass developed outstanding abilities in MOS circuit layout, a crucial skill dependent on creative spatial thinking rather than solid-state physics. He designed and built a twenty-bit shift register, not because there was a specific demand for it but to demonstrate the capabilities of MOS to potential customers. Wanlass held many meetings with prospective customers to promote MOS technology. At the 1964 Western Electronic Show and Convention, one of the industry's premier technical and trade conferences, GME rented a hotel room to demonstrate the part. Through Wanlass's work promoting MOS, Lee Boysel, a key figure in later MOS work at Fairchild, and A. R. DiPietro, a key advocate of MOS technology at IBM, received introductions to the technology.[23]

The government was GME's most important early MOS customer. Art Lowell, the founder of GME, was a retired Marine colonel who had previously headed the avionics branch of the Navy's Bureau of Weapons, and his connections seem to have been important in winning government contracts. This first MOS contract was for a chip with six to seven MOS transistors for NASA's Interplanetary Monitoring Platform satellite. MOS circuits provided low power and high density. Another important early GME customer was the National Security Agency, which was interested in MOS technology for the very high circuit density it allowed. Neither of these contracts was of a size to carry GME's MOS business, however.[24]

Lowell was a promoter with a touch of P.T. Barnum in him, and GME propagated an image of itself that at times far exceeded the reality. After he named the new corporation, he encouraged speculation that the company was backed by General Motors and claimed it would build twelve plants throughout the country within three years (GME was in fact financed by Pyle-National, a Chicago-based manufacturer of electronics equipment that was somewhat smaller than General Motors). For his part, Wanlass, in his zeal for MOS technology, was optimistic that all MOS's problems could be solved in short order. In mid-1964, with Lowell providing the business enthusiasm and Wanlass the technical enthusiasm, GME entered into an agreement with Victor Comptometer to provide a calculator made with MOS integrated circuits (figure 5.3). The calculator required twenty-nine MOS chips, each with complicated circuit designs involving a hundred or more transistors. Wanlass was extremely bright and might have been able to solve the problems of making an MOS calculator if he had been able to deal with each problem himself. But this project proved too large for one person, and in any case Wanlass left GME in December 1964, with the project still in its early stages.[25]

Although GME had a working prototype by the fall of 1965 and the calculator was then introduced at the Business Equipment Manufacturers Association trade show, things fell apart as GME tried to put the calculator into production. While GME tripled its floor space and its production staff, it was unable to make the chips in large quantities. On some chips the yields were zero. The engineers at GME had only a rudimentary knowledge of the chemistry and physics of MOS structures. Furthermore, the Victor calculator required a number of technical advances in MOS technology that proved to be greater than GME could handle on its allotted schedule. To put the required number of MOS transistors on a chip required extremely tight tolerances—tolerances that could not be maintained in production conditions. Although the photoresist purchased from Kodak had sufficed for use in smaller bipolar circuits, it proved unsuitable for use with the tight tolerances required in the MOS chips. Almost half of the chips had to be redesigned. Because each chip represented not just a single logic gate but many connected gates, GME had to develop new ways of testing chips to ensure that each worked completely. GME's problems were exacerbated by its acquisition by Philco-Ford in January 1966, which led to a large exodus of personnel.[26]

The fact that the Victor calculator was a system composed entirely of MOS technology made it an attractive product but ultimately was its downfall. For the MOS technology to prove itself viable, someone had to find an application in which the MOS transistor's physical characteristics translated into otherwise unachievable system characteristics. It was relatively easy for GME to produce MOS parts, such as shift registers, that contained many more transistors than were possible using bipolar technology. But it was not enough for MOS merely to be better: just because one could make larger shift registers with MOS technology than with bipolar technology did not mean there would be a demand for them. A calculator made totally from MOS circuits could be much smaller

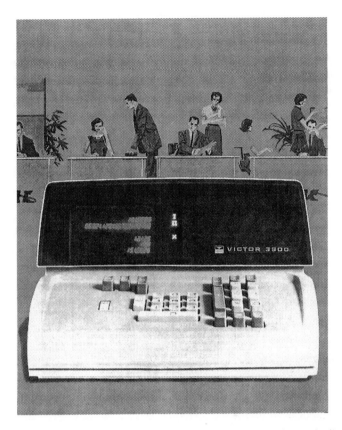

FIGURE 5.3 • THE VICTOR 3900 CALCULATOR. The calculator had a cathode ray tube display that was capable of displaying four lines containing twenty-two digits. The machine performed four functions and had two memories. Its initial price was $1,825, and it weighed twenty-five pounds. This figure, from the cover of a brochure, shows it was being marketed as a business machine. Courtesy Heinz Nixdorf MuseumsForum, Paderborn, Germany.

and much cheaper than one made with bipolar parts, but making even this small system proved to be beyond GME's capabilities. After spending several years fighting production problems, Philco-Ford stopped work on the calculator in June 1968.[27]

While GME was in the early phases of its struggle with the Victor calculator, Wanlass became impatient again. He made plans with a friend from graduate school and some of his colleagues at GME to start a new MOS firm that would be based near Santa Barbara. Before the deal could be consummated, Wanlass backed out, joined General Instrument (GI), and convinced several of his co-conspirators to do the same. In the early 1960s GI, a manufacturer of electrical and electronic equipment, decided to make a strong entry into the business of integrated circuits. It hired experienced semiconductor managers from Philco and IBM and, because it was a late entrant into the bipolar field, hired Wanlass at its Hicksville, New York, facility with the aim of getting into the nascent MOS business.[28]

At GI, Wanlass's first product was a twenty-one-bit shift register whose purpose was to wrest the title of largest shift register away from GME, which offered a Wanlass-designed twenty-bit part. (Obviously, there were difficulties in coming up with original products in MOS.) Later, GI introduced fifty- and ninety-bit shift registers. Shift registers were particularly attractive because they were serial. Although MOS technology made it possible to put hundreds of transistors on a chip, there had been no comparable breakthrough in the number of leads on a chip package, which frequently turned out to be the limiting factor in chip design. With a shift register, the number of pins required did not increase as the number of transistors inside (or the number of bits) increased.[29]

Another part of great interest to GI and those interested in MOS was the digital differential analyzer (DDA). Again, its serial architecture allowed it to have relatively few output pins, and shift registers were used to feed it data. Many saw it as a first step towards a computer on a chip, and it had potential applications in inertial guidance. GI introduced its first DDA in August 1966.[30]

By 1967 Wanlass had grown weary of Long Island. As GI struggled to move MOS into production, Wanlass became more and more unhappy with GI's unionized workforce, and he and a colleague proposed moving the entire MOS operation back to their home state of Utah, where they could be free of what they saw as union interference and corporate bureaucracy. Although GI executives initially agreed to the plan, they ultimately retained production on Long Island, allowing Wanlass to set up an R&D laboratory in Utah. Whatever other effects it had, such as lengthening the lines of communication between manufacturing and development, the lab in Salt Lake City (an inauspicious place for MOS work, given the chief culprit in MOS instabilities) kept Wanlass happy and at GI for several more years.[31]

In its 1969 review of the MOS business, *Electronic News* asserted that GI was the industry leader, with estimated sales in 1968 of $7 million. But Wanlass left GI in 1970, and GI quickly lost its position of leadership in the 1970s as MOS sales took off. The importance of Wanlass and his work at GME and GI came not from sales of MOS products but from the role GME and GI played as training grounds for engineers and centers for the diffusion of MOS technology. For all his technical genius, Wanlass had very little business savvy and was often more interested in developing an interesting new technology than in making money for the company he worked for.[32]

Many companies received significant inputs, either directly or indirectly, from Wanlass. Texas Instruments' first MOS part had been reverse-engineered from GI. Wanlass and others from GI gave IBM important counsel in the early days of its MOS program (see below). Even Fairchild, with scores of researchers, benefited from Wanlass's work after he left. Although Fairchild's scientific talent was extraordinary, it still had much to learn from Wanlass in terms of techniques. In 1966 Fairchild hired Lee Boysel from IBM to start an MOS circuit design group. While at IBM, Boysel served an apprenticeship under Wanlass at GI. In 1967 Fairchild hired Bob Cole, who had worked with Wanlass at GME, as a lead engineer in its MOS manufacturing operation. The June 1966 report of the Device Development Department at Fairchild R&D mentioned an MOS technique brought to Fairchild "based on information resulting from previous employment," apparently at either GME or GI, with Wanlass the source.[33]

Intel, which had the cream of Fairchild's R&D MOS talent, recognized Wanlass's abilities. Gordon Moore made an effort to hire Wanlass for Intel, but Wanlass turned him down. Within the first few months of its founding, Intel's lead MOS engineer attended a seminar held by GI at which Wanlass described GI's MOS work. Intel had already decided on a technological approach different from GI's, but hearing about GI's work offered Intel the opportunity to size up its plans against those of the industry's technological leader.[34]

Wanlass's role illustrates the importance of tacit knowledge. The design and production of MOS transistors had craft elements that could not be transmitted through papers. While the papers of Deal, Grove, and Snow were the fundamental citations of MOS technology, they were not sufficient to teach someone to design and build MOS circuits. Wanlass's skills, such as in designing and laying out integrated circuits, could be transmitted only personally, which meant that Wanlass was much less widely known than Deal, Grove, and Snow. In spite of Wanlass's absence from the elite professional journals, his abilities were very clear to industry insiders such as Moore.[35]

Interfirm Communication: The Case of IBM Research

At GI and GME, Wanlass met with representatives of other companies to promote MOS technology. In March 1965, IBM invited Wanlass and others to visit, ostensibly to discuss the possibility of GI developing MOS memory chips for IBM. For IBM the meeting had another purpose as well, for among those in attendance were key figures from Research's MOS effort. Wanlass's visit provided Research with the perfect opportunity to evaluate its program against GI's and pick Wanlass's brain.

Wanlass, as an MOS enthusiast more than a businessman, proved more open and candid than someone more concerned about his company's success would have been. Wanlass gave Research his candid assessment of its program. He was skeptical of the value of the programmable interconnect pattern approach, which used computers to wire up all the good circuits on a chip—an assessment that some in Research were coming to on their own. Wanlass thought IBM's plan to build MOS transistors with gate lengths of 50 microns was too conservative and suggested moving down to 12.5 microns instead. (The transistor's gate length is one of the key determinants of its speed.) As IBM Research started its MOS work it was not clear where it should be in terms of balancing what was possible against what would provide the optimal performance. Wanlass, with his greater experience with the technology, was able to provide guidance. A Research manager gave further details of the meeting: "G. Cheroff and I had lunch with Wanlass. We discussed broadly some of the fabrication procedures. Some of the key areas we have been investigating and which are consistent with the GME or GI production are: (1) the use of dry oxide, (2) the care taken in keeping a low level of sodium in the various solutions, etc., that the wafer is exposed to, (3) the effect of hydrogen and water."[36]

Wanlass's meeting with IBM was typical, not just for Wanlass or IBM, but for the industry as a whole. Each company involved in MOS research received information from other MOS programs in a variety of forms. While each firm had different networks of communication, the example of IBM Research shows these networks' extent and significance. Interfirm interactions took many forms, including informal conversations, intercompany meetings, confidential exchange agreements, and reverse engineering of artifacts. While there are some cases where it is possible to infer that a specific interaction had specific results (such as in Wanlass's meeting), more importantly these interactions transferred the industry's knowledge of MOS to IBM. As IBM Research considered potential directions in one area of its MOS program, it did so knowing what other companies were doing in that area. This did not necessarily mean that IBM Research would follow suit, but Research, which respected the work of others, could use this knowledge to assess the strength and weakness of its approach.

In analyzing IBM's intercompany contacts it is important to distinguish between IBM Research and IBM's Components Division (CD), the site of IBM's semiconductor production. Research was a much more astute observer of the industry and much more likely to adopt work done elsewhere. As mentioned previously, CD's position outside the semiconductor industry made it largely immune from competitive threats and allowed it to ignore what others were doing. Moreover, Research, working on projects aimed at realization in the more distant future, had the freedom to consider all possible alternatives, while CD, often working on a tight time schedule, felt more keenly the constraints imposed by its production technology. Research generally had a much more cosmopolitan outlook than CD, with more emphasis placed on publishing papers in journals and winning respect in the industry at large. In CD, recognition went to those who implemented a new technology that went into production, irrespective of whether an article was ever written about it or whether it was ever lauded in the industry at large.

One of the first interactions came with the formal announcement of the MOS transistor by RCA in February 1963. The week the announcement was made, J. Earl Thomas, a senior manager at CD who had made a wide assortment of contacts through previous stints at Bell Labs and MIT's Lincoln Labs, called his acquaintance Bill Webster, who was in charge of electronic device research at RCA's David Sarnoff Research Center. Webster told Thomas, "I think this is really a big thing" and suggested that RCA was launching a major program. Webster also gave indications that transistors made by researchers at RCA had only small variations from unit to unit, although he admitted the results from

RCA's production facility were somewhat worse. Thomas wrote a summary of his conversation and sent it to managers in both CD and Research.[37] Research was planning an MOS program at this time, and this knowledge of the broad outlines of another industrial research laboratory's program took away some uncertainty. If the technology ultimately proved intractable IBM Research management could point to RCA's failure alongside its own. And Research management could well imagine that once it became known that one of IBM's chief competitors was working in MOS, Research would be asked what it was doing in the area.

IBM had formal confidential exchange agreements with Bell Labs and Texas Instruments (TI), under which representatives of the companies met periodically to share information. Bell Labs, a late entrant to MOS transistor work, was pursuing a very different course from the rest of the industry, using an aluminum oxide–silicon dioxide double layer for the gate insulator, instead of silicon dioxide. While this promised to lead to a more reliable transistor, it also resulted in fabrication difficulties. Researchers from IBM left from a visit to Bell Labs' Allentown facilities convinced that Bell Labs was headed in the wrong direction. IBM did make use of a double-walled diffusion furnace design from Bell Labs, however. At IBM's meetings with TI, TI's engineers and scientists agreed that IBM Research's n-channel direction was promising, but they claimed they had been unable to win management support for their own n-channel program. Researchers from IBM called their colleagues from TI to check that measurements they had made on MOS structures were in agreement with TI's measurements. IBM and TI also exchanged information about methods of water purification.[38]

Much could be learned from actual devices that could not be gleaned from meetings. IBM bought transistors from all MOS producers and put them on long-term test, as a way of determining the stability levels other firms had been able to achieve. IBM also procured devices to reverse-engineer them and look for process changes. One member of the Research group had responsibility for keeping track of other firms' work on MOS stability. In September 1966, analysis of a recently purchased Fairchild integrated circuit showed a stability "equal to or better than anything seen on any devices to date," leading the IBM Research manager in charge of MOS devices to observe: "If this is being realized on any kind of yield basis in production, it is a significant accomplishment."[39]

That same month, analysis of GI's integrated circuits revealed changes in its process, which were verified through later conversations with GI personnel. In 1968, an IBM Research memo documented the specifications GI was using for laying out chips. While this did not necessarily imply that IBM would use the same specifications, it had a basis for comparing whatever specifications it developed.[40]

Because of the empirical nature of semiconductor processing, significant capabilities could be transferred between companies through the movement of technicians. In the mid-1960s chemists at RCA developed a rinsing technique for cleaning wafers used to make MOS integrated circuits. The key part of the technique was a recipe for the chemicals used in the rinse, which RCA did not publish until 1970. Around 1968, a technician moved from RCA to IBM and brought the details of the rinse with him. Empirical studies at IBM Research showed the rinse's superiority to IBM's existing technique. This rinse, known at IBM as the Huang Clean, after the technician who brought it to IBM, became the standard cleaning method used at IBM for many years. The rinse also became an industry standard, called the RCA Clean outside IBM.[41]

Taken as a whole, these external connections were crucial to the ultimate success of IBM's MOS program. These relations were vastly more important than relations with more basic research groups within IBM Research. IBM's MOS network meant that its MOS program was highly leveraged—the handful of people working on MOS at IBM could capitalize on the work of hundreds of others in the industry. IBM Research did not have to solve every problem itself. And even though MOS was a new technology, IBM could draw its bearings from others in the industry.

While the industrywide MOS effort helped Research implement the technology, it also helped Research sell the technology within IBM. Many of those in IBM's system development groups who were considering MOS technology and were in contact with the Research group had initially become interested in the technology through advertisements and articles in trade journals. As they made contact with an IBM purchasing agent to buy MOS parts, the purchasing agent informed them of the Research work.[42]

In some cases internalizing functions within a corporation so that administrative coordination replaced market coordination reduced transaction costs, but in this case the market provided a convoluted means of intrafirm communication that was more effective than any direct method IBM's managers could devise. The market was better able to disseminate information broadly within IBM, and the information transmitted by the market was considered more reliable than information coming directly from IBM Research. IBM was a very large and decentralized company with hundreds of development projects scattered over dozens of sites within the United States and Europe. It would have required an extraordinary effort on Research's part to communicate with every potential user of MOS. But one might imagine that almost all of these groups were reached by such trade periodicals as *Electronics* or *Electronic News,* which had news of the latest work in the semiconductor and computer industries. Research lacked credibility with others in the company because of its previous failures, so its assessment of MOS technology would have counted for little had it been alone in its advocacy of MOS. But Fairchild, the industry's technical leader and the pioneer in integrated circuits, was also working on MOS, and its work and the work of other companies gave Research's MOS program credibility.[43]

The industrywide work on MOS gave Research leverage as it tried to make the case for the new technology within IBM. In a 1966 memo, Sol Triebwasser, the manager in Research responsible for all MOS work, attempted to get CD to reconsider its decision not to use MOS technology. Triebwasser mentioned four recent developments bearing on MOS technology—two in Research and two in the industry at large. While some in IBM had questioned the reliability of MOS technology, Triebwasser claimed that an industrywide consensus had been reached that MOS was reliable: "The various industrial contacts we have been having at meetings and in random discussions indicates that [the] general industry attitude is that reliability is not a problem in the FET [MOS] technology. NASA had indicated that they have devices of this kind in orbit for some time with no failures." "The major monolithic manufacturers, T.I. and Fairchild, both now express very positive attitudes toward FET [MOS] technology," he added. "This was not clear six months ago."[44] As IBM Research fought to have MOS technology established within IBM, the work in the industry provided independent verification of Research's assessment of MOS.

While those in IBM's development and production organizations were not bound to follow the rest of the semiconductor industry, even they recognized its power. In 1965 Erich Bloch, the head of memory development within IBM's

Systems Development Division, wrote a memo arguing that it was premature to consider an MOS memory program. Among the questions that had to be answered before it could be determined if an MOS memory program made sense, he listed: "What is the effort the total industry is putting in pn junctions vs. MOS devices? It seems to me that this would have a great bearing on the progress one could foresee in the two technologies."[45] Bloch's question was an admission that the direction of the industry put constraints on even a company as large as IBM; although IBM had millions and millions of dollars to spend on research and development, it was advantageous to have programs consistent with industry trends.

The Tyranny of the Majority:
The Case of RCA and the Military

IBM Research's MOS program gained huge advantages by being within the scope of other industry programs; while it did not march in lockstep with them, it was close enough to receive collateral benefits. RCA's work on two complex variants of MOS technology shows the difficulties that could result from running too far ahead of the industry. Even though the military funded RCA's chosen technologies, they were shunned by the rest of the industry. Military funding kept the programs going—a mixed blessing—but it could not ensure a technology competitive with the simpler technology the industry was developing through continuous incremental improvements. The importance of the markets created by the rest of the industry ultimately forced RCA to buy the capability to make industry-standard MOS transistors from other companies.

Although some economists have written of the military's funding of technology development in the post–World War II period as though it were an unalloyed blessing, military support did not translate into commercial success for RCA's MOS program. RCA was the first company to pursue MOS technology vigorously, with much of its work done under government contract. While it continued as one of the leaders in MOS research throughout the 1960s, it ranked only thirteenth in MOS sales by 1971. Military funding reinforced tendencies that were present within RCA itself—tendencies that steered it outside of the technological mainstream.[46]

Both RCA and the Air Force, the principal funding agency for RCA's programs, had extremely long-range R&D outlooks that tended to favor revolutionary new technologies. The Air Force funded two of the most radical semiconductor programs in the 1950s and 1960s, gallium arsenide and molecular electronics. Although precisely what molecular electronics was remained a mystery to many in the industry, its premise was that one could eliminate traditional circuit elements by performing circuit functions directly at the molecular level. At a 1961 Air Force R&D conference, one of the Air Force's leading electronics experts scolded those in the industry for putting too much effort into research on silicon and germanium and not enough into looking for new phenomena.[47]

RCA Research's long-range outlook—up to twenty-five years ahead of the state of the art—came about because it had left near-term development work to the product divisions. This left Research able to pursue technologies, even when their commercial prospects were unclear. At the same time the military became an important ally for Research as a way to force RCA's product groups to show an interest in Research's technologies. Military funding also lowered costs in a period when RCA was increasingly investing in areas outside of electronics.[48]

The two complex strains of MOS technology that RCA Research pursued, complementary MOS (CMOS) and silicon on sapphire (SOS), had special resonances for RCA. In the early 1950s an engineer at RCA wrote a paper on the complementary symmetry properties of bipolar transistors. George Sziklai argued that because there were two different types of bipolar transistors (npn and pnp), it was possible to build powerful complementary circuits that had not been attainable with vacuum tubes. This principle of complementary symmetry became well established at RCA, so that shortly after Paul Weimer invented the TFT, he invented circuits using complementary symmetry, which were a close relative of CMOS. In 1963 Frank Wanlass, still at Fairchild, disclosed his CMOS circuit, which applied the principle of complementary symmetry to MOS transistor circuits. RCA researchers, after overcoming their initial annoyance at themselves for not having invented it before, embraced it fully. While complementary symmetry had led to some interesting bipolar circuits, in MOS it made possible power reductions of six orders of magnitude, making it very plausibly the ultimate low-power circuit. As chapter 3 discussed, two types of MOS transistors could be made: n channel or p channel. Most companies worked on *p* channel because it was simpler, while IBM worked on *n* channel because it was faster. CMOS consisted of putting both *n*-channel and *p*-channel transistors together on a single chip, creating something much more complicated than having all *n*-channel or all *p*-channel transistors on a chip and eliminating much of the simplicity that had been MOS's initial attraction. Initially, RCA was the only company that made any serious effort at developing CMOS. Fairchild, which had invented CMOS, looked for a technology that could be implemented sooner and largely ignored the possibilities of CMOS in favor of simpler alternatives.[49]

In 1965 RCA won a large contract from the Air Force for R&D on CMOS. RCA considered this contract, which called for building a computer with CMOS memory, crucial in establishing its position in digital integrated circuits. The Air Force's interest in RCA came from its CMOS work, whose low power offered advantages in avionics. With this contract RCA devoted almost its entire MOS program to CMOS. G. B. Herzog, a senior manager at RCA Research and the leading advocate of CMOS within RCA, asserted that CMOS was met with skepticism, if not hostility, by those in RCA outside of Research: "With a three-year contract from the Air Force, RCA had a legal obligation to pursue its work on CMOS. I am convinced that without this commitment stretching over an extended period, the CMOS array effort would have been killed at some point, as we experienced one failure after another. . . . Thus it was that the discipline of an outside agency provided the means for the Laboratories' [Research's] staff to keep alive a project in which they believed."[50]

Although Herzog viewed the commitment to CMOS positively, the Air Force's contract provided a stability that locked RCA out of other technological options very early in the development of MOS technology, when adaptability was key. In 1966 RCA's Defense Electronics Group needed MOS integrated circuits of a type that were widely available in the industry. Although RCA had a longer history of MOS work than any other company, RCA Research's pursuit of the more complex CMOS technology slowed down its MOS work such that it could not provide a usable technology in a timely manner. The Defense Electronics Group paid an MOS start-up company, American Microsystems Inc., for the right to its MOS process.[51]

RCA's CMOS efforts did bear some fruit. In 1967 RCA received further contracts for CMOS research from NASA, and the next year RCA introduced CMOS circuits as commercial products. Although CMOS was well suited for applications that required low power, its complexity restricted it to much simpler circuits than were possible using either *p*-channel MOS or *n*-channel MOS. RCA's concentration on CMOS effectively limited the company to a niche market and conceded the market for large-scale integration, such as memories, to other firms. In 1968, RCA was not in the top six in MOS sales. By 1971, it ranked behind such recent start-ups as Intel, Mostek, and Electronic Arrays. By 1974 it was clear even to RCA how large and important the market for MOS semiconductor memories was. To enter the field it had to pay a start-up company, Advanced Memory Systems, $750,000 for the rights to use that company's *n*-channel MOS memory technology.[52]

The other MOS technology that RCA Research favored was SOS, silicon on sapphire, an RCA invention. As the name implied, the MOS transistors were built on a layer of silicon grown on a sapphire substrate. SOS was to RCA a marriage of a standard MOS transistor with Paul Weimer's TFT—invented by RCA—and RCA hoped SOS would have the best features of both. It used silicon and so avoided the stability problems of the TFT, but the substrate was insulating, like the glass substrate on the TFT, so there were lower parasitic capacitances in an SOS transistor, making it faster. In 1965 the RCA annual research report asserted a belief that SOS would "give higher performance circuits, more flexibility, and possibly economic advantages."[53] The sapphire substrate made it more costly than single-crystal silicon and led to a number of technical problems. The Air Force backed SOS R&D for many years, but without giving it a large procurement contract. In the absence of such a contract RCA was unwilling to put SOS into production. RCA and Autonetics (later Rockwell), also with military support, were the only ones in the industry working on SOS in the 1960s.[54]

In 1970, Joe Burns, one of the leaders of RCA's SOS research effort, frustrated at RCA's reluctance to move the technology into manufacturing, left to form his own SOS company, Inselek. Inselek received a great deal of publicity in the trade press and offered a wide range of SOS products that had faster speeds than MOS. While the sapphire substrate led to speed advantages, it also made it much more costly. It proved uncompetitive for computer memories, a huge, cost-sensitive market. In 1974, L. J. Sevin, the president of Mostek, an MOS firm that had been dabbling in SOS, asserted: "I think for the next several years, if the substrate is not silicon, the part is in trouble. Silicon is the most cost effective substrate."[55] In the 1970s, as the large volumes of standard MOS integrated circuits led to lower and lower prices, the high prices of SOS confined it to an increasingly small market. In July 1975, Inselek, unable to get customers for this new and more expensive technology in the midst of an industrywide slowdown, filed for bankruptcy protection. RCA continued its work on SOS without ever making substantial inroads into the MOS market or finding significant new applications.[56]

While SOS languished, the irony of RCA's concentration on CMOS was that it was ultimately right. By the late 1980s CMOS became the only viable MOS technology, with its extremely low power dissipation making it possible to put millions of transistors on a chip without the chip melting. But RCA, which no longer existed as a corporation, was right on the wrong time scale. By immediately attempting to move to the technology's end point, RCA ignored how the technology evolved and locked itself out of lucrative markets. Other firms were able to take a simpler—one might imagine the RCA engineers would say inferior—technology; make profitable products; and, as they gained experience with the technology, add incremental improvements.[57] SOS ultimately fell before the onslaught of the silicon planar process, just as the TFT and germanium had before it. The steady influx of military funding made it difficult for RCA to face up to this reality, with RCA typically having to pay only 50 percent to 70 percent of the research costs for these programs.[58]

Conclusion

The industrywide commitment to the planar process and the communal nature of semiconductor R&D posed a dilemma for R&D managers. Their job was to develop new things and to consider what technology would come after the silicon planar process. Furthermore, it might be tempting to believe that a competitive advantage could be achieved by moving beyond existing practices or in directions different from those taken by the rest of the industry. Without suggesting that such inclinations never paid off, overall the industry exacted a severe penalty from those who strayed too far from its direction. As a company veered from the industry's trajectory, the industry's work became less applicable and the wayward firm was forced to pay a greater share of the development costs, both financial and intellectual. It took the further risk that it would miss out on whatever markets the industry was able to create. Although IBM's decision to work on *n*-channel MOS was costly, it could still take advantage of work done elsewhere on MOS. On the other hand, the TFT, IBM's germanium program, and RCA's SOS program all suffered from being unable to take advantage of the silicon planar technology. These technologies, with only limited numbers of adherents, did not offer compensating advantages that could overcome the high costs they suffered from being away from the industry's technical base. The steady improvements that came as the industry as a whole worked on the silicon planar technology might ultimately overwhelm a technology that seemed to have inherent technical characteristics superior to silicon.

In 1967 and 1968, the industry was devoting its vast resources to the production of silicon bipolar transistors. Even though the MOS transistor had much in common with the bipolar transistor and many adherents, there would still be substantial costs in introducing it as a new technology. The key question in 1967 and 1968 was whether the MOS community could come up with a compelling product that could not be made using bipolar technology and therefore justify those costs.

The End of Research

The year 1968 was pivotal in American history. The country was in almost constant turmoil for the first eight months of the year, from the Tet offensive beginning in January, to the assassination of Martin Luther King Jr. in April, to the assassination of Robert Kennedy in June, to the riots at the Democratic convention in August. The election of Richard Nixon marked the end of a period when Democrats had held the White House for 28 out of the previous 36 years, and it ushered in a period where Republicans would hold the presidency for 20 out of the next 24 years.[1]

While the general public did not realize it, 1968 was also a year of turmoil and reorientation in the semiconductor industry. The turmoil, part tragedy, part comedy, came through management upheaval at Fairchild that rippled through the industry. The year marked the end of Fairchild's domination of the semiconductor industry and the formation of Intel, which would remain the most important semiconductor firm up to the time of this writing.[2]

Intel's success was built on MOS technology, and so it is important to be precise about Intel's achievement. Intel was not the largest producer of MOS integrated circuits at any time until 1975, but Intel was the first to put into production the silicon gate MOS process, an extremely powerful method for making MOS transistors. And Intel was the leader in the most important market for MOS technology in the 1970s: semiconductor memories. In that area it achieved first mover advantages.[3] Although these advantages would last for only a single generation in that market, Intel was also extremely effective at creating a variety of new products that exploited the possibilities of the silicon gate MOS process. These new products, including the erasable programmable ROM and the microprocessor, formed the foundation of its later success.

Intel had an ambivalent relationship to its Fairchild past. Fairchild provided a large proportion of the people who worked at Intel, and it was at Fairchild that they had learned their skills. But the founders of Intel made a conscious attempt to transcend the organizational problems that had plagued Fairchild, most notably by not having a separate R&D lab, even though the first four people at Intel included the director and assistant director of Fairchild R&D. Intel's work on silicon gate MOS technology represented a synthesis of R&D and manufacturing approaches. Although the silicon gate process had been a project of Fairchild R&D, the key figure in developing Intel's silicon gate process was from manufacturing, and his experience in manufacturing proved crucial to the success of Intel's process.

Intel's early history is a classic story of a Silicon Valley start-up, serving as an exemplar of Gordon Moore's claim that start-ups are not better at creating new things than existing companies; instead, start-ups are better at exploiting new things. When Intel started, a great deal of uncertainty surrounded MOS technology—no one knew if there would ever be successful MOS products or what they would be. Flexibility was key. Intel was able to conduct a series of economic experiments by quickly producing a range of products and then pouring resources into those that succeeded. The resources available on the San Francisco Peninsula enabled Intel to move quickly and conserve capital.[4]

Moore's Law, the Disintegration of Fairchild, and the Founding of Intel

In 1965 Gordon Moore wrote an *Electronics* article that subsequently became famous as the original statement of Moore's Law. While this article on the future of integrated circuits included such optimistic predictions as home computers and personal communications equipment, it also implicitly recognized a daunting problem for companies that wanted to make money producing integrated circuits. The first eleven years of Fairchild's history revealed this ambivalence, for even though Fairchild had made a string of major innovations, it was unable to secure a lasting competitive advantage in the industry.[5]

A key section of Moore's 1965 article was based on an analysis of the cost of integrated circuit production. He claimed that at any given time there was a component density (the number of components per chip) that led to the lowest overall cost per component. Be too aggressive, by trying to put too many components on a chip, and the loss of yield would more than offset the increased component count. Being too conservative, making integrated circuits with many fewer components, and overhead items, such as packaging, would lead to higher costs on a per component basis. Moore asserted that this minimum cost point had been moving out in density by a factor of two per year (figure 6.1). So while an integrated circuit made with eight components yielded the lowest cost per component in 1962, the next year the minimum overall cost was achieved with an integrated circuit with sixteen components. Moore's analysis showed that as one moved out into chips with higher component counts, the increased density was almost offset by the decreased cost per component, making the overall cost of the integrated circuit roughly the same.

Moore's article (and Fairchild's history) tacitly suggested that a semiconductor company attempted to run up the down escalator, with continual increases in components per integrated circuit offsetting declining costs. A part that had been made using the greatest of ingenuity and engineering skill would within a few years be manufacturable by a firm of average capabilities. In an industry lacking barriers to keep competition out, if a firm did not move upwards into higher and higher densities, it could expect to face withering competition that would drive prices down close to costs. Implicit in Moore's article and Fairchild's history were two strategies for competition in the semiconductor industry. One could be a pioneer, pushing the frontier further out into higher and higher component counts on each integrated circuit. Or one could hold one's ground and try to outmanufacture competitors by having lower costs on existing products—but one would have to drive those costs very low in order to make a profit. Although the two were not mutually exclusive, the *Electronics* article left no doubt as to the preferred method of the head of R&D at Fairchild Semiconductor. While in either case costs on a per transistor basis would go

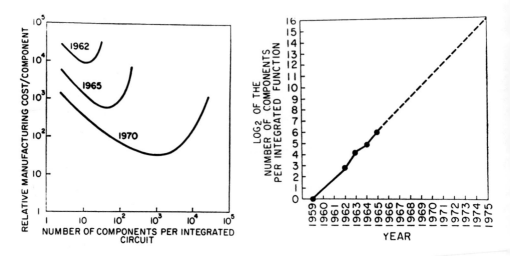

FIGURE 6.1 • GRAPHS FROM GORDON MOORE'S 1965 PAPER "CRAMMING MORE COMPO-
NENTS ONTO INTEGRATED CIRCUITS." The right graph is the original formulation of what is
now known as Moore's Law. It was derived from the left graph, Moore's analysis of integrated
circuit costs over time. He asserted that at any time there is a component density that gives the
lowest cost per component (the minimum on each of the three curves). He further claimed that
this minimum point had been moving out by a factor of two every year. On the right graph Moore
then plotted the number of components per integrated circuit at the minimum cost point as a
function of time, extrapolating out from 1965. He projected that in 1975 the semiconductor
industry would be producing integrated circuits with 64,000 components on them. From Gordon
E. Moore, "Cramming More Components onto Integrated Circuits," *Electronics*, 19 April 1965, 115,
116. Reprinted by permission of Penton Media, Inc.

down, the really big decreases would come only as one moved into higher and
higher densities.

One might see Moore's article as a commentary on the first eight years of
Fairchild's history. Its success had two elements: new products unique to Fairchild, and a strong manufacturing organization. By being the first to market a
product, Fairchild was able to set its own prices. When Fairchild began operations, it was the first to offer a silicon double-diffused transistor, which sold for
$150. In the early 1960s Fairchild developed a number of new transistor and
integrated circuit products that commanded premium prices.[6]

In an industry with low start-up costs, the potential profits from high-
margin new products presented a major temptation to Fairchild employees. The

start-up companies formed by Fairchild alumni between 1959 and 1963 were closely linked with Fairchild's new products. A typical scenario would be that as a product was introduced, a small group of the Fairchild development team would leave to start a separate company. For example, Fairchild announced availability of its first integrated circuit line in March 1961, and Signetics was formed in September of that year, led by David Allison, the former head of the Device Development Department at Fairchild. By March 1962, Signetics was offering its own integrated circuit products. In 1963, another group from Fairchild Device Development left Fairchild to start General Microelectronics.[7]

After the formation of General Microelectronics, no new start-ups were formed by defectors from Fairchild until 1967, when Charles Sporck, then the general manager of Fairchild Semiconductor, and a group from Fairchild took over a moribund National Semiconductor. The lack of start-ups during this period can be attributed to a growing product stability in the industry, with no blockbuster products at higher density levels coming out of R&D. During this time Fairchild attempted to enter new commercial markets by lowering its prices and by developing new, low-cost products.[8] Sporck's National represented a new kind of start-up, centered on manufacturing. The people Sporck took with him from Fairchild were primarily from the manufacturing side— none were from R&D. Although National had designers who developed new circuits (except for its linear circuits, these circuits were primarily minor modifications of circuits designed elsewhere), manufacturing was the heart of the firm. In the fall of 1967, National announced it was second-sourcing Texas Instruments' popular line of transistor-transistor-logic integrated circuits. This move gave National access to a potentially large market with minimal development expenses, but it put a premium on National's ability to manufacture effectively in an extremely competitive environment.[9]

A precipitating factor in Sporck's decision to leave Fairchild was the stinginess of the Long Island–based management of Fairchild Camera and Instrument (the parent company of Fairchild Semiconductor) in granting stock options to its employees. Sporck had been unable to hold Bob Widlar and Dave Talbert, Fairchild's talented linear design team, on this account. And as he took over National, he received a substantial equity interest in the firm that promised to reward him if it prospered.[10]

While R&D went through a period of relative stability in terms of personnel after 1963, Fairchild manufacturing continued to suffer major losses in the post-Sporck period. By offering stock options and higher pay, National was able to lure away various employees, from manufacturing engineers to equipment operators. One engineer working in production recalled that there was a going-away party for a departing employee almost every week.[11]

By 1967, lacking popular new products, which halted any movement up into higher densities, Fairchild's success depended on its ability to manufacture efficiently as its war of attrition with Texas Instruments and other competitors led to falling integrated circuit prices. In this war, Fairchild was hurt by the fact that many of its key soldiers had defected.[12] One tactic to provide at least some relief by substituting quantity for quality was developed by Fairchild's marketing head, Jerry Sanders. Fairchild would offer a new product every week for a year. Although none of the products might be blockbusters (and none were), for at least a short period of time Fairchild would be the only source of these new circuits and could therefore avoid commodity pricing.[13]

Fairchild Semiconductor's earnings plummeted in 1967. Fairchild Semiconductor's difficulties were acutely felt at its parent company, Fairchild Camera and Instrument, for despite its various efforts at diversification, Fairchild Semiconductor provided most of the profits to the whole company, with the press claiming that all the other divisions were money losers. In October 1967 John Carter, the chief executive officer of Fairchild Camera and Instrument, resigned after the company posted a 95 percent drop in earnings. After Fairchild's successful diversification into semiconductors, Carter was responsible for Fairchild's money-losing entries into duplicating machines, instruments, and vacuum tubes.[14]

With Carter's departure, Richard Hodgson served as Fairchild's chief executive, but when he was removed in the spring of 1968, Robert Noyce was the logical candidate to become the next chief executive of Fairchild Camera and Instrument. He had been a founder and general manager of Fairchild Semiconductor and was group vice president responsible for both the Semiconductor and Instrumentation Divisions, the most important components of the parent. But when Fairchild bypassed Noyce for the chief executive position, he quit. Gordon Moore, the head of Fairchild R&D, left with Noyce, out of a growing frustration over the difficulties in transferring products from R&D to manufacturing and a belief that any new head of Fairchild would likely undertake a major reorganization.[15]

Noyce and Moore, founders of Fairchild Semiconductor who had observed dozens of people start new semiconductor companies after them, resolved to start a new semiconductor company. This time they would keep control. They arranged financing through Arthur Rock, who had put together the original Fairchild deal, after agreeing to put up $245,000 each. Rock was quickly able to raise $2,500,000 by selling convertible bonds, including some to the other six founders of Fairchild Semiconductor. The Fairchild founders had each received $250,000 when Fairchild Camera and Instrument acquired the company, and although several were involved in firms that would be competing with Noyce and Moore's new venture, they were unwilling to pass up a promising investment. On 18 July 1968, Noyce and Moore incorporated their new venture.[16]

Fairchild still faced the problem of finding a new president. In August, Fairchild named Lester Hogan, the general manager of Motorola's semiconductor division, to be the president of the parent company, Fairchild Camera and Instrument. Hogan had served a short but noteworthy stint at Bell Labs, been a professor of physics at Harvard, and then led Motorola's semiconductor operations, making it one of the industry leaders. To lure Hogan from Motorola, Sherman Fairchild, the owner of Fairchild Camera and Instrument, paid a fabulous price: a salary of $120,000 a year and 10,000 shares of stock at a fraction of its market value, along with an interest-free loan of $5.4 million to exercise a stock option.[17]

Fairchild did not get just Hogan; it got a whole new management team. Before Hogan left Motorola, he persuaded a group of his senior managers at Motorola to come with him. These ex-Motorola managers, dubbed "Hogan's Heroes" by industry wags, formed an additional layer of management at Fairchild Semiconductor that mediated between Hogan and the previous Fairchild organization. This affront to the Fairchild employees turned the trickle of departures from Fairchild into a stream. After Hogan's arrival, *Electronic News* carried regular reports of the resignation of senior Fairchild personnel, often to form a new company. In December 1968, Jerry Sanders, formerly head of marketing at Fairchild, resigned, and in April of the next year he announced plans to start a new company, Advanced Micro Devices, with seven other Fairchild alumni. In November 1968, Lee Boysel, the leader of the MOS circuit design effort in Mountain View, resigned, and he later formed Four-Phase Systems. In March 1969, *Electronic News* reported that six groups of Fairchild employees were looking for funding to start new ventures.[18]

Choosing a Product at Intel: Semiconductor Memories

As turmoil enveloped Fairchild, Moore and Noyce began their new company. They immediately faced the issue of what product to make. As researchers at heart, they wanted to push integrated circuit technology to higher and higher component levels, not become yet another entrant in the commodity semiconductor business. By starting a new company, they were able to look at the work being done at Fairchild R&D and throughout the industry and choose the areas that seemed most promising. The challenge was to select a novel product and technology that would open new markets to Intel and become a viable business before they ran out of money.

Moore and Noyce seized on semiconductor memory as their area of concentration. It was not a new idea: Moore had directed work on it at Fairchild R&D, and Lee Boysel's group had worked on it at Mountain View. And it was becoming more popular: *Electronics* magazine carried articles in March and June 1968 describing work on semiconductor memories by other firms.[19]

Perhaps the most important sign that semiconductor memory's time was coming emanated from IBM. While semiconductor companies could embark on a semiconductor memory program, they still had to convince customers to buy what they had made. In 1965 engineers from IBM had described a bipolar memory containing sixteen bits used in an experimental scratch-pad memory. At the 1967 Electron Device Meeting, *Electronic News* reported (correctly) that IBM was working on a 128-bit bipolar memory and an MOS large-scale array. While these were small, titillating hints at what IBM was up to, if it became known that IBM, the dominant force in the world computer industry, was planning to use semiconductor memories in its computers, that knowledge could be expected to immediately establish a market for semiconductor memories as its competitors followed suit.[20]

Even at Fairchild, Moore had considered semiconductor memories the one innovation that was capable of supporting a new company. In April 1968, while still the head of Fairchild R&D, he had given an invited paper to the

International Conference on Magnetics on the prospects for semiconductor memories. Moore stated that while semiconductor memories would clearly be used in small, fast memories, the "interesting" question was the possibility of the use of semiconductor memory in place of magnetic cores for the main memory of computers. Moore claimed that it was possible to build semiconductor memories for a cost of less than one cent per bit, a figure that made them cost-competitive with magnetic cores.[21]

Moore anticipated that semiconductor memories would be cost-competitive, although they had not been earlier, because of the effect he had described in his 1965 article. Semiconductor memories offered Intel a way to continue to move up the down escalator and leave its competitors—such as Fairchild, National Semiconductor, and Texas Instruments—behind, at least for a time. In the preceding few years, even though it had been technically possible to produce integrated circuits with more and more components, Fairchild and the industry had hit a barren spot where they had been unable to find products that customers wanted. Semiconductor memories were a way to make higher-density integrated circuits that customers would buy.[22]

Semiconductor memory had several other attractions for Moore and Noyce. The computer-aided-design approach to semiconductor logic that Moore supported at Fairchild assumed that with large-scale integration each customer would require many part numbers. Fairchild R&D had developed a system using computers and programmers to make this economically feasible. Semiconductor memory required only a very few part numbers, common across different customers, thus allowing them to jettison the expensive infrastructure built up around the computer-aided-design approach. Furthermore, semiconductor memory would require only a very few chip designers and put a premium on devising a process to densely pack transistors onto a chip, which was compatible with Moore and Noyce's strengths in semiconductor process technology.[23]

As an exercise in counterfactual speculation, one might wonder what would have happened had Moore and Noyce remained at Fairchild. Although Moore was having doubts about the computer-aided-design approach, the project was not clearly headed for failure. Fairchild had spent many years developing the necessary tools to make it work. It was expected to become a profitable product within a few years, and other companies throughout the industry were adopting similar approaches. At that point it would have been much easier to keep the program going than to kill it. This is what Fairchild's new management group did, only shutting down the operation six years later in 1974 after it had still not proven financially successful. The limited resources of a start-up, as compared with Fairchild R&D, forced Moore and Noyce to concentrate on the most promising areas.[24]

Having made the decision to focus on semiconductor memories, Moore and Noyce then took an agnostic position on the MOS versus bipolar question; they would work on both. For each transistor technology, they chose a new approach that had been the subject of preliminary research by Fairchild R&D but not yet entered production. In 1968 the biggest problem affecting the speed of bipolar transistors was a phenomenon called saturation. As the transistor was turned on, a large capacitance was built up that had to be discharged before the transistor could be turned off again. (This capacitance can be thought of as a barrel full of water that had to be drained before the transistor was turned off.) Fairchild had developed a technique called gold doping that allowed the capacitor to discharge faster, decreasing the turnoff time of the transistor. (The gold doping technique can be thought of as punching holes in the barrel of water to allow it to drain faster.) This process was widely used in the industry, but even for an industry highly dependent on art, gold doping was an extremely tricky process.[25]

In 1967 a Fairchild researcher, Ted Jenkins, proposed a method for making a Schottky diode, a very fast type of diode that was compatible with integrated circuit processing. This allowed the creation of a circuit configuration with the Schottky diode in parallel with two terminals of the transistor. The Schottky diode acted as a clamp, keeping the transistor from entering saturation and the capacitance from building up. (The Schottky clamp can be thought of as a lid that allows only a tiny amount of water into the barrel.) This allowed the transistor to be turned off much more quickly than did a transistor that was saturated. The idea of clamping a transistor to prevent it from entering saturation was not new; a researcher from MIT's Lincoln Labs had proposed using a germanium diode to do the same thing in the 1950s. The beauty of the Schottky diode proposal was its compatibility with integrated circuit processing. While the germanium diode was totally incompatible with the planar process and required the addition of an extra discrete device, one could add the Schottky diode with almost no extra processing. Intel planned to use the Schottky process to build a sixty-four-bit memory that would be used for cache storage in computers.[26]

In large memory systems that served as the main memory for computers, MOS technology offered special advantages. Here speed was not a primary consideration, both because magnetic cores were not particularly fast by semiconductor standards and because of the growing acceptance of cache memories. But since main memories were large, up to eight megabytes in System/360 computers, density was important. The density advantage of MOS over bipolar translated into a significant cost savings and a reduction in the size of the overall computer.

In MOS, Intel chose to concentrate on silicon gate MOS, a process innovation previously disclosed by Bell Labs at the 1967 Metallurgical Society Meeting. Ed Snow of Fairchild R&D heard the paper, thought the idea was promising, and encouraged others at Fairchild to pick it up. Bell Labs had abandoned work on silicon gate shortly after describing it. Previous MOS transistors had had metal (mainly aluminum) as their gate electrodes. As its name implied, the silicon gate transistor substituted silicon, in this case polycrystalline silicon, for the aluminum gate. The substitution of silicon for aluminum gates allowed for reordering the entire MOS fabrication process. Unlike aluminum, silicon could withstand high-temperature processing, and so the silicon gate could be formed very early in the fabrication process and used to align the other two terminals of the MOS transistor. This feature, called self-alignment, made it possible to make MOS transistors that had much tighter tolerances than were possible previously, increasing the density of transistors on an integrated circuit. Furthermore, because the oxide was formed and covered early in the process, it was easier to keep free from contamination.[27]

Although Fairchild had done preliminary work on the silicon gate process in late 1967, sustained work began only in February 1968, after Federico Faggin arrived at Fairchild R&D from Fairchild's Italian affiliate for what was to be a period of six months. Faggin's work, classified "exploratory device technology," consisted of developing a process, building test structures, and fabricating an existing MOS part using the new process. By the beginning of July, when Moore and Noyce left Fairchild, the silicon gate MOS technology was in an intermediate stage of development, more than just a disembodied idea but still having problems to be solved before it could be manufactured. Faggin had brought silicon gate to a point where a competent manager could make judgments about what the key problems were and how difficult they would be to solve.[28]

Staffing Intel

While Moore and Noyce had an idea of what they wanted to do, they had to come up with the right people to work in the start-up. The respect with which Moore was held at R&D meant Intel could recruit almost anyone from R&D. When Andrew Grove, by this time the assistant director of the R&D lab at Fairchild, heard that Moore was leaving, he immediately volunteered to join the operation. Les Vadasz, the head of MOS development in R&D, also signed on quickly. Of course, one could not make a well-balanced firm of exclusively R&D people—people with other skills were needed.[29]

The fifth person on board was Gene Flath, the manager responsible for a large fraction of Fairchild's integrated circuit manufacturing, who became the manager of Intel's manufacturing efforts. Flath had taken the R&D semiconductor physics course and gotten to know Grove, and later he had dealt with Moore and Grove working on a serious manufacturing problem. When Flath heard that Moore and Noyce were starting a new company, he called to volunteer his services and was quickly hired.[30]

Flath proved critical in the staffing of Intel, for he provided the entry point into Fairchild manufacturing. People from R&D did not have the kinds of contacts or the credibility among manufacturing personnel that Flath had. He quickly hired two engineers from Fairchild manufacturing, Tom Rowe and Gary Hart. Two others who had worked for Flath at Fairchild, George Chiu and George Staudacher, came later. Those with experience in manufacturing would later hire Fairchild operators, although at times Intel would have moratoriums on this practice for fear of legal action by Fairchild.[31]

In its early phase, Intel primarily hired people with very specific skills that it needed to fill immediate needs. As a start-up, Intel could ill afford to allow workers a protracted learning period. This meant no hiring directly from colleges or universities, for universities did not teach the craft of semiconductor production. And it also put a limit on its hiring of engineers and scientists from

R&D, for although there were top-rate scientists from R&D who would have liked to have joined Intel, they either did not have the specific skills needed at the time, or duplicated the skills possessed by others already at Intel. Intel did not need people far out on the leading edge of technology; it needed people who were thoroughly grounded in the state of the art and could make use of existing equipment. A large portion of the first group of engineers and scientists hired from R&D served as managers at Intel. (Appendix figure A.4 shows the early engineering staff at Intel and their previous employers.)[32]

Intel quickly acquired the skills it needed, sometimes hiring people from Fairchild, sometimes going elsewhere. Perhaps the best example of this acquisition of skills is Flath's hiring of a complete set of manufacturing engineers, but there were other examples. As the silicon gate process was originally conceived, one of the key steps was evaporating silicon onto silicon dioxide to form the gate. Fairchild R&D had two technicians who had mastered this process. Intel attempted to hire one, and when he declined, hired the other, Larry Brown, who was at Intel by August 1968. Another case occurred in the bipolar process. Epitaxial growth was a key step in making bipolar transistors, and to do this work Intel hired Gus Skousen, an engineer who had worked on epitaxy at Signetics in Sunnyvale but had been laid off. As the work progressed, leaders of the MOS effort believed that the use of silicon nitride would be necessary to guarantee the device's reliability. To do the nitride work, Intel hired Bob Holmstrom, who worked at Sprague Electronics in North Adams, Massachusetts, and had helped set up a semiconductor facility at Berkeley as a student. Although he had no previous experience working with nitride, he was told that would be his job at Intel. At an interview with Gordon Moore in New York City, Holmstrom brought several nitride-covered wafers he and his technician at Sprague had prepared especially for the meeting.[33]

One area that was noticeably dominated by non-Fairchild alumni was MOS circuit design. The most experienced MOS circuit designers at Fairchild had been Lee Boysel and his group at Mountain View, but they had had antagonistic relations with R&D and an incompatible working style. Furthermore, they had left Fairchild to form their own start-up. Because of Fairchild R&D's work on computer-aided design of logic circuits, it had no one with experience designing semiconductor memories. For these skills Intel looked elsewhere. Its first

MOS designer, Joel Karp, had previously worked at Philco Microelectronics (previously called General Microelectronics). Taught MOS circuit design by the intellectual descendants of Frank Wanlass, Karp's last project had been the design of an MOS memory. When Philco moved its operations to the East Coast, Karp made plans to leave. He had heard press accounts about Moore and Noyce's new venture and sent them his resume. Shortly thereafter he was interviewed and hired. Intel's next MOS circuit designer was Bob Abbott, who had worked as a co-op student at Philco on MOS design before taking his first job at Motorola, where he had designed MOS shift registers and RAMs. In the spring of 1970 Intel hired John Reed, who had years of MOS experience at American Microsystems and Autonetics, to lead the work on the 1103 RAM.[34]

An important pointer to talent came from Andrew Grove's connections at Berkeley. Grove served as a visiting professor at Berkeley, where he taught the Fairchild R&D semiconductor physics course, which had been put into book form in 1967. Grove continued to teach even during Intel's early days. As Intel developed a need, he could use his contacts with faculty at Berkeley to find promising alumni in addition to those he knew through his course. Bob Holmstrom was hired through Grove's contacts with faculty, while Bob Abbott came to Intel's attention because he had been a student in Grove's course. (Neither was hired directly out of school.)[35]

Fairchild's problems put Intel in an advantageous spot in its search for talent in its first few years. Due to Hogan's demoralization of the staff, Fairchild was like a large passenger ship that was badly damaged but still seaworthy. While there was no need to abandon ship immediately, anyone with the chance to move to a more seaworthy vessel understood the advantages of doing so. Intel did not have to bring on board everybody it might want right away, but as a specific need came up, it would often look to Fairchild for the needed skills. For example, George Staudacher was a test engineer at Fairchild who had worked under Gene Flath. At the time Intel was formed, it had no need for a test engineer—it had nothing to test. Staudacher joined Intel in early 1969, but informal arrangements had been made for him to come on board several months earlier even as he continued to work at Fairchild. And while Intel had no need for a large contingent from R&D at the beginning, once Intel got established and began growing, places for these people opened up.[36]

Implementing the Silicon Gate Process

The silicon gate process was the foundation of both Intel's success in MOS technology and its success as a company, for it proved to be extremely manufacturable and adaptable to a wide range of products. In picking up the work done at Fairchild R&D, Intel took a process that was capable of yielding small numbers of laboratory devices and transformed it into a process that could consistently yield large numbers of reliable MOS integrated circuits. While Intel's work on silicon gate built on a knowledge base that had been established at Fairchild, it represented a different approach to making MOS transistors than Fairchild R&D had taken. Furthermore, Intel was able to avoid the organizational problems that had hampered MOS work at Fairchild.

Given Intel's success in the years since its founding, and the status Noyce, Moore, and Grove have attained as technical leaders and managers, it is easy to imagine that Intel was preordained to be a winner. But in a very real sense, a cloud hung over Noyce and Moore as they started Intel. One of their perceived legacies at Fairchild was an out-of-control research lab that did top-flight work but could not convert that work into profitable products. After Moore and Noyce's departure one Fairchild manager asserted that under Moore's direction the R&D lab was "becoming a university and a hobby shop."[37] A Motorola executive stated that Fairchild "seemed dedicated to technology for its own sake."[38] Charles Sporck, the head of rival National Semiconductor, told one person considering a job at Intel that while the company had some great R&D "guys" it would not be able to make money in the industry.[39]

Grove felt the pressures of Fairchild's failures in a more personal way. While he had done fundamental work on MOS stability at Fairchild, Fairchild's inability to make use of the work raised questions in his mind about the worth of that work. The Simon and Garfunkel song "Faking It" reminded Grove of his personal challenge, which could be addressed only by the successful production of MOS devices. At Fairchild there had always been others to blame, such as headquarters in Long Island or manufacturing in Mountain View; Intel would have no such easy scapegoats.[40]

Moore and Noyce recognized these problems themselves and put the highest priority on eliminating the problem of transferring technology between

R&D and manufacturing. At one level these problems were solved by the simple act of leaving Fairchild and forming a start-up. Inherently, the start-up would be small, without the numerous competing organizations that had existed at Fairchild. Development work would have to be done quickly to allow the introduction of a product.

But Moore and Noyce made conscious decisions affecting questions of technology transfer. In Moore's 1965 *Electronics* article on the future of integrated circuits, he asserted that to reach higher densities, what was needed was not fundamental research but engineering effort. Although Moore and Noyce had reflexively set up an R&D lab at Fairchild, their years of experience in the semiconductor industry had shown them that having a separate R&D laboratory greatly retarded the technology transfer process. Getting products into production quickly was essential in the semiconductor industry, and as they looked out into the future, they saw no immediate obstacles that would require a major research effort. Intel would not have the kind of research lab that Fairchild had had, one that had a separate identity and often produced only scientific papers.[41]

Another decision concerned the location of the company. By this time numerous semiconductor companies were located in the San Francisco Peninsula, and several of these firms were making transitions that gave Intel the opportunity to move into a building set up for semiconductor production. Union Carbide, a second-tier semiconductor producer, was moving its operations from Mountain View to San Diego. It had occupied two nondescript buildings only a few blocks away from Fairchild's main manufacturing facility and closer still to the Wagon Wheel Bar, a notorious watering hole for personnel from area semiconductor companies. Moore and Noyce chose the Carbide site. Intel's location might be seen as a symbolic positioning of the company not in the Stanford Industrial Park in Palo Alto (where Fairchild R&D had been) but in a manufacturing environment.[42]

Early Intel was also organized so as to minimize technology transfer problems. Grove was responsible for operations, with the MOS engineering, bipolar engineering, manufacturing, and quality assurance departments all reporting to him (appendix figure A.4). All the initial process development work was done in the manufacturing department, thus eliminating any technology transfer problems. This arrangement was made possible by the fact that Intel was not producing anything, freeing the manufacturing group to do process development.

As it developed the silicon gate process, Intel also developed its own style of dealing with the problem of making stable MOS transistors. Intel's style differed in significant ways from the style of Fairchild R&D. At Fairchild R&D, work had focused on the elimination of sodium, no matter where it was to be found, which followed naturally from the work of Snow, Deal, and Grove in showing that sodium played a key role in MOS instabilities. But at another level it was an artifact of the distance between R&D and manufacturing. At Fairchild, moving the MOS technology from research to manufacturing involved no less than two transfers, first from the Solid State Physics Department to the Digital Integrated Electronics Department at R&D, then from Digital Integrated Electronics to manufacturing. If Solid State Physics had been able to make stable MOS devices by eliminating sodium, then it was logical for R&D to assume that manufacturing could do the same. However, Solid State Physics was merely making MOS test structures, not actual transistors, and certainly not the complex integrated circuits that would be final products. Furthermore, Solid State Physics operated on a much smaller scale than manufacturing, which meant it had a much easier time adhering to the requisite standards of cleanliness. Intel adopted an approach that put much less emphasis on cleanliness and sodium elimination than Fairchild R&D had, and more on rendering sodium harmless. The process developers at Intel were from manufacturing and knew the standards that could be reasonably expected on a manufacturing line.[43]

The main responsibility for developing the silicon gate process at Intel rested with Tom Rowe, a young engineer who had received a bachelor's degree in metallurgy from MIT in 1966 and a master's degree in 1967. Rowe wanted a position in manufacturing and considered pursuing a Ph.D. but feared it would disqualify him from manufacturing work. With the type of background and intellect that would have made him ideal for R&D, he took a position at Fairchild's manufacturing facility in Mountain View. For someone with a deep intellectual curiosity, Fairchild was a perfect place to get started in the semiconductor industry. As a vertically integrated manufacturer, it offered the chance to see all phases of semiconductor production in one complex—from growing crystals of silicon to construction of packaging for finished chips. Rowe also served as the liaison to R&D for his area and had taken the R&D semiconductor physics course. Although Rowe took no written materials with him to Intel, he carried much in his head. For any straightforward process needed at Intel, he used procedures from Fairchild. And the basis for innovation frequently came from experience at Fairchild.[44]

The first challenge at Intel was to establish even the most rudimentary semiconductor processing facilities. Although Intel's building had been previously used for semiconductor production, it was in poor repair when Intel moved in. Its facilities for making deionized water, crucial for rinsing wafers and cleaning equipment, were inoperable. Rowe proposed a workaround using commercial bottled water. He picked up five-gallon bottles on the way to work and held the bottles over the items to rinse them. Intel purchased off-the-shelf production equipment, some literally off the floor of a major trade convention, and had fabrication capability by the fall of 1968.[45]

The next hurdle was to make stable MOS test structures. To motivate the people working on semiconductor processing, Grove proposed a bet requiring the achievement of three milestones, two related to bipolar processing and one related to MOS processing. The MOS milestone was to make an MOS capacitor that would have a voltage constant within one-tenth of a volt, a benchmark attainable in an R&D lab under good conditions, but not yet regularly achieved in a manufacturing environment. This benchmark would demonstrate standard MOS processing capability, but nothing unique to the planned silicon gate work. The two bipolar milestones were met relatively quickly, but as the deadline approached, with Intel unable to attain the cleanliness standards of an R&D lab, the MOS milestone had not been achieved. On the last day, the terms of the bet were changed so as to allow the use of phosphorus, which was known to have a stabilizing effect on MOS devices, but had been specifically excluded originally. Fairchild R&D had taken a strong stand against the use of phosphorus, claiming that stability was best achieved through maintaining high standards of cleanliness. Rowe, who had experience diffusing phosphorus for making bipolar transistors at Fairchild, added it, making a stable device. This demonstrated a flexibility on Intel's part and an ability to move beyond R&D standards. Instead of continuing to attempt to achieve cleaner conditions and meet the original conditions of the bet, Intel engineers went on to work on the silicon gate process.[46]

Whether a process was manufacturable was highly subjective, with the answer heavily dependent on whether the respondent was the one being asked to do the manufacturing. But there were some general principles involved. While much R&D work was carried out by technicians, in manufacturing the processing was done by less skilled operators, putting a premium on simplicity. Furthermore, each step had to have a high probability of initial success to ensure both yield and throughput. The process in manufacturing had to yield reliable devices that would last many years. As Intel started, a one-page description of the silicon gate MOS process was distributed, which one assumes represented the silicon gate process at Fairchild by this point. Intel managers seem to have believed that it was a workable process.[47]

But there were major problems. The principal one as work began was the difficulty in depositing the silicon gate onto the silicon dioxide. Fairchild R&D had used an electron beam evaporator, and Intel planned to use the same technique. Evaporated silicon would not consistently adhere to the silicon dioxide, and furthermore, the crucible that held the silicon as it was being evaporated had a tendency to explode. Both Brown (the technician hired from Fairchild to do the evaporation) and Rowe agreed that the process was too dangerous to be run by operators and did the silicon evaporation runs themselves.[48]

Gus Skousen, an engineer who had been hired to work on epitaxy (a step used only for bipolar transistors, not MOS transistors), proposed a solution. Skousen suggested that the same equipment used in bipolar epitaxy could be used to deposit polysilicon to form the silicon gate. The first such wafers were processed in December 1968 and showed excellent adherence to the silicon dioxide as well as much better uniformity than the evaporated silicon.[49]

In the early phase of the silicon gate work, the solution to one problem frequently created another problem, so that there were no immediate visible benefits overall. Had the development process been divided among different departments, shifting problems across departmental lines could have created tension. The localization of the work in one department (Flath's) and in essentially one person (Rowe) eliminated such difficulties.

Depositing polysilicon with an epitaxial reactor caused a problem of masking. In a later step, a photoresist would be applied to the polysilicon, with a pattern exposed to define the gates of the transistors. The photoresist adhered much more poorly to the polysilicon deposited with the epitaxial reactor than to the evaporated polysilicon and had the tendency to peel, threatening the performance of the transistors. The most important determinant of the MOS transistor's speed was the length of the gate. Peeling photoresist would lead to large tolerances on the gate and therefore also on the transistor's speed. Rowe proposed growing oxide on top of the polysilicon and then putting the photoresist on top of the silicon dioxide.[50]

This was again a solution that created problems. The very sharp delineation of the silicon gate made possible by the oxide mask created a steep step that was difficult to cover with metal, as required later in the process. Rowe had studied the problem of step coverage in bipolar circuits at Fairchild. Based on the results of those studies, Rowe addressed the step coverage problem for the silicon gate process by designing a special domed planetary jig for holding wafers while they were being covered with metal by an electron beam evaporator. The jig resulted in very good step coverage and had the added advantage of having a capacity of fifty wafers (existing jigs held only twenty). This was an important production consideration, particularly since the electron beam evaporator was one of the most expensive pieces of equipment used.[51]

While Fairchild had built much of its own production equipment, Moore and Noyce had made a decision at the outset that Intel would not. This decision reflected the belief that building production equipment was not an efficient way for a start-up to use its scarce resources, particularly since a significant semiconductor production equipment business had developed in the years after the establishment of Fairchild. Rowe approached Temescal, a manufacturer of electron beam evaporation systems, and it enthusiastically adopted his design.[52]

The silicon gate process created several other problems of step coverage. Traditional MOS integrated circuits had one layer of metal on top of the wafer. The silicon gate process had a layer of metal plus a layer of polysilicon, which meant it had more horizontal layers and therefore more vertical steps than a standard MOS process. Another steep step was created when a layer of silicon dioxide was deposited on top of the silicon gate. At an engineering staff meeting, Gordon Moore suggested that the problem could be solved if the silicon dioxide could be melted, thereby smoothing off the steep step, a suggestion that many of his colleagues met with skepticism. Rowe's time was already fully occupied during the day with other assignments, so during the evening he worked on attempting to melt the glass. Rowe added a small amount of phosphorus to the glass as a way of lowering its melting point. The ultimate solution was not to melt the glass but to flow it, softening it at a temperature below the glass's melting point. The use of phosphorus had an additional benefit, for its tendency to trap sodium was well known. Rowe doubted that it would be possible to maintain an environment free enough from sodium to make stable devices. The phosphorus on top of the gate would trap any sodium that was in the area and prevent sodium from reaching the gate oxide, where it would change the transistor's characteristics. Rowe's method would eliminate the need for applying phosphorus to the gate oxide itself, which Intel engineers believed to be too great a threat to stability to use.[53]

Although Rowe had responsibility for most of the silicon gate development work, Moore's suggestion shows the resources Rowe had to draw upon. This was visible at the weekly meeting of the engineering staff. The most prominent Fairchild alumni, Noyce, Moore, Grove, Flath, Rowe, and Vadasz, represented a wide range of organizations at Fairchild and would have rarely, if ever, met together. While Moore did work in the laboratory and was a frequent presence there, he and Noyce made their major contributions at these staff meetings, for they were talented freethinkers on technical problems. Grove, one of the leading theoreticians of the MOS transistor, tutored Rowe, keeping him informed of any articles he thought relevant. (Although Rowe read articles brought to his attention, he did not have the same detailed knowledge of the published literature that the people from R&D had.)[54]

By the late spring of 1969, after having made many technical breakthroughs that would have been unlikely to occur at Fairchild, Intel still did not have a working silicon gate process. The 1101 memory chip, the first silicon gate product, was a 256-bit RAM. Each wafer contained 180 chips, but for every 100 wafers processed, only 60 would typically avoid major failures and make it to final testing. And each wafer that made it to final testing typically would yield only five good chips, an overall yield of less than 2 percent, which made the part an economic disaster. By this time the bipolar process was already working, and control of the processing equipment had passed from the engineers to a group responsible for production, whose main priority was to build a large inventory of bipolar chips in preparation for their official introduction. Running MOS wafers became much less urgent. The fate of the silicon gate process was in doubt.[55]

One Saturday morning during this period, Rowe came in as usual to work on the silicon gate process. He was initially alone but was later joined by Grove. Rowe and Grove discussed the specific problem Rowe was working on and then moved into a general discussion of their discouragement and doubts about the process. Grove told Rowe not to hand him a live burial, meaning if the process were truly unworkable and dead it should be stopped, but if there were any signs of life, work must go on. They agreed that they had to continue work on the process.[56]

Just weeks later came the breakthrough that justified their perseverance. The semiconductor fabrication process consists of adding and removing layers of material. Although adding materials received most of the attention, the ability to effectively remove material was just as important. Material that was only partially removed could prevent the necessary adherence between layers. The main problem Rowe suspected this time was with residues. The answer again came out of the experience Rowe had had manufacturing bipolar transistors at Fairchild, where he had studied the effects of different chemical rinses on different materials. At Intel he used that knowledge to devise a new rinse. The morning Rowe introduced his new rinse, the wafers receiving it showed a tenfold yield improvement over previous wafers. These electrifying results brought people streaming out of their offices to see what was going on in the fabrication area. Les Vadasz, the manager of the MOS design area, who needed the silicon gate process to work if any of his products were to see the light of day, jumped up and down yelling with his Hungarian accent "It's a sooper dip," leading to Rowe's rinse being christened the super dip.[57]

But even after the yield had risen so dramatically, the question of reliability loomed large. How long would the parts work? MOS reliability was so poor that it was standard practice in the industry to burn in MOS integrated circuits by operating them at elevated power levels for an extended period of time prior to shipment so that any marginal parts would fail before they reached the customer. (Only bipolar parts requiring extremely high reliability, such as for military programs, underwent this procedure.) This was a costly step requiring the construction of a large number of test fixtures, since each part was typically burned in for twenty-four hours. A large failure rate at burn-in was potentially disastrous, for at this point, just prior to shipment, the assembled chips represented their greatest expense to Intel.[58]

The prime failure mechanism was a gate short, where a voltage applied to the gate of the transistor resulted in a short circuit through the normally insulating silicon dioxide. While a typical MOS transistor could withstand the voltages applied to it, defects introduced into the oxide during processing could dramatically decrease the maximum voltage a transistor could withstand. Intel's first MOS part, a 256-bit RAM, had roughly two thousand transistors, with these transistors having a distribution of voltages at which their gates would short. If even one gate on the chip shorted, the chip was unusable. During the development of the silicon gate process, the prevailing wisdom among Intel's management was that some additional step was necessary to prevent gate shorts and thereby burn-in fallout. Intel's plan was to add a thin layer of silicon nitride under the gate for this purpose, although the use of silicon nitride had the potential to reduce the stability of the transistor. Both Rowe and Desmond Fitzgerald, the quality assurance manager, opposed the use of nitride but, lacking an alternative, were forced to concur. After the basic process had been stabilized, Rowe developed a method using an extra oxidation step and chemical rinses that greatly increased the breakdown voltage of the gate oxide. With the enhanced breakdown voltage, not only were the burn-in losses greatly decreased, but the overall yield of the process soared, as some integrated circuits had had transistors that had shorted out even during initial testing. The use of nitride proved unnecessary, and ultimately Intel's silicon gate parts proved so hardy that they did not require burn-in.[59]

Although normal manufacturing and reliability problems would continue, by September 1969, Intel had developed a silicon gate process with good yield and reliability. In October 1969, Intel published an article on the process in the *IEEE Spectrum*. The choice of publication was revealing, for *Spectrum* was a general publication sent to all electrical engineers who were members of the IEEE. If Intel had planned to describe the technical features of its process, the natural venues would have been professional conferences such as the International Electron Device Meeting, or specialized professional journals such as the IEEE *Transactions on Electron Devices* or the *Journal of the Electrochemical Society*. Describing the silicon gate process at a professional conference would have exposed Intel to searching questions by practitioners familiar with the subtleties of semiconductor processing. The readers of the *Spectrum* were more likely to be customers than colleagues, and the *Spectrum* article functioned as advertising for Intel's upcoming products. While the silicon gate process was ostensibly described in some detail, crucial information was left out. None of the ten elements that Tom Rowe later described as crucial to Intel's silicon gate process were mentioned at all. One feature the article described was the use of silicon nitride, which had been removed from the process by the time Intel received the final proofs of the *Spectrum* article.[60]

Rowe epitomized the tendency towards secrecy at Intel. He had no inclination to present the work on silicon gate at a conference, not only because of secrecy concerns but also because he had too much else to do. As a result of Intel's secrecy, Rowe became the anonymous craftsman. Although his jig for use in depositing metal on wafers was used throughout the industry, no name was attached to it. Since he did not publish papers or give presentations at professional conferences, he did not get the recognition that the group working on MOS at Fairchild R&D had.[61]

An analysis of Intel's first silicon gate part by Fairchild R&D shows the success both of Intel's silicon gate process and Rowe's efforts to maintain secrecy. Rowe had attempted to make the process resistant to reverse engineering. Shortly after Intel introduced the part, Fairchild bought one to evaluate. The most important test of any MOS part was its stability—did the transistors' characteristics change over time? Fairchild R&D performed several stress tests on the Intel part and pronounced its characteristics "rock stable," a term never applied in Fairchild progress reports to a Fairchild device. In R&D's effort to reverse-engineer the Intel part, Fairchild engineers were unable to discern many key features of the manufacturing process.[62]

Designing MOS Memories

Intel's first planned MOS part, a 256-bit RAM called the 1101, was designed in parallel with the silicon gate process. The part was a static RAM, which meant that the memory cell retained data all the time power was applied to the chip. This made the circuit design straightforward. A dynamic RAM, while allowing for a higher capacity, would have been much more risky, for its functioning would have been more dependent on interactions between the silicon gate process and the circuit design. Intel concentrated on process innovation first. The 1101 ultimately served as a kind of test chip, proving to Intel that it could make silicon gate MOS.[63]

Intel's approach to MOS circuit design, while not involving the same amount of analysis that had been done at Fairchild, was nonetheless very methodical. Intel designed at least five test chips, both to improve the silicon gate process and to evaluate the circuits that would be used in the product. Initially, Intel designers did not use the computer-based circuit analysis programs that had been common at Fairchild R&D but relied primarily on graphical analysis. This required knowledge of each MOS transistor's current-voltage (I-V) characteristics, which were calculated based on the design parameters. One would plot a family of I-V curves and use these to design the circuits. To make these calculations by hand was possible but extraordinarily tedious. Intel's circuit designer, Joel Karp, did not know how to program a computer, but Grove taught him FORTRAN, and he then went down to the IBM Service Bureau in Palo Alto to key in and run his program for calculating and plotting MOS I-V curves.[64]

By January 1969 Intel had its 1101 designed, and in September, with the silicon gate process working, it introduced the 1101 as a product. The initial price was $150, dropping to $55 in large quantities. This represented a price of from twenty to sixty cents a bit—far above the penny-per-bit target Moore had set for semiconductor memories to displace cores. Intel engineers had designed the 1101 with little input from customers, and the part drew little interest.[65]

In its focus on memory, and the silicon gate and Schottky processes, Intel's strategy was to develop new technology and new products. The advantage of this strategy was that it freed Intel from the harsh world of commodity pricing that Fairchild faced. The disadvantage of this strategy was that insufficient demand might exist for the new products Intel developed, which proved to be the case with the 1101. When this happened, Intel had a temporizing plan. The early entrants in MOS, particularly General Microelectronics and General Instrument, had concentrated on shift registers, which provided serial rather than random access memories. By 1968 computer makers had found several uses for shift registers, so that they were the one standard part that existed in MOS. Although competition existed in this market, one could design a shift register that was compatible with the ones currently being offered and be reasonably sure that it would sell if it were priced competitively.

By mid-1970 Intel introduced its 1406 and 1407 shift registers, which were plug compatible with ones being sold by National Semiconductor. Because Intel used the silicon gate process and National did not, the Intel parts were not identical copies of the National ones, but would look the same to the user. National's parts had suffered from reliability problems, a frequent concern with MOS technology, providing an opening for Intel. The density advantage of Intel's silicon gate process over standard metal gate MOS allowed Intel to put the same function in a smaller area, an important advantage in a commodity market. Intel then designed a family of shift registers that had a larger capacity than anything then on the market, holding up to 1,024 bits of information.[66]

In spite of the tepid response the 1101 received, the atmosphere for semiconductor memories within the semiconductor and computer industries had markedly improved by late 1969. If in mid-1968 a careful observer of the semiconductor industry could have seen signs of work on semiconductor memory, by 1969 it had become one of the hottest areas in the semiconductor business. Primarily due to the growing awareness of the potential of both MOS technology and semiconductor memory, the period between 1968 and 1970 was an extraordinarily fertile time for semiconductor start-ups. In 1968 twelve new start-ups were established on the San Francisco Peninsula, with nine more the next year. Many more were established elsewhere in the country.[67]

A look at four of these new firms shows that while each had a different approach, there were broad similarities in what they were doing vis-à-vis semiconductor memories. In late 1968 defectors from IBM formed the Technology Division of Cogar Corporation, aimed at developing semiconductor memories. Cogar planned to develop complete memory systems that it would then sell to computer manufacturers. It first received widespread attention in July 1969 when IBM sued Cogar charging theft of trade secrets. The trade and business press subsequently wrote a number of favorable articles on Cogar.[68]

In October 1968, engineers from IBM, Fairchild, and Collins Radio formed Advanced Memory Systems (AMS), located in Sunnyvale, California. Like Cogar (perhaps due to the IBM influence), AMS planned to build complete memory systems, but instead of selling them to computer makers, it was focused on selling these systems directly to computer users through a third party. AMS also offered its memory chips, both bipolar and MOS, for sale individually.[69]

In 1969 three of the leaders of Texas Instruments' MOS development effort left to start Mostek, located near Dallas, Texas. As its name implies, Mostek concentrated exclusively on MOS technology. It made chips, not memory systems. It worked on MOS memory as well as other MOS parts, such as custom-designed chips for calculators. Mostek's main effort was in a new variation of MOS technology using a new fabrication technique called ion implantation.[70]

Another start-up with a large MOS effort but no intention of selling individual MOS chips was Four-Phase Systems. Four-Phase was founded in late 1968 by Lee Boysel, who led Fairchild's MOS circuit design effort in its Mountain View facility. Boysel recruited many of the engineers who had worked with him at Mountain View to join him in his new venture. Four-Phase aimed to make and sell computer systems that were composed almost entirely of MOS technology. Four-Phase designed almost all of its own chips and subcontracted the fabrication process. The Four-Phase computer was to have MOS semiconductor memory. By June 1969 Four-Phase had working hardware of a 1,024-bit dynamic random access memory (DRAM) it had designed. Boysel and his group had designed a smaller DRAM at Fairchild that served as a test vehicle for the Four-Phase part.[71]

Although Four-Phase did not publicize any of its work prior to the announcement of its computer system, Intel had access to that work. Robert Noyce, the president of Intel, served on the board of directors at Four-Phase. The first head of Four-Phase, Charles Sutcliffe, had known Noyce when they had both worked at Philco. Noyce saw the Four-Phase DRAM and brought back word of it to Intel. Although it is impossible to assert what the exact impact of the Four-Phase work was on Noyce and Intel, it certainly would have suggested that their 1101 part was not competitive and that DRAMs could be made successfully.[72]

The efforts of these start-ups and established companies resulted in substantial publicity for both MOS technology and semiconductor memory by 1969. Although the trade press had carried items on semiconductor memory previously, by 1969 it had come to the fore. *Electronic News* carried its report on the 1969 IEEE International Solid-State Circuits Conference under the headline "IBM Making Semicon Memory." The article quoted unnamed observers who asserted that IBM's use of semiconductor memories meant that the technology had reached its "taking off point," and the reporter claimed that semiconductor memory's "pace of development was gathering speed." *Electronic News's* report on the 1969 International Electron Device Conference carried the headline "IC Producers Jam Memory Cost Talks," and the article observed: "It appears that semiconductor companies are determined to pressure core memories before the next decade is half over." The focus on cost suggested that most of the technical problems were seen as solvable and that the adoption of semiconductor memory was now primarily a question of economics. The constant reports in the trade press on semiconductor memory created an environment that no computer maker could afford to ignore. For suppliers, the publicity would have reinforced their notion that they were in the right field.[73]

The spate of new start-ups also helped Intel in a more tangible way. In early 1969, Bob Cole, a Fairchild MOS manufacturing engineer, and Dan Borror, the head of mask making at Philco-Ford Microlectronics (formerly General Micro-electronics), started Cartesian. Cartesian was closely tied to Four-Phase Systems, but it offered an array of services, such as chip fabrication and mask making, to other companies. Cartesian provided the masks for Intel's early chips, freeing Intel from the need to dedicate time and resources to this essential part of chip making.[74]

The 1103: DRAM Takeoff

The impetus for Intel's second-generation RAM came from outside the company. Honeywell's group responsible for memory design for its line of mini-computers wanted to use semiconductor memories for its next machine. Because Honeywell, like the rest of IBM's competitors, did not have the capability to make its own semiconductor memories, it sent out a request for proposal (RFP) across the semiconductor industry. The RFP called for a 512-bit MOS DRAM and specified the memory cell to be used as well as the basic block diagram of the chip. Honeywell wanted several semiconductor producers to make this part to ensure a steady supply and a reasonable price. In the fall of 1969, Honeywell's engineers came to Intel. Although Intel already had an idea for an MOS DRAM, it halted that work in favor of the Honeywell proposal, doubling the size of the memory to 1,024 bits. The Intel engineer responsible for the design of the part kept in almost daily contact with Honeywell and in February 1970 gave a joint paper with a Honeywell engineer describing the RAM at the International Solid-State Circuits Conference. In early 1970, Honeywell had a computer populated with the Intel-built RAM.[75]

But the Honeywell-specified part had problems. Its slow speed limited its market outside of Honeywell. Furthermore, the cell structure Honeywell had requested resulted in several quirks that made the operation of the part marginal. For these reasons, Intel ultimately cancelled the Honeywell-specified part, which Intel called the 1102.[76]

Intel had another 1,024-bit RAM design, which it called the 1103, that used a different cell structure than the 1102, thereby avoiding some of its problems. The 1103 was faster, with the requirements of more demanding customers in mind. It was also smaller than the 1102, making it cheaper to build. Due to Intel's limited supply of MOS circuit designers, it had suspended work on the 1103, but by the second quarter of 1970, work on the 1103 resumed. By the summer of 1970, the first 1103 parts were built. After initial testing that showed the whole lot to be nonfunctional, a test engineer showed that the parts worked if they met an extremely severe timing requirement.[77]

In October 1970 Intel introduced the 1103 with a brash ad claiming "cores lose price war to new chip." The part was rushed out by a marketing group convinced that an enormous opportunity lay before Intel if it reached the market quickly, before many other firms introduced 1,024-bit RAMs. Some in engineering were not as eager for a quick introduction of the part, for they knew it needed a major redesign to eliminate the timing problems.[78]

Honeywell concurred with the switch to the 1103 and continued to work with Intel. Honeywell's engineers provided Intel with an understanding of how the part had to work in the computer system, and they helped Intel develop specifications for the part and characterize it. *Electronics* magazine wrote in 1973 that Honeywell "acted as the non-IBM-computer industry's semiconductor-memory steering committee" in its work with Intel.[79]

The 1103 was accepted as the industry standard within a remarkably short time. By June of 1971, seven semiconductor firms—including such industry leaders as Motorola, Fairchild, National, and Signetics—had all announced plans to second-source the 1103. This came at a time when Intel was still producing the 1103 in engineering sampling quantities. Intel announced that by the end of June 1971 it had shipped 52,000 pieces of 1103 memory, allowing Intel to say in its advertising that it had shipped millions of bits. But when translated into systems, this was a less impressive figure. A medium-sized mainframe computer that had 256 kilobytes of memory would require more than two thousand 1103s, while a minicomputer having 64 kilobytes of memory would require over five hundred. It seems likely that at this point none of the

1103s made had been shipped in computers; they were still being tested by computer makers.[80]

Intel's competitors played the key role in making the 1103 the industry-standard part. A long lead time existed between when a computer company might start evaluating a memory part and when it would be shipping the part in production computer systems, as each manufacturer had to design and evaluate prototype systems. Semiconductor manufacturers who were latecomers to the memory market had to place their bets on an existing part, for by the time their RAMs would be available they would have no realistic chance of getting a new standard established. Since Intel had one of the first RAMs on the market, its relationship with Honeywell was well known in the industry, and its founders were industry leaders, it made sense for the semiconductor industry to standardize its part. AMS, which had the first one-kilobit DRAM on the market, was not well known and had concentrated much of its efforts on selling add-on memory systems rather than chips. It attracted only one second source, Motorola, which had also announced plans to make Intel's 1103. By the end of 1972 Intel could state that the 1103 was used by fourteen out of eighteen mainframe computer manufacturers in the United States, Europe, and Japan and that it was the semiconductor part with the world's largest dollar volume.[81]

Intel established the manufacturing capacity to capitalize on the 1103's status as the industry standard. By 1974 *Electronic News* estimated that Intel controlled 70 percent of the 1103 market, down from a high of 85 percent. The only two effective competitors were Microsystems International Limited, a small Canadian firm that had licensed Intel's silicon gate process and 1103 design for $2 million, and American Microsystems, which was estimated to have 20 percent of the market. Ironically, many of the companies that announced their intention to produce the 1103 and helped make it the standard part, turned out to be nonfactors in 1103 production. The 1103 was a very difficult part to make, and they may well have given up as they started and saw what a large head start Intel had.[82]

Although the 1103 was a major source of revenue in Intel's early years, MOS semiconductor memory had a significance to Intel and the industry far beyond that of a single product: it became the product with a capacity for infinite expansion, which Gordon Moore's 1965 paper had implied that the industry needed. Even though Intel could command premium prices for its one-kilobit

RAM, as competitors entered the market and learned how to make it, prices fell dramatically. But computer memories provided the ideal growth path. Since the memory on a medium-sized to large computer ranged from 8 kilobytes to 2 megabytes, there was no question that a market existed for higher-density memory chips that would ultimately offer a lower price per bit (and Moore's 1965 paper asserted they would). Intel first described its four-kilobit memory chip in 1972.[83]

The experience of computer maker Digital Equipment Corporation shows how semiconductor memory interacted with the computer industry. In 1976 its top engineers gave a paper describing a fatal architectural flaw shortening the lifespan of their popular minicomputer family, the PDP-11. The terminal malady was the inability of the PDP-11 architecture, developed before the introduction of semiconductor memory, to accommodate continually increasing memory capacity. The PDP-11 architecture was capable of addressing only sixty-four kilobytes, and although this was a significant increase over Digital's previous minicomputer family, the PDP-8, it proved inadequate in the age of MOS memory. The Digital engineers' 1976 paper described a phenomenon where in the face of a continuing decline in prices of memories (by 26 percent to 41 percent a year), customers (no doubt encouraged by Digital's sales force) bought machines having the same price as their previous systems, but with much greater memory. A fundamental requirement for a long-lasting computer architecture became its ability to accept an ever-expanding memory. Digital's replacement for the PDP-11, the VAX, assumed that the system had to accommodate its memory expanding in a fashion similar to what Moore's 1965 paper predicted would happen with integrated circuits. While the VAX's ability to address 4.3 gigabytes might seem infinite compared to a system that could address only 64 kilobytes, Digital's engineers calculated that on the basis of the continued decrease in memory costs and the continued increase in memory sizes, the VAX architecture would run into that 4.3-gigabyte limit in 1999.[84]

With the acceptance of Moore's 1965 paper as a fundamental premise of computer memory design, it was as if the semiconductor industry was a virus that had successfully inserted its genetic code into the computer industry host so that henceforth the host would follow its dictates. In practical terms the computer companies had allocated an essentially fixed dollar amount of the price of their computers for the semiconductor memory, so that as the semiconductor industry produced higher and higher density MOS memory chips with lower and lower per bit costs, semiconductor firms would not be fighting among themselves for an ever-shrinking revenue per system, as they would have been if the memory size of computers had been held constant.

The Scope of MOS

In Tom Wolfe's classic 1983 *Esquire* article on Intel, he describes Andrew Grove teaching a seminar for employees on the Intel culture. According to Grove, the Intel approach was not about taking the ball yourself and running with it; instead, Grove explained, "At Intel you take the ball yourself and you let the air out and you fold the ball up and put it in your pocket. Then you take another ball and run with it and when you've crossed the goal you take the second ball out of your pocket and reinflate it and score twelve points instead of six."[85] Grove's definition aptly defines Intel's approach to the silicon gate technology. In a less colorful but more analytical way, Intel's approach could be described as fully realizing economies of scope, which Alfred Chandler has defined as "the use of processes within a single operating unit to produce or distribute more than one product." Intel used its basic silicon gate MOS technology to offer a wide range of complex and demanding large-scale integration circuits that offered the possibility of commanding premium prices. Intel was never content to rest on the success of the 1103. It introduced eighteen new products in 1971 and thirty-one more the next year. These new products were primarily implemented in silicon gate MOS technology, but they did not always have a clear market. They included an erasable programmable ROM, microprocessors, memory systems, and digital watches. Intel's willingness to improvise did not invariably lead to success, as its experience with digital watches shows. But Intel's experimental ventures started out on a relatively small scale so that any failure would not jeopardize the entire company. One of Intel's new products quickly became its most financially important, even while the traditional memory market attracted the most attention.[86]

Because the silicon gate MOS process was very dense, allowing the fabrication of many transistors on a chip, it was well suited for a variety of memory products as well as the microprocessor (to be discussed in chapter 8). In 1970

Intel introduced a 1,024-bit shift register, the largest shift register in production and a demanding part to build; it contained more transistors than the 1103. In 1972 Intel began introducing products using an n-channel silicon gate process that was significantly faster than its original p-channel process. At the beginning of the year it announced a two-kilobit shift register and later announced a one-kilobit static RAM. N-channel MOS parts were much easier to interface with bipolar circuitry than p channel, obviating the need for additional translation circuitry.[87]

While these parts were all obvious extrapolations of existing products, in 1971 Intel introduced a new type of memory made possible by the silicon gate process, but with an unclear market. The erasable programmable ROM, while not as well known as the microprocessor, interacted with the microprocessor and became Intel's most important and successful product in the 1970s. In 1969, Dov Frohman joined Intel from Fairchild R&D, where he had been involved in developing computer-aided-design tools for MOS design. He received a Ph.D. in electrical engineering from Berkeley in 1968, doing his dissertation on an MOS type of nonvolatile memory. (A nonvolatile memory is one that retains the information stored even when power is removed from the chip.) He then worked on this memory at Fairchild R&D and in 1969 joined Intel with the intention of continuing his Fairchild work. At Intel, Frohman also considered a reliability problem facing the silicon gate process. As he proposed a mechanism for the problem, he realized that this mechanism could be used to create a new type of ROM that was nonvolatile and erasable. Furthermore, Frohman's memory would use the silicon gate process essentially as Intel had developed it, while his earlier proposals required a unique process.[88]

ROMs were established products, used for the control stores in computers. Programmable ROMs (PROMs) were also established products. While the data stored in a ROM was determined in the manufacturing process and required a long lead time to change, PROMs were programmed by the user, allowing for immediate changes to microcode without going through the semiconductor manufacturer. But once the part was programmed, it could not be changed; new microcode required a new part. Intel introduced Frohman's EPROM after asking several customers if they had uses for such a device. Intel's initial expectation was that the EPROM would be used in small quantities for prototyping computer systems, when the computer system was in a state of flux, but then conventional PROMs or ROMs would be used when the system went into production.[89]

But Intel proved to be very wrong about how the device would be used. The first part, a 2,048-bit design, was introduced in 1971, but instead of being used only for prototyping, customers used it much more for production than Intel had anticipated. The microcode for systems remained in flux longer than Intel had thought, and customers put a high value on the ability to reprogram a part. Intel had priced the chip very high, expecting companies to buy only a few, but instead they bought many. Here, unlike in DRAM, Intel had no competition. The EPROM was more difficult to make than the DRAM and did not attract the same attention in the press. Meanwhile, Intel made an active effort to hide the success of its EPROM—Intel truly hid the football in its shirt. Intel raised the price of EPROMs until they cost ten times more than an equivalent ROM. The EPROM was Intel's most profitable product throughout the 1970s, even while DRAMs and microprocessors got much more publicity.[90]

Another way Intel took advantage of economies of scope was to develop higher-level assemblies using its memories. Intel's efforts to sell semiconductor memory for use in computers and peripherals ran into several problems. The first had to do with their complexity. Intel's 1103 DRAM was an extremely difficult part to use. Ted Hoff wrote a twenty-eight-page application note describing how to use the chip and avoid potentially dangerous conditions. For large computer makers, this complexity was not insurmountable; in fact, since they typically had large organizations to design and debug magnetic memories, the complexities of the 1103 gave these organizations an outlet for their skills, perhaps making the new technology less threatening to them. But for a small company considering putting semiconductor memory into a peripheral such as a terminal or printer, the complexities of semiconductor memory were potentially overwhelming.[91]

This was one impetus for the founding of Intel's Memory Systems Division in 1971. The other was the fact that IBM's production of its own memory chips excluded Intel from the largest segment of the memory business. To begin the work, Intel hired three engineers from Honeywell who had worked designing the 1103 into Honeywell computers. The new division's first product was a small memory meant to be sold directly to the manufacturers of computer equipment for incorporation into their systems. In contrast to firms like Cogar, which focused exclusively on selling memory systems, Intel offered its customers a choice. Those who had the capability to design their own systems could buy memory chips only, while those who lacked that capability could buy a memory board or system. Later, Intel designed a series of add-on memory systems for IBM System/360 and System/370 computer systems to be sold directly to the computer user.[92]

Intel's memory system work offered ancillary advantages to the company. It gave Intel a group of engineers who understood how memories were used in computer systems, which proved helpful as Intel developed later generations of semiconductor memory. Intel's memory systems business also allowed Intel to make use of a large number of memory chips that would have otherwise been unsalable. Among the largest yield detractors in memory chips were random defects that affected only a single bit of memory. Even though 1,023 out of 1,024 memory cells worked, the part could not be sold to a customer: most customers typically believed that any failure in the chip meant that the entire chip was compromised. Intel designed memory systems to use partially good chips, by addressing only those parts of the chip that were good, a technique invisible to a user of an add-on memory system.[93]

After semiconductor memory and calculators, one of the most popular applications for MOS integrated circuits was digital watches. Complementary MOS (CMOS) used very little power and allowed for battery operation. In 1970 RCA won a million-dollar contract for CMOS chips to be used in Hamilton Watch Co.'s $1,500 Pulsar watches. In 1971 Intersil, a Silicon Valley start-up, announced a multimillion-dollar contract with Seiko to supply CMOS chips for a $650 watch. By 1972 the trade paper *Electronic News* was regularly announcing new entrants into the MOS watch business.[94]

In June 1972, Intel made its entry into the field with the acquisition of Microma, a Silicon Valley start-up that made watches retailing for $150. But the large number of semiconductor manufacturers participating in the field, coupled with the effect Moore noted in his 1965 *Electronics* article, drove prices down. In contrast to semiconductor memories, no growth path appeared that would allow for larger and more complicated chips to keep revenues up. By 1975 forty firms were selling electronic watches. By 1976 Texas Instruments was selling watches for twenty dollars, a price it halved the next year. While

Texas Instruments was experienced at selling consumer electronics and willing to battle it out in a commodity pricing environment, Intel was not. In 1978 it sold its Microma division and exited the watch business poorer, but wiser.[95]

A Unidirectional Information Machine

Both the semiconductor industry and Silicon Valley were based on a relatively free flow of information and people between companies that might be considered competitors. Intel made a major effort to make the process unidirectional. This marked a significant difference from Fairchild, which had been one of the leading sources of information for the entire industry, through the licenses it granted for the planar process, the work of its R&D labs, and the movement of its people. Intel was much more aggressive in trying to discourage people from leaving for competitors and much more parsimonious in what it disclosed. By not having an R&D lab, Intel generated less information that could be transferred.

Fairchild had licensed its most important invention, the planar process, to any semiconductor company willing to pay the fee. While this brought in handsome revenues, it enabled the competition. Ultimately, Fairchild lost control of the markets the planar process made possible, as other companies proved to be much more effective manufacturers. Intel relied much more on trade secrets than on patents to protect its semiconductor work. Fairchild's planar process patent had been effective because it resulted in a definite structure that could be easily identified through reverse engineering, so that infringement could be detected. Intel did not have the patent on the silicon gate process, and its achievement had been to make it manufacturable. The key process steps, such as the exact composition of a rinse, would not have been evident to someone reverse-engineering the part. By patenting its version of the silicon gate process, Intel would simply be revealing what made the process work while receiving marginal protection, since infringement could not be inferred from the artifact. Intel had a policy of not patenting inventions that were not visible on the integrated circuit itself and so generated relatively few patents.[96]

Intel's reliance on trade secrecy for protection made stopping the flow of informal, unauthorized transfers of information much more important. The center for these transfers in the Santa Clara Valley, through engineers both swapping secrets and changing jobs, was the Wagon Wheel Bar, located just blocks from Intel's facility. Intel management made the Wagon Wheel off-limits to Intel personnel.[97]

As Intel grew it was able to take advantage of the experienced pool of talent in the Silicon Valley area. Fairchild proved to be a particularly attractive target. Although Fairchild probably could not have stopped any individual defections, had it responded aggressively to these raids, with legal action or threats of legal action, it would have made its employees much less attractive to Intel. Fairchild did not respond, and Intel continued to look to it for employees.

Intel knew how to discourage defections. In late 1975 one of Intel's top MOS RAM designers, Jimmy Koo, left to join AMS, a local competitor. Ed Gelbach, an Intel senior vice president, wrote to AMS demanding that it not obtain proprietary information from Koo and that it go beyond verbal assurances and provide a method to guarantee compliance. Gelbach further insisted that AMS agree not to hire other Intel employees having proprietary knowledge of Intel's RAM technology for two years.[98]

But behind Gelbach's bluster, Intel had a problem: over the preceding three years it had hired away twenty-one AMS employees, very often into positions similar to ones they had held at AMS where there might be reasonable questions about whether they had brought over proprietary AMS knowledge to Intel. The head of AMS asked for assurances that Intel had not appropriated any of its trade secrets. The incident had a predictable resolution. Both sides made professions of their honorable behavior, while Intel hired Fairchild's best remaining MOS designer to take Koo's place. When an AMS executive challenged Gelbach about Intel's raids, Gelbach replied, "We didn't think you cared." While Fairchild truly seemed not to care, no one could labor under that misapprehension with Intel.[99]

As Intel grew and wanted to develop new products and move into new areas, the veteran employees from the area were a key asset. For example, nineteen of the twenty-one employees from AMS had worked on memory systems, enabling Intel to move quickly into an area where its founders and original employees had little experience. Although those coming from Fairchild often worked in new areas at Intel, they had semiconductor experience. Frequently

Intel turned to Fairchild R&D for people like Noyce, Moore, and Grove—Ph.D.'s with backgrounds in semiconductor processing. They provided important managerial talent for Intel.[100]

Intel limited formal transfers of information to the rest of the semiconductor community through conferences and papers, a process that can be seen by examining Intel's participation in the International Electron Device Meeting (IEDM), the profession's premier device conference, a common site for introducing innovations. Fairchild had presented its original planar process here in 1960, and each IEDM typically had a number of papers by Fairchild researchers. Between 1968 and 1975, Gordon Moore was the only Intel employee who presented a paper at an IEDM, and the papers he presented in 1968 and 1975 were at plenary sessions where he discussed in general terms the state and future of semiconductor technology, not specific Intel work.[101]

Its absence from the IEDM sessions may be explained during Intel's first years by the demands of starting a new company, but as it grew one might have expected that to change. Even as it became an established semiconductor producer, Intel had no time for this type of work and wanted to keep its process details secret. Between 1969 and 1975, Intel hired three Ph.D. researchers from Fairchild R&D who had been contributors to the IEDM and might have been expected to continue to do so at Intel. At a rapidly growing Intel, they were never heard from again at IEDM sessions. Gordon Moore has said Intel operated according to what he has called "the Noyce principle of minimum information," giving the following description of its operation: "One guesses what the answer to a problem is and goes as far as one can in an heuristic way. If this does not solve the problem, one goes back and learns enough to try something else. Thus rather than mount research efforts aimed at truly understanding problems and producing publishable technological solutions, Intel tries to get by with as little information as possible."[102]

This is not to say that Intel found the IEDM irrelevant. Every year between 1968 and 1975, Intel had representatives on the program committees. Intel would continue to take in the work from the leading industrial research labs, whose tenets practically required that they present and publish their work. But Intel, with no research lab, would no longer do the work in the more academic style often presented at these meetings. And Intel would closely guard the work that was directly related to the processes it used.[103]

Although Intel did not present papers when they served only the vague purpose of diffusing information within the industry or impressing others with its work, Intel did present papers when they served some clear and specific purpose for the company. For example, at the IEEE Reliability Physics Conference, a typical paper might discuss new techniques for testing the reliability of transistors, or the physics of some failure mechanism. Between 1969 and 1975 Intel presented three papers at the Reliability Physics Conference. Each was of the same form: presenting a reliability study of a new semiconductor technology used in an Intel product. For example, in 1971 three Intel engineers gave a paper studying the reliability of silicon gate MOS products. It concluded, "The Si-gate technology provides excellent reliability." Since silicon gate was new, it was reasonable for customers to wonder about its reliability, and this paper, in showing that Intel had done the studies to prove it reliable, helped to market Intel products.[104]

Conclusion

By 1974, its sixth year of operation, Intel had established itself as a major semiconductor producer, with sales of $134 million and profits of nearly $20 million. By 1975 Intel was the world's largest producer of MOS integrated circuits and one of the ten largest semiconductor producers. Intel was clearly an MOS company by this time, with its aggressiveness and energy in developing new MOS products standing in marked contrast to its conscious decision not to compete in most areas of bipolar technology. And in spite of its diverse product line, including microprocessors and watches, Intel was predominantly a semiconductor memory company.[105]

While Intel rose, Fairchild fell. Fairchild suffered through such a multitude of problems in the years after 1968, many of them self-inflicted, that even cataloging them is difficult. Fundamentally, after the departure of Noyce and Moore and the arrival of Lester Hogan and his lieutenants, Fairchild lost its integrity as a company. While it had too much strength to collapse, it never regained a basic soundness. Hogan came to Fairchild as the head of the parent company, and he moved the corporate headquarters from Long Island to Mountain View. Ironically, while Fairchild was now in fact a West Coast company, it began looking more like an East Coast vertically integrated company.

As Fairchild employees left, frustrated by the arrival of Hogan's Heroes and the implicit lack of trust shown in the existing employees, Hogan frequently filled in the gaps by hiring managers from Bell Labs or other large East Coast firms. One former Fairchild employee spoke of Hogan bringing with him "Mahogany Row," well-appointed executive offices isolated from engineers and workers. In October 1970, *Electronic News* reported that many disgruntled Fairchild employees had adopted as their underground slogan "We couldn't care less, because Les doesn't care anymore." Fairchild continually lost its best technical people to smaller firms in the area and tried to fill in by hiring people from elsewhere, often from East Coast companies. Frequently, these people would use Fairchild as a way station, getting acquainted with the area, then moving on to a company with better prospects. The flow was almost always from East Coast firms to Fairchild and then from Fairchild to Intel or other start-ups. One did not leave Intel for Fairchild.[106]

Fairchild MOS management suffered from a continuous churning process, bringing in what one Fairchild veteran called the "VP of the year" to reorganize the MOS operations, typically with a new strategy and ideas about new products. However, the senior management typically got dissatisfied before the new plans had time to show results, the vice president was fired, and the process began again. Fairchild essentially sat on the sidelines while Intel developed the first generation of memory products, and it was a fast-paced game where there were no timeouts to allow a sluggish Fairchild to amble back onto the field.[107]

While Intel had conceded the bipolar market and concentrated on MOS, Fairchild developed both, keeping its emphasis on bipolar. While MOS technology had established a position below bipolar, at a higher density and lower cost, in the early 1970s Fairchild attempted to squeeze MOS from above and below. On the high end it worked on developing bipolar technologies that would offer densities similar to MOS at bipolar speeds. Basing their assessment on an innovative bipolar technology Fairchild had recently announced, in 1971 senior Fairchild managers confidently predicted that bipolar memory would split the market roughly evenly with MOS. Obviously, such analysis made it less urgent that Fairchild establish a position in MOS. But it was one thing to make such an assertion; it was quite another to carry it out. Intel had established a solid position in computer memories with its 1103, and Fairchild lacked the vigor to unseat Intel. Throughout the 1970s, the consulting firm Integrated Circuit

Engineering estimated the split in semiconductor memories to be roughly three-quarters MOS and one-quarter bipolar.[108]

From below MOS, Fairchild worked on a new semiconductor technology, charge-coupled devices (CCDs), which could be used for memory, offering higher densities than MOS. However, CCDs were so much slower than MOS memories typically used in computers (by a factor of 300), Fairchild had great difficulty finding a market for the new technology. Fairchild's flanking efforts not only failed but spread out its forces, making its MOS program much weaker. And most new applications in the semiconductor industry came not at the high end, where bipolar dominated, or the very low end, where Fairchild tried to establish CCDs, but at the middle and low ends, where MOS was satisfactory. While the bipolar market was still profitable and growing, it grew at a slower rate than MOS. Figure 6.2 shows the comparative breakdown of bipolar and MOS business at Fairchild and Intel as estimated by Integrated Circuit Engineering.[109]

The final blow to Fairchild came in 1979, when it was bought by the French oil services company Schlumberger. While Silicon Valley entrepreneurs were establishing firms under local control, Fairchild was moving in the opposite direction—further and further east. Schlumberger assigned one of its own executives, who had no experience in the semiconductor industry, to manage Fairchild. While Schlumberger brought a new set of problems, Fairchild continued its long love affair with bipolar technology. In the 1980s, when the balance between bipolar and MOS was shifting decisively in favor of MOS, Integrated Circuit Engineering estimated that Fairchild's MOS sales were falling dramatically. In 1986 Fairchild announced the development of a new bipolar technology, which it claimed offered the circuit densities of CMOS. The engineering manager in charge of Fairchild's sixty-person bipolar development group, a former IBMer, bravely claimed the resurgence of bipolar technology. The next year National Semiconductor bought Fairchild, paying less for it than the value of Fairchild's real estate and buildings.[110]

Intel owed a great debt to Fairchild, but it was also a recrystallization of many people from Fairchild into an entirely different structure. The new structure lacked the polarities of R&D versus manufacturing and (initially) MOS versus bipolar. This depolarization can best be seen in two people, Andrew Grove and Tom Rowe. Grove's main accomplishment at Fairchild had been as the theoretician of the MOS transistor, and he became famous in the profession

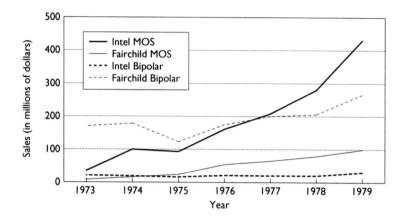

FIGURE 6.2 • MOS AND BIPOLAR INTEGRATED CIRCUIT SALES AT FAIRCHILD AND INTEL AS ESTIMATED BY INTEGRATED CIRCUIT ENGINEERING, 1973–1979. The cyclical nature of the semiconductor industry can be seen in the sharp downturn in 1975. *Sources:* Integrated Circuit Engineering, *Status 1975: A Report on the Integrated Circuit Industry* (Scottsdale, Ariz.: Integrated Circuit Engineering, 1975), 24 [1973 numbers]; Integrated Circuit Engineering, *Status 1976: A Report on the Integrated Circuit Industry* (Scottsdale, Ariz.: Integrated Circuit Engineering, 1976), 24 [1974 numbers]; Integrated Circuit Engineering, *Status 1977: A Report on the Integrated Circuit Industry* (Scottsdale, Ariz., Integrated Circuit Engineering, 1977), 3–2 [1975 numbers]; Integrated Circuit Engineering, *Status 1978: A Report on the Integrated Circuit Industry* (Scottsdale, Ariz., Integrated Circuit Engineering, 1978), 3–2 [1976 numbers]; Integrated Circuit Engineering, *Status 1979: A Report on the Integrated Circuit Industry* (Scottsdale, Ariz.: Integrated Circuit Engineering, 1979), 5–2 [1977 numbers]; Integrated Circuit Engineering, *Status 1980: A Report on the Integrated Circuit Industry* (Scottsdale, Ariz.: Integrated Circuit Engineering, 1980), 55 [1978 numbers]; Integrated Circuit Engineering, *Status 1981: A Report on the Integrated Circuit Industry* (Scottsdale, Ariz.: Integrated Circuit Engineering, 1981), 62 [1979 numbers].

for debates at professional conferences over the causes of MOS instability. As chief of operations at Intel, he was the driver, instilling discipline and making sure schedules were met. While he had been a prolific author of technical papers at Fairchild, he next achieved recognition as an author in the 1980s, writing management books.

Rowe, a manufacturing engineer whose experience had exclusively been with bipolar transistors at Fairchild, had played the crucial role in developing Intel's MOS technology. None of the R&D people had as much firsthand experience with manufacturing problems. In an organization committed to making MOS technology work, his bipolar experience proved valuable, for he believed that MOS technology should have the robustness of the bipolar technologies then in manufacturing. At Fairchild, Rowe never would have had a chance to do what he did at Intel, for the organizational barriers were too rigid.

While the operational side of a technology-based company typically had research, development, and manufacturing organizations, Intel lopped off the research part. This was a calculated gamble based on the large amount of data accumulated about silicon semiconductor technology and the large amount of work put into the public domain by other research groups. As might be expected, Intel came up with nothing that caught the rest of the industry by surprise or was based on some deep scientific knowledge. Instead, Intel took a broad range of product concepts that were in the minds of knowledgeable people in the industry and quickly converted them into real products.

IBM

MOS AND THE VISIBLE HAND, 1967–1975

Intel, a small start-up, had a very tight organization with very few technology transfer problems. Its MOS process was developed in the same run-down Mountain View building where it would be manufactured. At IBM, the establishment of MOS memory technology involved at least five different sites and the transfer of people and technologies across thousands of miles. IBM's MOS memory program achieved high-volume production at roughly the same time as Intel's, but the path to that point was completely different.

IBM is a classic case of the oligopolistic, vertically integrated company analyzed by Alfred Chandler. In IBM's semiconductor operations, management coordination replaced market coordination. IBM's semiconductor work had no direct contact with markets, allowing managers an extraordinary degree of freedom to direct work as they saw fit. Intel developed both bipolar and MOS technologies; it had no idea beforehand whether either would be successful. The success of MOS in the marketplace dictated that Intel would be primarily an MOS company. At IBM, the MOS versus bipolar decision was one made by a manager. Once IBM made the decision to use MOS memories in its computers, the success of the program was virtually guaranteed. IBM would pour whatever resources necessary to make the program a success and would become one of the world's largest producers of MOS memories.[1]

At IBM, questions of coordination loomed large. It had to manage the transition from the older magnetic core memory technology to semiconductor memory without jeopardizing its enormous revenues. Managers had to determine how to marshal the various organizations and resources at their disposal. IBM did not use its resources with perfect efficiency and had many duplicative efforts and false starts, but even these served a purpose, by helping IBM to manage the risks of moving to the new technology and satisfying various constituencies within the company. When it was over, Burlington, Vermont, had emerged as the center of MOS work in IBM.

Although IBM's MOS program could be called a success, that success was qualified in two important ways. First, IBM's philosophy of semiconductor production could only be practiced within the IBM environment, as shown by the failure of a start-up formed by IBM engineers. Second, IBM's MOS success was confined to memory. MOS memory had clear relevance to IBM's most profitable line of business, mainframe computers. But MOS logic was too slow for mainframe computer applications, and while IBM had many smaller systems where it could have been productively used, bipolar technology proved to be a satisfactory alternative.

The Joint Program

In 1967 managers from IBM's Components Division (CD) and Research signed an agreement to conduct a joint program in MOS memory development, involving participants from three different divisions and four different sites within IBM. The main work in the program would be carried out by those outside the centers of power of IBM's semiconductor work. Research was in this position, for it had made practically no contribution to the semiconductor technology currently being manufactured by IBM. It was joined by another group with a marginal mission, located in IBM's Boeblingen, West Germany, facility.

IBM Boeblingen had been established as a development site in the early 1950s to work on solid-state electronic devices. Germany was chosen to take advantage of native physics talent, Boeblingen because of its proximity to an IBM manufacturing facility. The history of IBM Boeblingen's electronics efforts up through 1963 paralleled IBM Research's, with an unsuccessful foray into gallium arsenide.[2]

In 1963 IBM Boeblingen transferred the silicon planar technology used in solid logic technology (SLT) from CD in Poughkeepsie (later to become East Fishkill), New York. Boeblingen then used that base to develop products meeting local needs for variant semiconductor technologies, while the most critical and most widely used semiconductor products still came from IBM East Fishkill. Boeblingen developed a semiconductor crosspoint switch, to be used in telephone communications systems.[3]

Wolfgang Liebmann, who had led the Boeblingen effort to transfer the silicon planar technology from East Fishkill, came to Yorktown to lead the MOS joint development program. Liebmann had received a doctorate in engineering at the Berlin Technical University in 1958, held a Fulbright Fellowship at the University of Pennsylvania, and then worked at RCA's David Sarnoff Research Center from 1960 to 1963. Because of his previous work in the United States, Liebmann had the credentials that would allow him to travel to the United States on short notice, and so he became Boeblingen's chief agent of technology transfer. American managers recognized Liebmann and the Boeblingen group as being extremely capable but lacking a mission. Liebmann brought with him a group of four people, which later grew to ten. Many had Ph.D.'s and would have worked in Research if IBM had a research lab in Germany.[4]

With the joint program, IBM's MOS work entered a more structured phase. CD had a formally defined eight-stage product development cycle designed to guarantee the integrity of programs as they neared high-volume manufacturing. In the joint program, Research would be responsible for bringing the program through the fourth phase (design), which proved the feasibility of the design and allowed for an assessment of the technical risk in going forward.[5] This phase required the development of preliminary performance specifications, the description of a fabrication process, the forecasting of process yields, and the forecasting of demand. Suffice it to say that all these requirements represented a new way of operation for Research, which was not used to such formal procedures. IBM lacked a formal means for transferring products from Research.

Liebmann took over technical leadership of the program, particularly on the process side. Although he had no previous experience with MOS technology, Boeblingen was scheduled to assume development responsibility for MOS after the program's conclusion, and Research managers believed that in order to give the receiving location a stake in the success of a new technology, the receiving location had to be given control of the program. Liebmann formulated a plan that would allow for learning and a rational evaluation of the technical alternatives. First, a fabrication process with relatively relaxed tolerances would be made, which would then be scaled to yield one that could be used to make a competitive product. A key contribution of the Boeblingen group was to add discipline to the MOS process. Research had made frequent modifications to its fabrication process, but the Boeblingen group fixed it at a point and documented it, requiring further changes to be made formally.[6]

Liebmann reopened the question of whether to do n- or p-channel MOS. His group fabricated and evaluated both transistor types with the relaxed process. A critical factor in the final decision proved to be which type held up better to IBM's packaging techniques. While a bare chip might work in the laboratory, IBM covered its chips with a layer of quartz to ensure that they would work over many years in customers' offices. Many doubted whether this packaging process would prove satisfactory for MOS transistors. Tests showed that while both held up reasonably well, the n-channel devices did better than the p channel. Research had originally chosen n-channel MOS for its speed. While Liebmann confirmed the decision, his primary criterion was not speed but compatibility with IBM's manufacturing processes.[7]

In addition to Liebmann's group, there was another newcomer to MOS: a small memory development department in IBM Poughkeepsie, managed by Dick Gladu. The department had previously worked on capacitive memory for read-only storage. While most of the memory development groups at IBM had been working on magnetic core memories, this group had not. Gladu's managers gave him the MOS assignment because he had people available and explicitly told him that they did not consider the work promising. Gladu's managers further expressed a preference that his people not directly involve themselves in the work but confine themselves to evaluating the work being done. Gladu gave two young engineers, Dick Linton and George Sonoda, the task of working on the MOS memory technology. Just as it had for Boeblingen, MOS memory represented an opportunity for Gladu and his group. They had no conflicting allegiances, either to magnetic cores or to bipolar semiconductors, and if MOS memory was to take off, their careers might be expected to follow it.[8]

Just as the program began, low-level Research management had to come to terms with an invention that had the potential to totally upset the program as it had been constituted. In late 1966, Robert Dennard, a member of the Research MOS program, was inspired during an conference reviewing all Research projects. As Dennard heard a presentation on IBM's magnetic memory program, he began to think of how analogies to the magnetic program might lead to simpler semiconductor memory structures. At the time Research's MOS memory program was based on a memory cell consisting of six MOS transistors. After several weeks of work, Dennard came up with a memory cell that consisted of only a single MOS transistor. Dennard's cell resulted in a dynamic memory in which the electrical charge essential to hold the information gradually leaked off the capacitor, requiring periodic restoration.[9]

Those in Research's MOS program recognized the significance of Dennard's work immediately, for it quite plausibly represented the ultimate MOS memory cell. But it was so far ahead of the existing state of Research's MOS program that it was incompatible with it. Before it could be implemented, it would require a whole series of innovations in peripheral circuit design and in MOS processing technology. Dennard and Dale Critchlow, Dennard's manager, decided not to use Dennard's invention immediately, which would delay the program as they pursued the additional required innovations. Instead, they proceeded with the existing memory cell design. While the members of Research's MOS program were criticized by some in CD for being too academic, they clearly cared about making a contribution to IBM's business and were not overly concerned with personal recognition.[10]

Dennard's memory cell further suffered from being so far ahead of its time that its implications were not fully comprehended. MOS memory was new, and few understood it. At first, IBM Research's patent attorney declined to file a patent on it. In April 1967, Dennard presented his memory cell at an internal conference of development personnel working in semiconductors and magnetic memory. Dennard showed that his memory design yielded a fivefold improvement in density over the approach then being pursued by IBM Research, but instead of being electrified at being among the first to learn of one of the most important inventions in the history of the MOS transistor, the audience sat silent, not comprehending the significance of Dennard's work. His talk prompted no questions or discussion. Although Dennard received a patent for his invention, he never published a paper on it.[11]

Dennard and other Research engineers worked closely with Gladu's group. And while the Research group trained Gladu's engineers, they all worked together as one team, with little distinction between the two. They designed four different memory chips in the course of the joint program. They first designed p- and n-channel memory chips using the relaxed process. After this they used the more competitive process to design a cross section of a 512-bit memory chip, the expected size of the first MOS memory product. The cross-section chip allowed thorough testing and evaluation of the internal circuitry, which would not be possible with a complete memory chip. Finally, they designed a 512-bit memory chip.[12]

Because IBM's ultimate concern was the performance of computer systems, the design and construction of an MOS memory system was a crucial milestone. One of the earlier memory chip designs was used to build a memory system that had one thousand sixteen-bit words. The system proved that all the necessary components had reached a level of maturity so that IBM could proceed with confidence. Research had previously done little work on packaging its memory chips, and the feasibility model required packaged chips, such as would be used in an actual product. Five peripheral chips had to be successfully built for the system to operate. The feasibility model allowed IBM engineers to test for interactions between components. In addition, the group completed a paper design showing what could be expected of an actual production MOS memory system.[13]

At the end of the joint program, Research management ordered the Research MOS group to disband and move into other areas. Research had transferred MOS technology to the development group, and therefore management considered Research's work finished. Research management seems to have continued on in a belief that it had held early in Research's history that Research and CD would not work on the same semiconductor technology at the same time. Therefore, the two groups would cooperate only briefly—during the transfer of the technology. Most of the Research MOS engineering group moved into communications work, and so the most experienced people in MOS technology in IBM were largely unavailable to the corporation during the most critical period of its development.[14]

Research's departure from MOS work caused anxiety among a number of members of the MOS group. They worried that in their absence the MOS technology they had developed would not survive, giving special significance to what they thought was to be their last contribution to the MOS technology, the publication of a design manual. Titled "Design of Insulated-Gate Field-Effect [MOS] Transistors and Large Scale Integrated Circuit Chips for Memory and Logic Applications," it defined and fixed the MOS technology as the Research-developed technology, in obvious and subtle ways. The manual included a description of the MOS process as developed by Research, with its rules for laying out circuits.[15]

The design manual worked through the disciplinary boundaries of circuit designers and process developers to fix the technology in a more substantial way. Process developers, who could be chemists, physicists, metallurgists, or electrical engineers, developed the fabrication processes for making integrated circuits. Circuit designers, typically electrical engineers, put together individual transistors to perform some needed function. Circuit designers would not need to know the exact process used to make a transistor, but they would need to know the electrical characteristics of a transistor made by that process, for these played a critical role in determining the performance of the overall circuit.

The manual contained pages and pages of electrical characteristics of MOS devices, which could be used by a circuit designer working with MOS technology. There was one hitch in using them, however. The device characteristics described in the design manual were obtained by using the Research-developed MOS process. Any other process could be expected to yield significantly different results. The Yorktown process was embedded within the characterization curves, even though the curves did not explicitly mention the process, and so they both had to be taken together.[16]

A designer would have ample reason to want to use the characterization curves. Nothing was available, either internally or outside IBM, that was as practical and design oriented as the design manual. Research published thousands of copies of it and distributed them throughout the company. In distributing the manual, they were distributing their version of the technology, making it harder for anyone else to offer a different version. Perhaps most important was the acceptance of the manual's basic tenets by those who would be carrying on the MOS work within IBM's development organizations, namely Liebmann and Gladu. Gladu's memory designs would have to undergo substantial changes if the process and the device characteristics were changed, a costly and tedious process he would resist.[17]

The design manual contained much practical information, available nowhere else, that would help a designer get going. It contained definitions for standard tests on MOS transistors. It provided several techniques for protecting MOS transistors from electrostatic discharge, a phenomenon that was much more troublesome in MOS transistors than in bipolar ones. It contained samples of masks Research had used to make various parts and discussed methods of defect analysis. All of this could have been duplicated independently, but the costs and time required would have been enormous.[18]

By any measure the joint program was a success. It brought groups from widely varied backgrounds and organizations to work together on a common goal, which would benefit all. By the time the program ended in the summer of 1968, the MOS technology had solidified and reached a point not too far from where it needed to be for it to go into large-scale production. But large-scale production was still years away.[19]

Commitment to Semiconductor Memory and the Search for an Orderly Transition

The fate of MOS memory depended on business decisions that had little to do with the success of the joint program; the joint program was necessary but not sufficient to get MOS established within IBM. The key executive responsible for the introduction of MOS memories was Ed Davis, who had responsibility for IBM's main memory development programs. Davis had a Ph.D. in electrical engineering from Stanford and had served a stint as administrative assistant to Frank Cary, a top IBM executive who would become president in 1971. Davis served as an important link between the highly technical world of semiconductor technology and the nontechnical world of top IBM management.[20]

Initially, Davis had responsibility for only the magnetic memory programs, but in 1967 he also gained responsibility for IBM's exploratory semiconductor memory work. Davis had spent the majority of his career working in semiconductors, giving him an inclination to favor semiconductor memory, and he soon killed most of the magnetic programs. In addition to Davis's predilections, several other factors proved important in these decisions. The exploratory bipolar memory program removed some of the risk in moving to semiconductor memories. Magnetic core memory was reaching a point of saturation, where increasing complexity was required to squeeze out improvements in cost and performance for the next generation. IBM's size and commitment to automated production techniques made manufacturing extremely capital-intensive. If IBM was to move to semiconductor memories in the near future, large investments in capital equipment for one more generation of magnetic memory were not prudent.[21]

To sell these decisions to top management, Davis made a masterful presentation demonstrating the complexities of the magnetic core under development and the difficulties it would present for maintenance. Frank Cary, at this time the head of the data processing group, with responsibility for development, manufacturing, and marketing of IBM's large computers, then requested support for his (and Davis's) decision to cancel a magnetic core memory development effort that IBM had already spent $37 million on and move to bipolar semiconductor memory. Initially, IBM limited its plans for semiconductor memory to two System/360 follow-ons, code named NS1 and NS2, that would use bipolar memory. To help win support for the bipolar chip, Davis called the bipolar memory chip that would replace cores ϕ2i (pronounced "phase two i"). IBM already had a functioning memory chip called ϕ2, and Davis picked the name of the new chip to suggest to top managers (who were not technically trained) that there were only slight differences between ϕ2i and ϕ2 and that therefore it represented only minimal risk. In fact, the two programs were only distantly related.[22]

Davis ran into trouble when he presented his plan to the corporate staff, which argued for the immediate use of MOS technology. Davis, whose career rested on the successful implementation of his plan, had taken considerable risk by committing to develop semiconductor memory and was loathe to take on any more. The corporate staff had the role of the loyal opposition, pushing development groups to be as aggressive as possible, but Davis would take the blame for failures of execution. Davis got the staff to agree that the prime consideration in whether MOS memory represented an acceptable alternative was whether it would be possible to maintain the schedules for two systems using bipolar memories if MOS memories were substituted. In January 1968 Davis and the staff convened a group to study the question. That group included the main constituencies involved: the corporate staff, Research, CD, and the Systems Development Division.[23]

Davis orchestrated the study to arrive at his desired answer. The study concluded that although schedules of the MOS and bipolar memory program were nominally the same, pilot line data were lacking to provide sufficient confidence that the MOS program would meet its dates; at the same levels of confidence it was actually six months behind. Supporting this conclusion was the fact that only one hundred finished MOS memory wafers had been made and ten MOS memory chips tested, while five thousand bipolar memories had been made.[24]

But the report also provided the most powerful support to date for the advantages of an MOS memory program. The simplicity and density of MOS would make possible a significant reduction in costs over bipolar. Through the work of planning departments, IBM had estimates of how many systems it could expect to place, and thus how many memory bits were required. While over 6,400 wafer starts per day would be required to produce the necessary bipolar memory chips, an MOS design would require only 2,000 wafer starts per day to make these same bits. The curves of wafer starts per day versus time ramped up quickly in the first two years to the 6,400 figure, then fell down quickly, due to the higher yields that would come with increased experience. The number of wafer starts per day was a key determinant of the size of the factory and the amount of capital equipment in it. Sixty-four hundred wafer starts per day was an enormous figure that would require one of the largest semiconductor factories in the world. The cost advantages of MOS were abundantly clear, and the report concluded that IBM should take steps to have specific systems use MOS memory.[25]

Davis's decision to cancel the ferrite memory development programs practically meant that IBM would use MOS memory technology. The initial decision to use $\phi 2i$ had been limited to two systems in IBM's follow-on to System/360, but there were many systems in the family, which led to an overwhelming demand for memory. The advantages of MOS technology over bipolar for memory were too large to overlook.[26]

A key factor making MOS satisfactory for mainframe computers was cache memory, which obviated one of the main advantages of bipolar memory technology, its speed. A cache memory is a small, fast memory interposed between the processing unit and the main memory. With a well-designed system, most of the information the processor required would come from the faster cache, rather than the main memory, partially decoupling the performance of the processor from the speed of the main memory. In 1966 IBM began development work on what would be the first cache memory in a commercial computing system.[27]

While Davis himself may have known that future systems would use MOS memory, it had not yet been established as the official IBM strategy. Within IBM the key to successful development of a semiconductor product was a committed user. CD could choose to develop something, but large amounts of money would flow to a program only when it was backed by a committed product or system inside the company. However, a user would be hesitant to commit to using a semiconductor technology unless there was some certainty that it could be developed and manufactured in a timely way.[28]

Shortly after the study group met, Davis agreed to launch an MOS memory program. The MOS program was scheduled to be a year later than the bipolar program, to cost half as much, and to have equal or better reliability. While the MOS memory program had no product commitments, Davis could ensure that it would be used because he had control over all the alternatives. (The stated plan was to have most of the computers being developed use existing core memory systems.)[29]

Davis tightly linked the MOS memory program to the bipolar program, claiming that the MOS process and thus the MOS manufacturing line would be as close to the bipolar process as possible. Davis would not be asking the company to make investments in two large manufacturing facilities, only one. MOS memory would come after the bipolar program had reached its peak wafer starts/day and so would take advantage of the extra capacity available.[30]

This arrangement had the added benefit of minimizing what Davis considered a major danger of the program: that he could be forced to stop the bipolar program and proceed immediately to MOS. He believed that the bipolar program was much less risky and could definitely be successfully implemented. If he gave significant resources to the MOS program, early progress could lead to added pressure to accelerate it, even though many potential problems lay ahead. Davis had to apply the brakes to the MOS work. While the study made in January stated that the two programs were six months apart, the MOS program Davis agreed to launch in February was to be one year behind the bipolar. Davis did not want an MOS versus bipolar competition. His chief strategist asserted that to be successful, the MOS and bipolar memory programs had to be seen as a single integrated program, where each would benefit from the total volume. Davis had arranged it so that the bipolar and MOS programs would share both the pilot line and manufacturing facilities. Since the MOS program would enter the pilot line only after the bipolar memory work had been completed, it provided the guaranteed separation Davis wanted.[31]

Davis wanted to ensure an orderly technical transition. In his mind one of the biggest hindrances to solid evolutionary progress in IBM came from those who tried to press for too much at once. After IBM's announcement of System/360 with SLT, SLT's perceived inferiority to integrated circuits led senior management to back a major leap into integrated circuits. After fighting problems for several years, the program was cancelled in favor of a more gradual transition to integrated circuits from SLT, something Davis had advocated all along. Davis believed IBM's nontechnical senior management was vulnerable to engineers who could make impressive promises. (Davis himself was no mean salesman.)[32]

MOS work would enter a holding pattern and would not enter East Fishkill, IBM's main semiconductor development facility, for quite some time after the end of the Research joint program. The Research pilot line kept running under the direction of East Fishkill personnel. Wolfgang Liebmann and his group went back to Boeblingen and took the MOS process with them. But little was done in East Fishkill.

There were many feints and false starts. The corporate staff continued to put pressure on CD to do MOS memory immediately and in parallel with bipolar memory. CD's lack of work on MOS memory was, to the corporate staff, a sign of insincerity. In the spring of 1968, Davis and Don Rosenheim, a senior manager at Yorktown, discussed moving people from Research to CD to speed the program along. Rosenheim sent Davis a list of forty-four people who might have been appropriate to hire, including all the key people in Research's MOS effort. In August, Davis asked for permission to hire eight people. No one was actually hired.[33] In 1968, Hank Bridgers from General Instruments, the leading MOS firm, was hired into East Fishkill ostensibly to get MOS technology into production, but he was soon shunted into a position with less responsibility. These actions seem to have been at least partially designed to deflect criticism as Davis implemented his plan.[34]

At the same time the Boeblingen group continued its work on MOS technology. Boeblingen had the best pilot line in the company, with excellent equipment and a top-flight and disciplined group of engineers. In spite of Boeblingen's earlier charter to develop MOS memory, CD pulled it back to East Fishkill, partly because as it grew in significance, it required more resources than Boeblingen had available, but partly so that CD could control the progress of the program.[35]

IBM: Sui Generis

In any discussion of IBM, it is difficult to overestimate its sheer size and unique position in the American economy. In 1969, while Intel had barely half a million dollars in revenues, IBM had revenues of $7.2 billion, and although definitions of the term *market* were contested, the Department of Justice claimed in its antitrust suit that IBM had between 71 percent and 83 percent of the market for computers. Since IBM made its own semiconductors, it dwarfed all other producers. In 1976 an IBM researcher estimated that between 1965 and 1967 IBM had consumed 80 percent of all the digital logic made in the United States. The fact that IBM would be using semiconductor memory meant that it would be manufacturing it on an unparalleled scale. In 1971, before IBM had announced its first computer using MOS memory, an IBM executive claimed that IBM had produced more MOS memory for use in system test (testing hardware and software prior to the introduction of products) than the rest of the world had made for all purposes. Gordon Moore estimated that in the 1970s IBM accounted for 60 percent to 70 percent of the market for computer memories.[36]

IBM's size and its position in the computer industry rather than the semiconductor industry meant that IBM was not bound by semiconductor industry standards or practices. IBM's semiconductors would be used only within IBM and sold only as part of IBM computer systems. This led to a distinctiveness that could be seen in the most basic matters. When the MOS transistor was first announced in the early 1960s, different companies called it different names. By the last third of the 1960s, the merchant semiconductor industry had settled on "MOS transistor." Each prospective MOS producer had an interest in a standard name as a basic requirement for clarity in communication with potential buyers. IBM continued to use a variant not used in the merchant industry, IGFET (insulated-gate field-effect transistor) or just FET (field-effect transistor). While a former researcher from Fairchild visiting IBM found this terminology confusing, industry standards were irrelevant to IBM's semiconductor operations, which had only internal customers. Similarly, while the rest of the industry settled on standard schematic symbols for the bipolar transistor and the MOS transistor, IBM adopted its own symbols (figure 7.1). IBM had a proprietary language.[37]

Although IBM's semiconductor devices operated according to the same physical principles as those produced by the semiconductor industry, the processes used to make them were dramatically different. IBM's processes were more capital-intensive and more automated. IBM's electromechanical heritage gave it the mechanical engineering skills necessary to build its own tooling. Although the semiconductor industry was famous for its mobility, after 1964 no one came to a position of responsibility in IBM's semiconductor work from another firm, so there were few advocates for following industry practices within IBM. IBM's semiconductor operations existed as a closed system.[38]

The fundamental sign of IBM's distinctiveness in semiconductor processing was the size of the wafers it used. For many years, they differed from those used by the rest of the industry by one-quarter of an inch. Much of the tooling was based on the wafer size, but since IBM designed and built much of its own production equipment, it did not need to follow industry standards.[39]

Three specific examples of IBM custom-developed processes were its methods of capsule diffusions, glass passivation, and chip packaging. While the semiconductor industry diffused dopants into its wafers by putting them in open tubes, IBM used tubes that had both ends sealed with a blowtorch. In the IBM process, powdered silicon doped to the appropriate level was put in the tube along with the wafers; then it was evacuated and sealed. This process involved the development of a special technology to open the tubes at the end of the process and special laboratories to prepare the powdered silicon. It did not have any significant advantages over the open-tube process.[40]

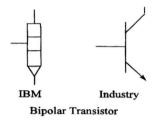

IBM   **Industry**

Bipolar Transistor

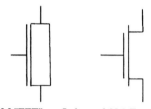

IBM "FET"   **Industry MOS Transistor**

FIGURE 7.1 • TRANSISTOR SCHEMATIC SYMBOLS, IBM VERSUS THE SEMICONDUCTOR INDUS-
TRY. Because IBM's semiconductors were used internally, IBM did not have to conform to semicon-
ductor industry standards. What IBM called the FET (field-effect transistor), the semiconductor
industry called the MOS transistor. The symbols on the left were still in use at IBM in the 1980s.
Examples of IBM's symbols can be seen in D. L. Critchlow, R. H. Dennard, and S. E. Schuster,
"Design and Characteristics of n-Channel Insulated Gate Field-Effect Transistors," *IBM Journal of
Research and Development* 17 (September 1973): 431; George Deuber, "Microminiature Circuits
Used for Compatible IBM 360 System," *Electronic News*, 13 April 1964, 5. Examples of industry
symbols can be seen in Robert J. Widlar, "New Developments in IC Voltage Regulation," *IEEE
Journal of Solid-State Circuits* 6 (February 1971): 2; *Fairchild Semiconductor Integrated Circuits* (Moun-
tain View, Calif.: Fairchild Semiconductor, 1966), 55.

Several of the major differences between IBM's processes and the semiconductor industry's came after the chip was made. To protect the chip from contamination after it was made, standard practice in the industry was to her-
metically seal the integrated circuit. IBM covered its integrated circuits with a
layer of glass, which obviated the need for hermetic sealing.[41]

By the late 1960s, IBM had developed a flip-chip method of connecting a
chip to its package. By this time most of the rest of the industry had settled on
wire bonding, in which a wire was essentially welded to the chip and then onto
the package. Each wire was individually welded, with the entire process frequently done by a poorly paid overseas worker. In the late 1960s, IBM devel-

oped an evolutionary extension of the method it had used in SLT, what it called its C4 process for attaching chips to their carriers. On each integrated circuit, solder balls were formed at every location where a signal was to be connected to the chip carrier. The chip carrier had pads corresponding to the locations of the solder balls. The chip would be placed, face down, on top of the chip carrier, and the assembly would be heated, reflowing the solder and connecting the chip with the carrier. This process had the advantages of making all the connections between the chip and the carrier simultaneously and of allowing assembly-line-style production, where a continuous number of chips and carriers could be sent down a conveyor into a furnace. However, this was an extremely complex process to develop, requiring a large effort to find solutions to a number of critical problems and expensive tooling to implement. As one example, the height of each solder ball had to be closely controlled so that the chip was supported by all solder balls equally. This brilliant piece of engineering was too complex and too expensive for the rest of the industry.[42]

IBM's position in the computer industry shaped the way it packaged its chips. Firms in the semiconductor industry were most concerned with the silicon part of the chip and devoted relatively little attention to packaging. They had every incentive to keep most of the value in the semiconductor portion, which they made, rather than in the package, which they often purchased from an outside vendor. IBM, with its ultimate concern the performance of the overall system, was willing to spend great amounts in packaging if it helped with that goal. In both its bipolar and MOS programs, IBM developed a package containing more than one chip (figure 7.2), allowing greater density at the system level than the industry-standard techniques. In 1972 IBM's group of seven hundred working on developing new packaging techniques dwarfed the effort of any one semiconductor firm and probably the industry as a whole.[43]

IBM could afford the extra costs of developing unique technologies not only because of its size, but also because it was not in a commodity business as the semiconductor industry was. Because IBM sold only complete systems and these systems had a much higher profit margin than semiconductors, it was not of the utmost importance that IBM be the low-cost semiconductor producer; the higher costs would be lost in the system profits. The freedom IBM had in its semiconductor work was a double-edged sword. It had the freedom to tailor its memory chips to its computer systems, rather than accept the more generic

FIGURE 7.2 • IBM'S MULTICHIP MODULES. While the semiconductor industry packaged its RAMs in inexpensive dual inline packages (one chip to a package), IBM developed a sophisticated multichip module packaging technology for its RAMs. The modules to the left and to the rear are stacked and contain two of IBM's two-kilobit RAMs on each of their two levels. The large module on the right contains four sixty-four-kilobit RAMs on each of two levels. The two-kilobit chip, named Riesling, was introduced in IBM products in 1973. The sixty-four-kilobit chip, named Common, was introduced in IBM products in 1978. Courtesy IBM Corporation.

solutions offered by the semiconductor manufacturers. It also had the freedom to ignore work done elsewhere and pursue its own approaches, even if other work was superior. Through the sheer magnitude of the resources it could bring to bear on a program, IBM could will an approach to work, where all others would have given up.[44]

Given the nearly limitless number of solutions to any problem, IBM's engineers and managers could almost always find a rationale for doing things differently than the rest of the industry. IBM could afford to pay the price of distinctiveness; the question was whether the benefits to IBM were worth the price. From this distance it is impossible to do cost/benefit analysis on every decision made, but the preference for unique solutions makes one suspect that at times managers chose difference for the sake of difference. Although they doubtless had rationalizations, new and unique processes would give added challenge to their creative people, and more people for them to manage.[45]

The IBM manager operated in an environment where it was essentially impossible for him (and all those considered here were male) to do something that would result in an unsatisfactory condition for IBM. The organization (or what John Kenneth Galbraith has called the technostructure) could mitigate the effects of any failure. A development program might not meet its deadlines, necessitating a delay in introducing and shipping a new product. In such a case, IBM would have temporizing plans to extend the life of older products. A program might greatly exceed its estimated costs. While this was not to be applauded, IBM's profit margins were such that cost overruns disappeared in a sea of black ink. While an IBM manager leading a project in either of these situations might be in for some uncomfortable moments, he would not necessarily lose his position.[46]

There was, however, one unforgivable sin: the failure of a manufacturing manager to meet production goals. This was unfailingly dealt with by removing the offending manager. Even an IBM salesman could not sell (or lease) what had not been made. Production shortfalls had more dire consequences for IBM than for a semiconductor company. A semiconductor company was denied the revenue it would have received from the part; IBM was denied the revenue from an entire computer system, rendering impotent the efforts of IBM's development, manufacturing, and marketing organizations. The importance of an assured supply was one of the prime reasons IBM manufactured its own semiconductors: while it could never be absolutely sure that a supplier would always come through, with its unlimited resources, IBM could guarantee its semiconductor supply.

IBM's MOS memory program had a much greater degree of stability than Intel's: while the key technical details of IBM's program changed very little after it left Research, Intel's MOS memory program went through a number of iterations that improved its characteristics. One explanation for the difference could be IBM's vertical integration, where having the computer developers and chip developers in the same company allowed it to choose the optimal design immediately, while Intel floundered, trying to learn what computer manufacturers wanted. But far more important were the environments the two operated in.

Intel put a premium on flexibility, looking for technological innovations that would give it an advantage over its competitors. IBM's mass production systems made changes difficult. But at IBM there was little need for such changes; IBM's technological choices could always be made satisfactory. While Intel's managers and engineers knew that their memory technologies would be tested in the competitive fires, IBM engineers had no such worry.[47]

Development: The East Fishkill Pilot Line

In late 1969, after the ϕ2i memory was transferred into manufacturing, the East Fishkill pilot line started work on MOS memory. It was run by Paul Castrucci, who had joined IBM in 1956 after receiving a B.S. in physics from Union College. He had led the development efforts for IBM's first bipolar memory chip and every one thereafter. Castrucci, like many other engineers in development, was less oriented to the semiconductor work done outside IBM than those in Research were. All of Castrucci's previous work had been done in bipolar technology, and Castrucci had not been an advocate of MOS, at one point going so far as to say that he would rather throw salt in his pilot line than run MOS transistors in it.[48]

Castrucci's group had little contact with the MOS work in Research, both because of doubts about its applicability and because Research had largely halted its MOS work. Castrucci had a pilot line that had been built to make bipolar circuits, with processes that had been shown to work in large volumes in bipolar technology. The combined ϕ2i/MOS memory program had been sold to IBM's senior management partly on the basis that the two technologies could be made on the same line, both in development and manufacturing (although Davis and several of his managers had suspicions that this would not be so). Since Castrucci would be making MOS memory for only a short time and did not have the responsibility for meeting large production goals, he had the incentive to keep the process as close to the bipolar ϕ2i process as possible.[49]

A major part of the development process was adapting Research's work to IBM's unique production processes. Research, with a small pilot line, used either production equipment of its own design, or equipment bought from vendors, not the high-volume production equipment used in East Fishkill. Furthermore, Research had dealt only minimally with questions of protecting and packaging the chip after it was made. The pilot line also had the job of scaling up production to the point where it could yield meaningful data on the process. The Yorktown pilot line had started on the order of a few wafers per day; East Fishkill would start fifty. The East Fishkill line would produce wafers that would be used to ensure the reliability of MOS memory as well as wafers to meet early production requirements.[50]

Early work suggested that making MOS transistors presented no problems. While the bipolar memory devices were extremely difficult to make and had had yields of 3 percent for an extended period, the first runs of MOS memories had a yield of 20 percent. But by April 1970, the IBM MOS program faced serious trouble. IBM had established various checkpoints and milestones in the development process to ensure the integrity of products as they entered manufacturing. The MOS program failed the critical T2 checkpoint. Jack Riseman, one of East Fishkill's technical leaders, audited the program and found three major problems: gate shorts, threshold instabilities, and charges on the glass passivation. East Fishkill managers formed three groups from the senior staff to address the problems. Although East Fishkill did not have experience making MOS devices, through its enormous bipolar programs it had people with unmatched experience in all areas of semiconductor technology. The veterans of the MOS program in Yorktown would not be brought in to save the day.[51]

The problem with the glass passivation was one of adapting a process developed for bipolar technology for use on MOS transistors. To protect its chips from the environment, East Fishkill covered them with a layer of glass, which was applied by radio frequency (RF) sputtering. In this process, conducted in a vacuum chamber, an RF electric field was applied to a glass, creating a plasma deposited on top of the semiconductor; the electric field had charged the plasma. Because the central problem of MOS transistors was controlling charges on the surfaces, the RF sputtering process was not what one would have chosen had one been developing a process specifically for use on MOS transistors. But because it existed and was part of East Fishkill's repertoire, it was used. East Fishkill developed a new sputtering process for MOS work that minimized charge on the glass.[52]

The biggest problem turned out to be gate shorts. When a voltage was applied to the gate of an MOS transistor, at times it would create a short circuit through the silicon dioxide, destroying the individual transistor and potentially rendering the entire memory chip unusable. This was a relatively rare problem that had not been addressed by Research, but given the volumes of memory chips IBM would be using, even a rare problem could be disastrous. No one knew the exact mechanism by which gate shorts occurred. IBM engineers found that a key element in reducing gate shorts was the cleanliness of the environment in which the wafer was processed, a factor that had not been as important for bipolar circuits. A simpler way to obviate the problem (although slowing the chip down) was to increase the thickness of the oxide. Ultimately, the thickness of the oxide was increased from 500 angstroms to 700 angstroms, allowing the previously failed T2 to be achieved in December 1970, roughly nine months late.[53]

The difficulties the MOS program faced made comparisons with the rest of the semiconductor industry inevitable. Intel had been selling MOS memory since 1969 and had described a 1,024-bit chip at the 1970 International Solid-State Circuits Conference. If a small start-up could make MOS memory, why couldn't IBM? These difficulties also showed how important MOS had become to the company as a whole. And they created a window where IBM's MOS program could have been significantly reformed.

One option was to buy MOS memories from other firms. In March 1970 Thomas J. Watson Jr., IBM's chairman and chief executive officer, asked its president, T. V. Learson, to see about having an independent evaluation made of IBM's monolithic memories in comparison with outside firms'. That same month Learson wrote to the head of the corporate staff, noting that four new systems were waiting for FET (MOS) memory. "It is quite obvious," Learson said, "that FET's are the pacing item in our whole machine marketing strategy." Learson wondered why IBM could not buy MOS memories from outside vendors, claiming that the volumes were large enough that IBM could still maintain its own internal manufacturing. The corporate staff recommended that MOS not be bought from outside vendors, first because the volumes were so large no vendor could provide the necessary quantities, and then because using components from outside IBM would involve disclosing proprietary information. Whether Learson agreed with the recommendation or not, with CD obviously against the use of outside vendors and the corporate staff concurring, he had no one else to turn to.[54]

Another option was to bring in an outsider to run the MOS operations. In

November 1970, Erich Bloch, the vice president for operations of CD, sent a note to A. R. DiPietro of the corporate staff asking for the names of those outside the company who had been successful in getting MOS into manufacturing. DiPietro, who had long been a voice calling for IBM to follow industry-standard practices, assumed that Bloch was looking for someone to run IBM's MOS work and provided four names. But even DiPietro was cautious in believing that any of these people could successfully run IBM's MOS program: "My best guess is that his experience would not be much help in practicing our IGFET [MOS] technology in manufacturing. The uniqueness and difficulty of our approach, coupled with his lack of familiarity with our internal operations, could actually make it impossible for him to contribute. On the other hand if he were able to control development as well as manufacturing and were allowed to make appropriate changes in our approach we could materially improve our chances for a successful IGFET [MOS] program."[55] No such outsider was ever hired.

In the spring of 1970, in response to an audit of the MOS program, IBM launched a new program that was much closer to industry standards, developing a *p*-channel silicon gate MOS memory. While twenty-five people worked on the program, its main goal seems to have been to deflect criticism; since it was over three years behind the established work, it would be unlikely to yield something that would be ready for production sooner unless the existing program ran into catastrophic problems. Whatever problems IBM's MOS program had, ultimately they were not great enough to lead to a reformation of IBM's semiconductor work.[56]

Cogar: The Failure of an IBM Start-up

As IBM engineers battled to produce MOS memories in East Fishkill, a similar struggle was under way just nine miles north in the town of Wappingers Falls, New York. There a group of sixty-six former IBMers, the heart of the technology division of the Cogar Corporation, tried to apply IBM's methods in a start-up company. They failed spectacularly, going through $27 million before producing significant quantities of anything, and demonstrated that IBM's semiconductor operations were an organism that could thrive only in the IBM environment. A comparison of Cogar and Intel, which was almost the perfect antithesis of Cogar, highlights a number of factors that were critical in Intel's success, for they were strikingly absent at Cogar.

The story of Cogar provides a hint of what would have happened had a group of Soviet semiconductor engineers formed a start-up in the United States. While Cogar's ex-IBM engineers were technically very talented, almost everything they had learned at IBM was exactly the wrong lesson for a start-up, but they failed to understand that the rules they had previously operated under no longer applied. Cogar was unable to make the adaptations necessary to develop a new technology in an environment of competitive capitalism.

In November 1968, Robert Markle, the IBM manager in East Fishkill who had not long before received responsibility for the MOS program, and his top lieutenant, Ray Pecararo, left IBM to become the founding members of the Technology Division of the Cogar Corporation. The head of the corporation and its namesake, George Cogar, was a high school dropout who with a technical education received in the Air Force became a senior computer designer for Univac. In 1964 Cogar left Univac to become one of the founding members of Mohawk Data Systems, a start-up based in Utica, New York, whose initial product was a keypunch that could record to magnetic tape. Although Mohawk succeeded, with sales of $9 million by 1967, Cogar left after a dispute with the other founders. Cogar formed the company bearing his name in January 1968 with the aid of a group of investors in the Utica area. Later that year he recruited Markle to head the Technology Division, which was to be the centerpiece of the company. Cogar himself had no previous experience in the semiconductor industry, but he subscribed to the common belief that with the advent of integrated circuits and large-scale integration in particular, it was essential for a computer company to develop its own semiconductor technology.[57]

While Cogar and his ex-IBM managers would produce semiconductor components, they did not want to sell components as such. It was not uncommon for those in the computer industry to think that selling semiconductors was a lousy way to make money. They saw price wars and low profit margins and compared them to the higher profit margins in the computer industry. Cogar believed that profits lay not in components but in systems. Cogar's plan was to sell complete memory systems or printed circuit cards containing both memory chips and supporting circuitry, with the expectation that these products would be immune from the price pressures common in the semiconductor industry. Cogar would sell these systems to computer manufacturers, in effect reorganizing the process of computer design by seizing the memory design function from the manufacturer.[58]

If a company could be said to have genes, all of Cogar's on the operational side came not just from IBM but from IBM East Fishkill. Robert Markle, the head of Cogar's Technology Division, had worked at IBM since 1951; Ray Pecararo, the head of engineering, had worked for IBM since 1956; Howard Geller, the head of marketing, had been with IBM since 1955. All had worked at East Fishkill since its founding, and none had significant experience outside IBM.[59]

Cogar's leaders, though accomplished, reflected their experience at IBM and IBM's approach to semiconductors. Markle had risen steadily since joining IBM, being sent to Harvard Business School by IBM, then serving on the staff of T.V. Learson, at the time an IBM vice president and group executive and later IBM's chief executive officer. Markle's most significant accomplishment at IBM had been directing the final development of SLT—managing the solution of the final technical problems in a technology that had already been defined. Markle had never participated in any of the major semiconductor professional conferences. While at IBM, Markle had very little contact with semiconductor professionals from other companies. Pecararo had served as a manager in the SLT program and later became the pilot line manager for semiconductor memory at East Fishkill. Although George Cogar had no experience with any phase of the semi-conductor industry and no knowledge of how the industry worked, he pro-vided the vision behind the operation and was the chief long-term thinker.[60]

Based on George Cogar's proven track record and the IBM management team he had assembled, Cogar Corporation had no trouble getting funding. In 1969 it raised $9.6 million through a sold-out stock offering and borrowed another $2.7 million. In the same year it began work on a new $3.6 million semiconductor plant in Wappingers Falls, about nine miles from East Fishkill. Cogar's advertising called the plant "the industry's newest, most automated facility."[61] Money flowed freely in the early days, and Cogar's managers had budgets so large they might have forgotten they were no longer with IBM.

The former IBMers did not have to move because Cogar put its operations right in the heart not of a company town but a company county. IBM had first established a facility in Poughkeepsie in 1941 to produce ordnance for use in World War II. A Works Progress Administration guide to the county published just prior to IBM's entry into the area had noted that no single industry dominated the area, with the largest manufacturers being a maker of cream separators and a maker of ball bearings. IBM's Poughkeepsie plant grew with IBM, and in 1962 IBM began construction of its East Fishkill facility for producing semiconductors. By 1972, when the county had 220,000 residents, IBM employed roughly 10,000 people in Poughkeepsie and 8,000 more in East Fishkill—a quarter of the county's workforce. IBM's employees, encouraged to stay by a lifetime employment policy and welfare capitalism, were much less likely to change employers than their Silicon Valley counterparts. The county lacked any university or other technology-based company that might have provided a counterweight to IBM.[62]

Cogar's location in Wappingers Falls showed its intent to operate as part of IBM's closed system and poach IBM employees from East Fishkill and Poughkeepsie. By July 1969 Cogar had hired sixty-six former IBM engineers. In spite of this apparent success in recruiting IBMers, there were significant limitations in Cogar's former IBM recruits. No one from IBM Research's MOS program joined Cogar. This may have reflected the tensions existing between East Fishkill and Research, where Cogar's East Fishkill managers (with their vast amount of semiconductor experience derived from work on SLT) denigrated Research's work as impractical and not applicable. Nor did anyone from the development side of the joint program join Cogar. Cogar began its efforts at MOS memory production without any engineers experienced in MOS technology.[63]

In theory, Cogar had a number of options as to what sort of semiconductor process it could use. It signed cross-licensing agreements with General Instrument, one of the recognized leaders in MOS production, as well as Texas Instruments and Fairchild. Instead of using processes from these firms, Cogar tried to implement the IBM Research–developed MOS process, which while containing some familiar elements was by far and away the most complex and difficult MOS process among Cogar's alternatives. While the semiconductor industry was working on p-channel MOS with the technical descendants of Frank Wanlass scattered abroad, Cogar, isolated from the rest of the industry, chose to follow IBM down the n-channel road.[64]

Cogar also tried to emulate IBM's production philosophy, which meant that it could not simply rely on buying standard off-the-shelf production equipment—it had to either design its own or copy IBM's. Apparently it did both.

Cogar's 1970 annual report spoke of "sophisticated tooling" and "automated equipment" for the production of semiconductors that were "principally designed by the company." Of the sixty-six patents issued to Cogar before 1975, twelve were for semiconductor production equipment. For the sake of comparison, Intel, which had made a decision not to make its own production equipment, received twenty-five patents between 1969 and 1975, and only one was for semiconductor production equipment. Cogar's employees appear to have taken the drawings for some of IBM's semiconductor production equipment and then attempted to have them duplicated by outside companies.[65]

Cogar also followed IBM in doing all semiconductor production work in-house, unlike Intel, which contracted out some jobs, such as mask fabrication and packaging. Cogar had to spend time, manpower, and capital developing each step of the process, where Intel could choose where to put its resources for best effect. Intel packaged its chips in the industry-standard dual inline package (DIP) and contracted out the assembly of its chips onto DIPs. Cogar developed its own version of IBM's extremely complex C4 process and then packaged several chips onto the distinctive square ceramic modules that immediately betrayed the IBM heritage. While the IBM/Cogar approach was technically superior in some ways, it cost Cogar valuable time and money. One former employee asserted that Cogar spent its first year making production equipment. Cogar, like IBM, internalized almost every function, with an in-house patent attorney, an in-house cafeteria, and an in-house library (appendix figure A.5). While Intel had a spartan lunch room, with a few vending machines where an employee could buy a sandwich, Cogar had a 150-seat company-subsidized cafeteria, which periodically served filet mignon.[66]

In July 1969 IBM brought suit against Cogar Corporation and the sixty-six former IBMers who worked for Cogar, charging theft of trade secrets. The suit had the immediate effect of halting the flow of employees from IBM, as Cogar tried to show it was not just trying to grab talent from IBM and potential recruits realized that they too would face legal action should they join Cogar. But the suit won Cogar free publicity and placed IBM, itself being sued for antitrust violations by the Department of Justice, in the uncomfortable position of looking like a giant trying to stamp out a small competitor. The *Wall Street Journal* quoted George Cogar as claiming that IBM was trying to "create and enforce servitude on IBM technical personnel."[67] The suit was finally settled out of court in March 1970, with IBM receiving the right to examine Cogar's facility and Cogar receiving the right to use IBM's patents. Many of IBM's claims against Cogar appear to have been well founded.[68]

Whatever industrial espionage and theft of trade secrets there were should be seen as part of Cogar's continuing dependence on IBM and IBM's continuing dominance of Cogar's thinking. Articles published by Cogar employees in technical and trade journals were peppered with references to IBM's work and appear to have been part of an effort to indoctrinate the rest of the industry into IBM's methods of memory system design. Cogar's semiconductor production line was arranged in a fashion similar to that at East Fishkill, with the diffusion furnaces labeled in the same manner as the furnaces at IBM Research.[69]

IBM had one of the strongest corporate cultures of any firm in the nation, and the strength of its hold on Cogar can be seen in such seemingly trivial matters as interior decoration and status markers. While Tom Wolfe has made Robert Noyce's secondhand metal desk at Intel legendary, Cogar's managers sat in offices that were essentially replicas of IBM managers' offices, with wood desks, matching water pitchers and coffee sets, and hard-wood flip-chart cabinets. However, Cogar did break away from IBM traditions in one area—in spite of Thomas Watson Sr.'s legendary antipathy towards alcohol, some managers had liquor cabinets in their offices.[70]

Cogar was a Wall Street favorite, and analysts frequently lumped Cogar and Intel together without appreciating their fundamental differences. *Business Week*'s articles on new semiconductor firms gave Cogar prominent mention, calling George Cogar a "computer technology wizard" and suggesting that it and Intel were among the entrants most likely to prosper.[71] A 1970 assessment of efforts in MOS technology by the consulting firm Quantum Science considered Cogar, along with Intel, one of the new firms most likely to succeed. While it acknowledged that Cogar needed orders, it asserted that the firm had sufficient capital and was ready to begin volume shipments. Quantum Science's overall ranking of the firms noted Cogar's strengths in design and manufacturing and accorded the firm a ranking matched only by Texas Instruments. A 1970 analysis of Cogar within IBM stated: "The Cogar people are IBM trained, have IBM processing skill and must be able to produce a comparable product." Another claimed that Cogar had "the ability to grow as a business."[72] In early

1970 Cogar's stock rose to a price of ninety, over four times the price of the initial stock offering.[73]

Cogar employees' detailed knowledge of IBM business plans gave them an enormous competitive advantage. They knew that IBM was planning to use semiconductor memories in its next line of computers, and they counted on the rest of the industry following suit. Cogar's strategy was to position itself as the savior of the rest of the computer industry in the face of IBM's move to semiconductor memories. It would bring IBM technology to the dwarfs of the computer industry that could not afford a major technology development effort of their own. In a full-page ad in *Electronic News,* Cogar asked the industry, "How will you sell against the New Systems?"—a thinly veiled reference to IBM's planned follow-on to System/360, which bore the internal code reference NS. The campaign also included a billboard outside the headquarters of computer maker Honeywell asking the same question. Cogar sent industry executives packages that when opened revealed the same menacing query.[74]

At IBM East Fishkill there had been no need for outside marketing of semiconductors, and Cogar was trying to sell to IBM competitors that Cogar's ex-IBM employees never would have talked to before. Almost all members of the computer industry had been customers of Fairchild, so they knew Robert Noyce and Gordon Moore. While Intel and Honeywell engineers worked closely together, Cogar could get in the door at Honeywell only through the mail.

Not surprisingly, Cogar's methods failed to win it business. While they do show the problems Cogar had in making contact with its customers, there was a more substantial problem Cogar faced in trying to sell memory systems. In designing a complete memory system for a computer, Cogar was competing with memory system organizations within the firms it was hoping to sell to. Furthermore, Cogar was asking these firms to cede control of a subsystem critical to the overall system's performance without any compelling economic reasons to do so.

As Cogar began to run short on funds, it invited Robert Noyce, the cofounder of Intel, to visit with the hopes that the two firms could conduct some sort of joint program. Noyce was horrified at Cogar's strategy of selling memory systems to computer manufacturers and, believing it would not work, refused to have anything to do with Cogar's operation.[75]

The resistance Cogar experienced from computer makers bore some similarities to the resistance that Fairchild faced when it tried to sell integrated circuits to computer manufacturers in the early 1960s. Computer firms did not want to cede control of a function to a semiconductor manufacturer. Fairchild had won the day because its integrated circuits could offer advantages of cost, reliability, and miniaturization. But the same situation did not apply here. Computer manufacturers like Honeywell and Burroughs could (and did) buy individual memory chips from Intel or other semiconductor manufacturers, keeping both the control and the profits that came from designing and building the memory system themselves.[76]

The problem with Cogar was not so much Cogar's managers' unfamiliarity with marketing as their unfamiliarity with markets. During their stints at IBM, Cogar's managers had never had to deal directly with a customer; they had dealt with other IBM divisions. IBM was very successful in controlling its environment and minimizing market forces. But the managers who went to Cogar after having been sheltered from the market for many years found themselves in a situation where they were now fully exposed to the market, and they were poorly equipped to deal with it.[77]

Computer manufacturers refused to let Cogar take over their memory system design, and Cogar could find no customers. But Cogar adapted only slowly to the failure of its initial strategy. In Cogar's 1970 annual report, George Cogar was still expressing confidence in Cogar's ability to win contracts from computer manufacturers for memory systems. It was not until January 1971 that Cogar introduced an alternate product, an add-on memory for IBM System/360 computers to be sold to end users. Cogar did not undertake to sell these directly but signed a contract with Potter Instruments, a manufacturer of IBM plug-compatible disk drives and tape systems, to sell Cogar memories. Cogar still adamantly refused to sell individual memory chips.[78]

By July 1971 Cogar's semiconductor memory effort had required $26 million in capital and Cogar was looking to raise $9 million more. At the same time, it had been unable to manufacture MOS memories at anything near the levels needed for production, with yields from March through June at a fraction of a percent. Later in July Cogar experienced a breakthrough in its manufacturing process, and yields rose enormously. But just as Cogar's hopes rose, disaster struck. Cogar's memory chips, which had passed initial tests, been assembled into systems, and been delivered to Potter Instruments, experienced massive reliability problems whereby memory systems would fail after only hours of use. The problem was faulty solder joints in Cogar's memory cards. Potter cancelled its agreement with Cogar to sell its systems. With no customers, Cogar's bank refused to extend its line of credit, and in April 1972 Cogar's memory operations were forced to shut down.[79]

With Cogar's demise, its engineers scattered to the winds. After their defections, IBM would not rehire Cogar employees. Markle took a job in California and Pecararo in Florida; General Instrument on Long Island hired a group of former Cogar engineers, and Fairchild hired another group.[80]

In July 1972 Cogar's facilities and equipment were put up for auction. Two engineers from Intel came over to put in bids and were astounded by what they saw. While Intel had gotten by with one Reichart microscope, a premium-quality instrument that cost five to ten times more than other models, Cogar had them everywhere. One of the Intel engineers, Ted Jenkins, said of Cogar: "It looked like someone had just opened the purse strings on the capital and let them buy whatever they want, with no regard to ever making a business out of the thing." Much of Cogar's equipment was custom-built in the IBM style, essentially useless to anyone who did not adopt the entire IBM philosophy of production. Few firms had any interest in this equipment (or the IBM approach), and bids were very low.[81]

Burlington, Vermont: MOS Finds a Home

Although the formal pilot line work on MOS memories was done at East Fishkill, the MOS technology as developed in East Fishkill had in some ways been forced to fit into a bipolar mold. MOS found an abiding home at an IBM plant in Burlington, Vermont, hundreds of miles from the centers of power in IBM. Except for its leader, the group responsible for MOS manufacturing at Burlington was by and large composed of newcomers to semiconductor technology within IBM. Burlington had both bipolar and MOS technologies in manufacturing, and it was an environment where neither dominated. Learning that had occurred in bipolar memory manufacturing proved to be crucial for the MOS work. The importance and distinctiveness of MOS technology were formally recognized in 1972, when the site was divided into two organizations, one bipolar and one MOS.

With the decision to use semiconductor memories in IBM systems, IBM managers selected Burlington as the site for their manufacture. It would require a large plant to make these memories, and East Fishkill was already one of the largest semiconductor facilities in the world. IBM's plant in Burlington had been established in 1957, the same year that across the country a group of engineers had left Shockley Semiconductor to form Fairchild Semiconductor. IBM's chief executive, Thomas Watson Jr., had made connections in Vermont through the ski lodge he had in the state. A group of local leaders concerned with the lack of jobs in the area had formed the Greater Burlington Industrial Corporation to promote business in the area, and IBM initially occupied a Greater Burlington Industrial Corporation–built facility to manufacture reed relays. Burlington then became a site for the production of SLT modules over and above what East Fishkill could provide. It had a small development laboratory, which worked on magnetic memory programs but no semiconductor development work. With the advent of semiconductor memory, Burlington's manufacturing mission was defined as semiconductor memory, while East Fishkill would build logic. East Fishkill would retain its role as the center of IBM's semiconductor work, and it would develop and pilot memories before they were transferred to Burlington for manufacture.[82]

The manager in charge of MOS manufacturing in Burlington was Bill MacGeorge. MacGeorge had received a B.S. in mathematics and physics from Ursinus College, working for six years as a semiconductor engineer at Philco before joining IBM in 1962. At Philco, MacGeorge had developed the habit of closely following (and copying) developments in the rest of the industry, but he came to IBM just as it was closing itself off to outside influences. MacGeorge progressed up to a senior manufacturing management position in East Fishkill before a desire to unload his house led him to transfer to Burlington.[83]

Putting MOS technology into manufacturing involved scaling production up by a factor of forty over the development line and progressing down the learning curve to meet production requirements. The move of the technology from research to development involved taking away a level of abstraction and making it more concrete, and the move from development to manufacturing was a further step in this direction. The completion of development work did not mean that high-volume manufacturing was a trivial task. IBM East Fishkill completed piloting work on φ2i in late 1969, and the system using it, System/370 Model 145, was announced in September 1970. At the time IBM was starting six thousand φ2i wafers per day in Burlington and the projected yields were 18 percent, but the actual yields were stuck around 6 percent. This meant that the costs would be three times higher than estimated and that IBM was two factories short of meeting the required demand. It was working to establish those two factories by building a new facility in Manassas, Virginia, and by contracting with Motorola to process six thousand wafers per day. While it was worth paying almost any price for IBM to meet shipment schedules, these schedules could also be met by raising yields to the projected levels.[84]

In October 1970 Paul Low, who had previously been the laboratory director in East Fishkill, was named the assistant plant manager in Burlington, with responsibility for fixing the φ2i yield problems. (The previous senior manufacturing management in Burlington was removed.) At the same time, MacGeorge, who had been an industrial engineering manager, moved over to φ2i. MacGeorge took a yield management system that he and a group of engineers had developed on a smaller scale and implemented it for φ2i. This system (see below) allowed the yield problem to be broken down and each part of the problem to be addressed separately.[85]

After the φ2i yields had reached acceptable levels, MacGeorge moved over to MOS work. Other than MacGeorge, who came over from the φ2i program, most of the experienced semiconductor manufacturing people remained with φ2i. Burlington had a pool of engineering talent available in those who had worked on the magnetic memory programs that Ed Davis had killed, and many of MacGeorge's people came from those programs.[86]

MacGeorge had closer relations with Yorktown than the East Fishkill group had. After receiving the assignment to manufacture MOS, he spent a week there, finding out what they had done. He thoroughly went through Yorktown's MOS manual, which was to prove to be an important reference for all Burlington MOS engineers. MacGeorge brought to MOS manufacturing a willingness to pick up and adopt work done elsewhere, a trait not always common within IBM. Burlington manufacturing engineers also kept in close contact with Yorktown.[87]

In 1971, at roughly the same time Burlington began MOS manufacturing, IBM established a semiconductor memory development group in Burlington. While Dick Gladu and his group of MOS circuit designers moved up from East Fishkill, most of the group was made up of people who had worked on the cancelled magnetic memory development programs. The formation of this group in Burlington provided the critical mass of semiconductor talent needed to allow the site to move outside the influence of East Fishkill and become more than a satellite plant. For many people in Burlington, East Fishkill served as the negative reference point. East Fishkill had control over what was done in Burlington, and Burlington personnel had to follow East Fishkill's direction even when they disagreed with it. Gladu and his group worked closely with Mac-George to get MOS memory into manufacturing.[88]

MacGeorge had begun his MOS line in Burlington with what he called "pots and pans." All available manufacturing space was being used up by ϕ2i, so MacGeorge took some nonclean space to set up the processes unique to the MOS line. For ϕ2i, MacGeorge had set up three separate manufacturing lines, each with a capacity of two thousand wafer starts per day. This allowed comparisons between the lines and a measure of protection should disaster strike one line. When the yield had risen to the point where demand could be met with only two of the three lines, one was converted to MOS. As MacGeorge and his team planned for a full-scale production line, they did not replicate the process developed in East Fishkill exactly; they saw certain areas where they believed changes would be needed and planned for additional steps. This was particularly true with regard to wafer cleaning, an area that was much more crucial in MOS than in bipolar. MacGeorge had a measure of freedom to go beyond the process East Fishkill had transmitted to him because of his proven track record and his contacts with senior managers in East Fishkill. In spite of earlier claims that it would be, the Burlington MOS manufacturing line was not interchangeable with the bipolar line.[89]

At the same time, important learning that carried over from ϕ2i was crucial to the success of the MOS memory, the most important being the yield management system. The semiconductor manufacturing process was extremely long and complex, consisting of between seventeen and thirty operations carried out on a wafer over a period of a month or more. If final test was the only measure of the quality of the process, one might not know about problems until long after they had occurred, with no certainty that the offending process step had not changed dramatically again in the meantime.

The yield management system included a model that broke the overall yield into component pieces. Data on individual steps were then taken and fed into the model, which would predict the overall yield. The yield management system allowed engineers to monitor the manufacturing process in close to real time and see how changing one process step affected the overall yield. This system became a powerful tool for driving up yields, by identifying the areas that would offer the greatest benefit.[90]

The yield management system was not at all specific to MOS; in fact, it was first used for bipolar, but it was unique to Burlington. Several factors account for its not being used in East Fishkill. The individual transistors that East Fishkill had manufactured for SLT modules provided less incentive for pushing yields, but the memory chips manufactured in Burlington were larger and more complicated than anything previously made in East Fishkill. The area of the 2,048-bit MOS memory was thirty-four times the area of an SLT transistor chip. A tiny random defect on an SLT wafer would ruin only one transistor; almost a thousand good ones would be left on the wafer. On the 2,048-bit chip, a random defect could make over twelve thousand transistors worthless. Costs (or at least prices) for MOS memory could be quantified and compared much more easily than they could for any type of logic, since many other firms offered MOS memory circuits for sale. Although memory systems were not a commodity, the existence of competitive add-on memory systems encouraged IBM to pay attention to costs. Any sort of outside comparison of IBM's logic semiconductor products was much more difficult, because no one offered anything similar and IBM did not face the same type of unit-for-unit competition it had in memory.[91]

While the yield model applied to either bipolar or MOS technology, it was particularly well suited to MOS technology, because the parameters that affected MOS yield could be more readily measured and controlled. One of the most critical parameters in MOS was defect density, the number of defects of a given size in a particular chip area. Because MOS circuits were denser than bipolar, they were more sensitive to an improvement (or declension) in defect density. MOS circuits were also very sensitive to contamination (as measured by changes in threshold voltages).

Implementing such a system was expensive but consistent with IBM's capital-intensive style of semiconductor production. The process control system required several mainframe computers, as well as professionals to write software. IBM devoted substantial engineering resources to characterizing and monitoring its manufacturing lines. IBM had a team of engineers that would regularly do a thorough failure analysis on parts that had fallen out at wafer test, to determine what the key yield detractors were. This was in contrast to a leaner style of manufacturing typical in Silicon Valley.[92]

Another crucial factor in Burlington's success manufacturing MOS memories was a sister plant in Sindelfingen, West Germany. Strictly speaking, the relationship was a hierarchical one, where the American plant controlled the German plant. Normally, Sindelfingen would be expected to implement the Burlington process without deviation, but MacGeorge allowed it the freedom to make modifications as long as all differences were documented. The German plant had a very talented workforce and access to a different set of tooling. Results across the plants were continuously monitored, using computers linking the sites. The two plants had a continual competition to see which one could achieve the higher yields. In an extremely empirical process, the existence of another plant with some degree of freedom allowed engineers to compare each step, choose the better approach, and use it in both plants.[93]

The East Fishkill MOS pilot line was operated for a period of time to produce parts under the assumption that East Fishkill's superior engineering talent would speed up early production. But shortly after the Burlington MOS line started up, it exceeded the East Fishkill yields. The Burlington line was optimized for both manufacturing and MOS, while the East Fishkill line was optimized for neither.[94]

Burlington's success in manufacturing MOS memories allowed it to move quickly to higher densities. While the 512-bit chip was the starting point, Gladu's group had designed a 1,024-bit chip and a 2,048-bit chip, while the Boeblingen lab had designed its own 2,048-bit chip. As a product worked its way down the learning curve, at a certain point it became more efficient to switch to a denser product rather than to fight for relatively small increases in yield at the asymptote. Within an eighteen-month period Burlington went from manufacturing 512-bit chips to 2,048-bit chips. (The German design, called Riesling, was ultimately chosen.)[95]

But a manufacturing manager could never get overconfident. Disaster could strike quickly even at a well-run, finely tuned semiconductor plant, dropping yields significantly. Because of the microscopic quantities of materials involved in the semiconductor manufacturing process, even the most trivial changes anywhere in the plant could have catastrophic results. One might think of the output of a semiconductor plant as being a function of an almost infinite number of variables, some of them, such as the putative process and the equipment used, fairly well understood and controlled. But practically every atom of matter in the plant was potentially significant if it got into the wrong place.

In 1972 Burlington lost the process for making MOS devices, as the threshold voltages on test sites went out of control. Analysis of the test sites showed the existence of microscopic amounts of titanium on the transistors, which were finally traced back to a seemingly trivial action. A senior (nontechnical) IBM executive had visited Burlington and raised questions about spots on the floor. In response, MacGeorge had the floors regularly buffed, inadvertently putting particulates in the air and thus into the rinse tanks.[96]

In August 1972 IBM introduced two new large computer systems that used MOS technology for their main memory. The System 370 Models 158 and 168 both used the 1,024-bit MOS memory, packaged four chips to a half-inch module. *Electronic News* carried a story on the announcement's impact on the semiconductor industry, quoting industry executives who self-interestedly proclaimed IBM's course an endorsement of their own memory programs. Robert Noyce of Intel cited IBM's program as an endorsement of *n*-channel MOS and stated it might mean the end of new investments in magnetic core memories. Robert Lloyd of AMS stated: "There's nothing like a blessing from the Vatican."[97] The next year IBM announced that it was moving to 2,048-bit chips for the System 370 Model 168, again putting four chips into a module. (At the time most semiconductor manufacturers were committed to making the step from 1,024-bit memory chips to 4,096-bit chips.)[98]

The distinctiveness of MOS technology was formally recognized in a 1972 reorganization of the Burlington site, shortly after MOS memory had been introduced in IBM's System/370 Models 158 and 168. Traditionally, IBM sites were fundamentally divided between plant (manufacturing) and lab (development). Burlington had responsibilities for semiconductor memory, bipolar and MOS, manufacturing and development. The site's general manager, Paul Low, needed both technologies but found that some of his senior managers were more interested in one or the other technology. He divided the site into a bipolar organization and an MOS organization, with development and manufacturing in each. At this time, close cooperation between development and manufacturing groups was particularly important as the clear first priority was to ensure that Burlington met the production requirements for IBM's new systems. Two years later, Burlington moved back into the traditional plant/lab organization, with MOS technology clearly dominant.[99]

MOS Logic: Incomplete Establishment

IBM's MOS memory program strongly resonated with IBM's strengths. It was a very large—one might say monolithic—program, with clear implications for IBM's central business line, mainframe computers. But the establishment of MOS logic, which had no such implications, proved to be much more problematic. Although IBM ultimately established an MOS logic program, many of the systems that could have used it stayed with bipolar technology. An MOS logic program that would have confined bipolar technology to a small niche within IBM ultimately failed. The problems that the MOS logic program would face could be seen back in Dan Doody's 1965 semiconductor strategy. MOS technology was directly competitive with bipolar, and while all indications were that MOS technology would be significantly cheaper, it did not allow one to do anything that was not achievable in bipolar. Doody had recommended that since IBM needed the bipolar technology for the high speeds it offered in mainframe computers, it should not develop MOS but use bipolar exclusively. Although there would be some savings from using MOS, this would be balanced by not having to develop a second technology. (For more on Doody's efforts to halt MOS work in Research, see chapter 2.)

The corporate staff had advocated MOS logic early on. While the staff saw the interest in MOS logic that was shown by those outside IBM, it had a serious problem. No IBM system that would require a large number of parts had expressed a firm desire for MOS logic, and CD had expressed no interest in making it. At one meeting where the corporate staff was pushing for MOS logic, the representative from CD asked the representative from the Systems Development Division (SDD) if it had any need for MOS logic. (As its name implies, SDD was the division responsible for developing computer systems in IBM.) The SDD representative said it did not, allowing CD to say it made no sense to develop a technology that no one wanted. With this the corporate staff worked to revise the charter of CD, clearly stating that not only did it have responsibility for developing technologies that systems groups wanted, it also had responsibility for "injecting new technologies into the product divisions" regardless of whether systems groups required them.[100] Of course, the deep-seated reasons for the lack of interest in MOS logic could not be completely remedied by a change in mission statement.

In 1968 East Fishkill hired George Cheroff, who had been a manager in the Research MOS program, ostensibly to start an MOS logic program. In fact, Cheroff's major role was to deflect criticism that CD had no effort under way in MOS logic. He had no pilot line, only five people, and a very small budget. Cheroff had a simple, straightforward desire to implement MOS logic in a product. That was not what East Fishkill managers wanted, and he ultimately left in frustration.[101]

IBM's position as a vertically integrated producer of computer systems complicated matters. For a company that had to buy its semiconductors from an outside source, MOS logic circuits, although low speed, would be cheaper than bipolar circuits. A buyer would see the price differential between bipolar and MOS technologies and have every incentive to switch to MOS logic for low-speed applications. But IBM was different. It needed bipolar technology for its mainframe computers. Using bipolar technology on the low end would spare it an extra development effort and an extra manufacturing line. The design tools for use in bipolar circuits were also superior and more familiar to systems designers. Its complicated system of charging internal customers for semiconductors based on costs rather than setting prices muddied the waters further. While the ability to use bipolar technologies where others used MOS might be seen as an economy of scope to IBM, there was a danger to IBM as well. If MOS advanced more quickly than bipolar technology or allowed users to do things not possible in bipolar, sticking with bipolar could prove costly.

The staff's lack of success in pushing MOS logic among either CD or potential users led it to consider other ways to advance its cause. In June 1969 E.R. Piore, IBM's chief scientist, wrote to Frank Cary, the head of the group responsible for data processing development and manufacturing, suggesting that it might be appropriate to issue an edict requiring IBM's low-end systems to use MOS technology. Piore noted that edicts had been used previously within IBM to good effect, such as when Thomas Watson Jr. had decreed that all future systems would use transistors.[102]

In 1971, four years after the corporate staff had revised CD's charter, very little had happened with MOS logic. IBM's Corporate Technical Committee, an advisory group whose members included some of IBM's top engineers and scientists, issued a warning about the pace of MOS logic work. It stated that it was "alarmed at the lack of real and solid commitment across the laboratories" to the use of MOS logic.[103] It noted that the only system committed to using MOS logic was itself in jeopardy. It also suggested that an edict be issued. Cary eventually issued an edict on the use of MOS logic, but it was not effective. Compared with the vacuum tube/transistor issue, the bipolar/MOS situation was much more subtle and complex. Bipolar was not clearly an obsolescent technology, and there were many reasons why a designer would want to use it.[104]

The problems MOS logic found can be seen in looking at some of the low-performance systems IBM introduced in the mid-1970s that could have but did not use it. For example, the Model 5100 computer, an early personal computer, made extensive use of East Fishkill's bipolar technologies but did not use MOS for logic. The Model 3800 printer used older East Fishkill bipolar technologies as well as components bought from outside IBM. The Model 3270 terminal exclusively used components bought outside of IBM. One of the most popular uses of IBM's MOS logic family was for read-only storage, an application where bipolar technology provided no satisfactory alternatives.[105]

In the early 1970s, IBM had another program that would have brought MOS logic to the heart of the corporation. In conjunction with FS, a new mainframe architecture that would replace System/360 and System/370, East Fishkill launched the Beacon complementary MOS (CMOS) logic program. Beacon was a very fast version of CMOS, able to meet the needs of all but the fastest FS systems, which would still use bipolar technology.[106]

The Beacon program had few connections to MOS work done previously in the company. Research made no contribution to the definition of the program and found out about it only after the program had been formulated. Most of the key managers and engineers had worked only in bipolar technology. One wing of the program had come from another MOS program, but its leader opposed the direction Beacon took.[107]

The bipolar-oriented Beacon group developed a technology that was in spirit much like bipolar. It was faster than all but the fastest bipolar circuits. But a price was paid for this speed. CMOS was much more complicated to make than the MOS devices used in the memory program. More significantly, the Beacon group tried to shrink the dimensions of the transistors so much to improve speed that they ran into a new reliability problem, hot electron trapping, where electrons were injected into the silicon dioxide, changing the transistor's threshold voltage. The use of CMOS led to another major difficulty, called latchup, that had the potential to destroy the chip. The engineers in the Beacon program were unable to produce functioning hardware, and the program was ultimately cancelled, with IBM beating a path back to the more proven bipolar technology for its mainframes. The lesson that East Fishkill took from Beacon was not that there had been flaws in the program's objectives, staffing, and relations with other organizations, but that CMOS (and MOS for logic) was not a promising technology.[108]

Conclusion

By 1975 IBM's MOS memory program was clearly among the largest in the world. Although the path it had followed from research to production had hardly been direct, it had made the transition from magnetic memory to MOS memory less risky for the company and had satisfied the various constituencies within the company. IBM could not feel market forces in the semiconductor industry, so it was up to managers to understand these forces and generate analogues useful for a computer company. This did not happen with regard to MOS logic.

In 1975, IBM reorganized its semiconductor work so that all MOS technology, both memory and logic, was located in Burlington. Previously, IBM's semiconductor work had been organized by the function it performed in the computer, logic or memory. With this reorganization, IBM recognized MOS as a distinct technology bearing on both memory and logic.[109] Unfortunately, the recognition came too late. Based on its previous memory charter, Burlington was already well along in developing an MOS technology that was utterly unsuited for use in logic. Furthermore, many of IBM's systems had already made commitments to use bipolar logic. There would be no economies of scope in MOS extending to logic.

With the semiconductor industry increasingly concentrating its work in the West and Southwest, IBM's facility in the small city of Burlington, Vermont, was symbolic of the differences between IBM and the semiconductor industry. While a complex local ecology accounted for the proliferation of semiconductor firms on the San Francisco Peninsula, it would be hard to argue that Burlington had any inherent advantages for semiconductor development and manufacture. IBM's plant there existed because of management fiat. Since the fiat was made by managers at IBM, they had the power to do whatever was necessary to make Burlington one of the largest semiconductor facilities in the world. IBM did things its own way. Cogar showed what happened to a company that tried to act like IBM without having IBM's resources.

The Logic of MOS

Like IBM, Intel had problems developing MOS logic products. Intel finally circumvented those problems, and the emergence of what is now known as the microprocessor stands as one of the milestones in the history of the MOS transistor. While memory was the largest early market for MOS technology, here it replaced magnetic core memories in established computer families. The microprocessor opened up entirely new applications to semiconductor electronics. The microprocessor proved to be the key engine of growth, with new applications appearing with each of the endless rounds of price cuts and performance improvements. From the two main lines leading from Intel's initial work, the microcontroller and the microprocessor, digital electronics has proliferated into almost every area of American life, through the small unseen computers that do their work in appliances and automobiles as well as through personal computers.

The history of the microprocessor reveals the complexities of invention. Many in the industry had recognized that at some point it would be possible to put most of a central processing unit of a computer on an integrated circuit. The 4004, generally seen as the first microprocessor, was preceded by another chip that was in significant ways more sophisticated than the 4004 but was seen in a different light because its designers were in the computer business. Intel typically did not call its 4004 a microprocessor. At the time of Intel's initial work, definitions were far from clear, but Intel proved to be quite adaptable in finding new markets for its new products, whatever they were called.

The microprocessor is best understood as an economy of scope of MOS memory. Intel had chosen the silicon gate MOS technology for its density in memories, but it turned out to be extremely well suited for microprocessors. However, the market for a microprocessor was too small to support even a start-up at that time. Because Intel had developed a successful MOS memory business, it could afford to devote resources to the more speculative microcomputer work. And one of the early goals of selling microcomputers was to increase the sales of memories. But Intel had to transform itself to make this economy of scope possible.

Foreshadowings

In the late 1960s and early 1970s, the idea of putting a subfunction of a computer or even an entire computer on a chip was not new. Soon after integrated circuits became viable, some engineers understood that a logical future step would be putting an entire computer on a chip. In 1964 Westinghouse engineers published an article in the *Proceedings of the IEEE* titled "The Evolution of the Concept of a Computer on a Slice." (By "slice," they meant an entire silicon wafer.) They asserted that it was "numerically possible to produce an entire computer subfunction on a single slice of silicon."[1] In 1966 Glen Madland, an industry consultant, wrote: "The utilization of MOS (metal-oxide-silicon) structures and other concepts which introduce the 'computer-on-a-silicon-slice' technology are perhaps much closer to practice than the average engineer realizes."[2] And in 1968, at the end of an MOS handbook for internal users, IBM Research engineers wrote: "Hopefully the day of the 'computer on a slice' is nearly dawning."[3]

The keyword came from the title of the Westinghouse article: concept. These were prophecies given at the beginning or end of a larger technical discussion and lacked specifics about what a computer on a chip would look like or how it would be implemented. No one discussed what sort of architecture a computer on a chip might have or precisely defined what constituted a computer on a chip. The yields then achievable in processing semiconductor wafers made the computer on a chip such a distant possibility that it was pointless to speculate on questions of implementation.

There was one exception: digital differential analyzers (DDAs). Several companies worked on DDAs using MOS technology as a way to produce a complex MOS circuit with minimal input/output pin requirements. Because of its serial nature, a DDA required fewer transistors to implement than a parallel computer. In 1966 General Instrument introduced a DDA that *Electronics* called "a move toward the development of a computer on a chip—a goal still well beyond the current state of the art." A General Instrument spokesman claimed that with just a few chips it would be possible to build a complete DDA computer capable of tens of thousands of problem solutions a second. While one could build an integrator or an analog computer from the DDA, one could not make a general-purpose digital computer. DDAs had a very small market.[4]

Viatron, Four-Phase Systems, and the Microprocessors That Weren't

Two important predecessors to Intel's work came from the computer start-ups Viatron and Four-Phase Systems. Viatron had something it called a microprocessor prior to Intel, but its microprocessor consisted of the primary components of what would now be called a personal computer. While Viatron received a great deal of publicity and shook up the computer industry, its plan proved to be more than it could execute, leading to a spectacular collapse. Four-Phase Systems produced a chip that could have been called a microprocessor according to a modern definition of the term, but Four-Phase never called it that. Four-Phase's perception of the chip, the AL1, and the development path it followed were shaped by Four-Phase's position in the computer industry rather than the semiconductor industry.

In October 1968, Viatron, the most publicized computer start-up of the late 1960s, announced its plans to lease its System 21 small computer for forty dollars a month. Viatron's announcement electrified both the computer industry and the investment community. The Computer Research Bureau headlined its 21 October newsletter for investors and stockbrokers "Viatron Blitzkriegs the Industry with the Development of the Decade: The Computer Market Will

Never Be the Same." The newsletter went on to say that Viatron "literally had the computer industry reeling" and asserted that Viatron could be "the Xerox, the IBM, or the UCC story of the next decade."[5] *Business Week* asserted that Viatron was trying to "crack wide open the market for electronic data processing equipment with low-priced mass-produced EDP [electronic data processing] hardware" and stated that Viatron had the potential to upset the pricing structure of the broadest portion of the EDP market.[6]

The inexpensive price of the Viatron System 21 made its market seem unlimited. The Computer Research Bureau wrote: "The Viatron breakthrough is so significant that the prospect of a computer on every secretary's, clerk's, and manager's desk is easy to envision." It also implied that the System 21 might be suited to home use, as did an *Electronics* article that noted that the lease price was less than many home phone bills.[7]

At the heart of System 21 was what Viatron called its microprocessor. Although Viatron never precisely defined what the term "microprocessor" meant, its material described it as a microprogrammed computing element, including both logic and memory on a number of chips. For example, the Model 2101 microprocessor had 400 characters of read-write memory (now called RAM) and 512 words 12 bits wide of read-only storage (now called ROM).[8]

Viatron had been founded in 1967 by a group of engineers who had left MITRE, a nonprofit organization spun off from MIT to do the systems engineering work for the SAGE air defense system. The president of Viatron, Edward Bennett, had managed a data management system that had been developed for use in Vietnam. Viatron's leaders had little experience in computer hardware, semiconductors, or any profit-making enterprise. Bennett, with a Ph.D. in psychology, was a particularly strong promoter of his vision of Viatron. Initially, Viatron planned to make nothing itself, contracting out both production and assembly, a strategy suggesting that Viatron's founders believed that R&D, not manufacturing, was the crucial skill needed in a computer company.[9]

Viatron would buy its MOS circuits from General Instrument and other suppliers. Although MOS large-scale integration (LSI) circuitry was the linchpin of the Viatron plan, the prototype it demonstrated in the fall of 1968 made little or no use of such circuitry. The seven hundred integrated circuits in the prototype meant it could never be profitable at the quoted price; Viatron hoped to get the number of chips down to fifty using MOS technology. Viatron's plan also depended on receiving massive orders for its System 21, with the economies of mass production making the low prices profitable. While Viatron received many orders, no semiconductor company had demonstrated the ability to mass-produce MOS integrated circuits in the volume Viatron was projecting. By 1969 Viatron had contracts with nine semiconductor companies to produce its MOS circuits.[10]

Viatron shared with Cogar the belief that it could start out as a large company. It raised $40 million in a year and a half, received fifty thousand letters of intent, and placed orders for $70 million in integrated circuits. In 1969 Bennett, not one to think small, claimed that by 1972 it would have delivered more "digital machines" than had "previously been installed by all computer makers." In contrast to the staid ex-IBMers at Cogar, who set out on a path and relentlessly continued down it, Bennett was very colorful and adaptable, changing his plans frequently and selling each new strategy with as much conviction and enthusiasm as the previous ones.[11]

Viatron's success was based on everything going exactly as planned; its managers had left no margin for error. Viatron needed to be able to manufacture large volumes of its systems to achieve the low lease price it had quoted, which had attracted so many orders. The limited ability of the semiconductor industry to produce MOS circuits forced Viatron to make the investments necessary to make some of its own integrated circuits. But it could not achieve anywhere near the production goals it had set for itself of five thousand to six thousand systems per month. These deficiencies in production hurt Viatron in two ways, driving up unit costs and leading frustrated customers to cancel orders. By July 1970 the other board members had removed Bennett, and by March 1971 Viatron was in Chapter 11 bankruptcy proceedings.[12]

Although Viatron had managerial problems, its main technical problems were ones that would soon be obviated by the semiconductor industry's continued progress along the course described by Gordon Moore's 1965 *Electronics* article. While MOS chips of the density Viatron needed were hard to make and in short supply in 1970, in a few years they would be commonplace. If Viatron had waited a few years, it would not have had to establish its own semiconductor fabrication capability.[13] But by the summer of 1970 Bennett's brief fling with the computer and semiconductor industries had ended. Undaunted, he had a new company and a new direction (energy machinery) by the fall and was promoting it with the same vigor he had Viatron.[14]

Lee Boysel founded a computer company that did not get as much attention as Viatron but had much more success. Boysel, the leader of Fairchild's MOS circuit design group at Mountain View, had long been interested in starting a company that would build computers, and in October 1968 he did so. While many were fleeing Fairchild because of the arrival of Lester Hogan and Hogan's Heroes, Boysel's work determined the timing of his departure. Boysel had used Fairchild as the unwitting funder of the development work necessary to start his company, and by the fall of 1968, with the fabrication of a dynamic RAM, he had in place all the components needed to build a computer using MOS technology.[15]

Boysel's company started as a partnership with two other engineers from his group at Fairchild and several others. They worked in a rented dentist's office using laboratory equipment from Boysel's home lab. No one was paid while Boysel sought financial backing. Four-Phase Systems was incorporated in February 1969, with $2 million in long-term notes. One early backer was Corning Glass Works, which had also supported the semiconductor manufacturer Signetics. The name hinted at Boysel's technological enthusiasm, for it was an abstruse type of MOS circuitry. After Four-Phase could start paying salaries, other members of Boysel's group at Fairchild joined him.[16]

Boysel planned to build the computer he had described in his design manifesto in September 1967, using MOS technology for the RAM, the ROM, and the processing unit. Through the use of four-phase circuitry, which made possible the use of very small MOS transistors, Boysel was able to build much denser chips than he had envisioned in his 1967 manifesto. The computer was to be equivalent in power to a medium-sized IBM System/360.[17]

Boysel was unwittingly entering the same market as Viatron, but he had a fundamentally different approach. His computer would be used to control terminals. Although it could be used for data processing, Boysel envisioned its typical application would be to interface between a mainframe computer and terminals. This way Boysel could satisfy his investors that his company would be a credible competitor of Viatron, with its lease price of forty dollars per terminal per month. Under Viatron's plan each terminal would have enough electronics to essentially make it a small computer, but Boysel's terminals would consist of little more than a cathode ray tube and a keyboard, making them cheaper. The electronics in the computer/terminal controller would be shared across many terminals. Boysel later refined his strategy so that his system would offer plug-compatible replacement for IBM terminals and terminal controllers. This provided Boysel access to an established market and lowered risk for potential customers, for if his system proved unsatisfactory, they could quickly replace it with an IBM system.[18]

In contrast to Viatron, Four-Phase was centered around engineers with experience in MOS circuit design. Boysel and his engineers had designed prototypes of each key MOS circuit Four-Phase would need to build its computer (ROM, RAM, and arithmetical logic unit) before they had left Fairchild. Boysel early on recognized the importance of reducing the number of separate chip part numbers required to build his computer, and he partitioned the system so that it could be built from roughly a dozen different chip designs, many less than Viatron's system required.[19]

While Four-Phase would not build its own integrated circuits, Boysel had made arrangements to ensure its supply. At the time Boysel left, Bob Cole, who had worked in MOS manufacturing at Fairchild and had been Boysel's ally in his battles with Fairchild R&D, also left to start a company. Cole's company, Cartesian, would process MOS wafers for computer companies, such as Four-Phase, that had the capability to design their own circuits. Cartesian would implement the MOS process that had been used at Fairchild manufacturing, which Boysel and his group had used in all their previous designs. While Cartesian was independent of Four-Phase, the two formed a dyad, with Boysel arranging financing for Cartesian along with Four-Phase.[20]

By the spring of 1970, Four-Phase had an engineering-level system operating, and it publicly introduced its system that fall. As the sassy start-up made its debut at the Fall Joint Computer Conference, it was inadvertently aided by IBM in the quest for publicity. In late September, shortly after Four-Phase had introduced its system, IBM introduced its System/370 Model 145, which used semiconductor memory. IBM launched a series of advertisements claiming the Model 145 was the first computer with semiconductor main memory. Four-Phase then had large copies of the ads put up around its booth, with the text modified to assert its primacy in semiconductor memory.[21]

By June of 1971, Four-Phase systems were in operation at Eastern Airlines, United Airlines, Bankers Trust, and McDonnell-Douglas. One user asserted that the cost of the Four-Phase system was roughly half of what equivalent IBM hardware would have cost. All users gave positive reports on their experiences with the system. By March 1973, Four-Phase had shipped 347 systems with 3,929 terminals to 131 different customers.[22]

The heart of the computer was the AL1 chip, which Boysel had conceived and designed while still at Fairchild. The AL1 contained an eight-bit arithmetic unit and eight eight-bit registers (including the program counter). It was an extremely complex chip, with over a thousand logic gates in an area of 130 by 120 mils (figure 8.1).[23]

The AL1 was the tacit subject of an April 1970 article in *Computer Design* by Boysel and Joe Murphy, one of his colleagues. The article discussed computer design using MOS technology and four-phase circuitry, specifically how a mini-computer had to be repartitioned to take advantage of the density of MOS. Boysel and Murphy had no simple term to describe what their chip did, alternately calling it "the main LSI block of a low-cost fourth-generation commercial computer system," an "eight-bit computer slice," and an "arithmetic logic block."[24]

The story of the AL1 is the story of the dog that did not bark: Four-Phase made no claims that the AL1 represented an invention. In April 1970, a time when Federico Faggin was just beginning the design of what would be called the 4004, Boysel and Murphy had great freedom to call the AL1 whatever they wanted. Although the AL1 was an outstanding piece of engineering and central to the success of his overall computer, Boysel did not see it as an innovation in and of itself. Boysel had done two previous designs of adders or arithmetic units at Fairchild, and the AL1 represented simply a continuation of that work; it was a change in degree, not a change in kind. To Boysel, the AL1 was not a computer on a chip or a processor on a chip. First of all, because of the system's twenty-four-bit width, three AL1s were required per system. Furthermore, in Four-Phase's nomenclature the central processing unit (CPU) circuit board was made up of three MOS integrated circuits (the AL1, a ROM chip, and a chip called the random logic chip) replicated three times. The CPU included the control store (ROM). The AL1 was not a microprocessor, because the definition of the term at that time meant a microprogrammed processor including both ROM and RAM.[25]

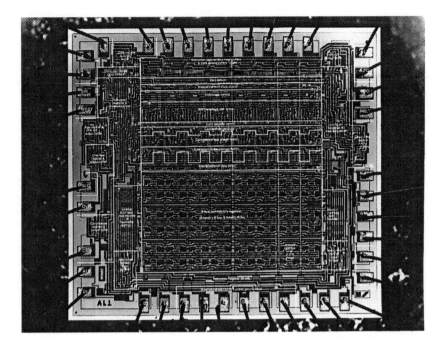

FIGURE 8.1 • THE AL1. Courtesy Lee Boysel.

Boysel, as an entrant into the computer business, was interested in the characteristics of the overall system, such as performance or cost. Unless it offered some such system benefit, he had no reason to claim his AL1 represented something new (figure 8.2). Boysel's nonchalant attitude toward putting a computer on a chip can be seen in his statements made for an article published in *Electronic Design* in February 1970. At this time what would become the 4004 and 8008 microprocessors from Intel existed as block diagrams but had not been implemented or made public. However, Four-Phase had a working computer system built around the AL1 chip. Boysel said: "The computer on a chip is no big deal. It's almost here now. We're down to nine chips and we're not even pushing the state of the art. I've no doubt that the whole computer will be on one chip within five years."[26]

In fact, being in the computer business gave Boysel reason to be quiet about the AL1. While there was no reason to believe that a customer's decision to acquire a Four-Phase system would hinge on exactly how much of the CPU was

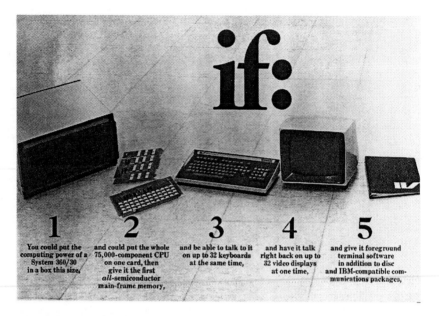

FIGURE 8.2 • EARLY AD FOR FOUR-PHASE SYSTEMS MODEL IV/70, CIRCA 1970. The following page states: "THEN: You'd Have System IV/70—the data base access system of the 70s." Note that the advertisement asserts only that the IV/70 had "the whole CPU 75,000 component on one card," rather than making any claims about a computer on a chip. The top circuit card above "2" is the CPU card with the AL1 chips, random logic chips, and ROM chips. The lower circuit card is the memory card, which used Four-Phase's one-kilobit RAM chips. Courtesy Lee Boysel.

put on a single chip, the AL1 was one of Four-Phase's most valuable pieces of intellectual property, and if other firms copied it, they could conceivably offer a competitive product. Four-Phase refused to sell the AL1 to a manufacturer of computer terminals, believing that such a sale would jeopardize its computer business, which it considered more profitable than the semiconductor business.[27]

Boysel's position in the computer industry also shaped how the part was packaged. A traditional constraint in the packaging of integrated circuits was in the number of inputs and outputs the chip had. At this time low-cost dual inline packages (DIPs) were available with sixteen or eighteen pins. For an eight-bit processor (with eight inputs/outputs and nine lines for control) such packages would mean that numerous lines would have to be multiplexed together, slowing down the system. The alternative was to use a very expensive forty-pin package. The cost for such a package (around five dollars each) was greater than the cost of making the integrated circuit itself. A semiconductor producer might be expected to be highly resistant to this arrangement, which put a majority of the cost of the component outside of its control. But for a computer systems company that was building a computer with a purchase price of ten thousand dollars or more, the increased performance more than made up for the few extra dollars paid for high-pincount packages.[28]

Four-Phase's position in the computer industry shaped the further development of the AL1 chip. In the early 1970s, while the semiconductor industry was not capital-intensive, the computer industry was, due to the way medium- to large-sized computers were acquired. Following a pattern that had been set in the precomputer office equipment era, most computers were leased, not bought. This meant that every computer that Four-Phase made was a capital item that would only gradually pay for its costs over the course of its lease. Four-Phase had to raise funds, either debt or equity, to pay for every computer it made. Leasing acted as a rein on the growth of a start-up, as the capital requirements would strangle the company if it grew too quickly. By 1974, the company's first full year of profitability, Boysel had had to raise $27 million to keep it going.[29]

Leasing made Boysel, at heart a technological radical, more conservative. Each new model that Boysel introduced threatened his installed lease base. Four-Phase's conservatism can be seen in its use of semiconductor technology. Four-Phase designed its own chips and then contracted with several foundries to have them built. The yields Four-Phase was able to achieve on its relatively simple, forgiving process were extraordinary, as high as 50 percent. Had these chips been products made by the semiconductor industry, their introduction would have been followed by a period of incremental improvements in both yield and performance, leading to cheaper, faster chips and larger markets. Four-Phase was able to quickly produce enough chips to meet its requirements for chips into the foreseeable future. Production was halted, and the chips were stored for later use. Such action would have been inconceivable in the semiconductor industry, where a part's value only went down over time. But at Four-Phase the value of the chips was related to the lease price Four-Phase could get for its systems, which remained relatively constant over time. While the semiconductor industry had a highly elastic market for chips, Four-Phase had an inelastic market for its chips; even if they could be produced for nothing, the intricacies of lease financing would determine how many systems Four-Phase would build.[30]

Four-Phase's position in the computer industry made the AL1 an unknown element in a black box to its customers. Few of Four-Phase's customers knew (or cared) that such a thing as the AL1 existed. They could see the computer system itself and think of new ways to use it but could not deal with the AL1 directly. Any innovation in how the AL1 was to be used had to come from Boysel and his company.

Although it never became the size of IBM, or even Digital Equipment, Four-Phase did achieve a substantial level of success with its approach. Its systems were widely used by hospitals to handle billing as well as by government agencies for data entry. Although the Four-Phase annual reports made constant reference to the continual need for capital (i.e., bank loans) required in the computer leasing business, Four-Phase stayed in the good graces of the banks and financial markets. In 1979 the firm had revenues of $178 million and net income of $16 million. As Four-Phase faced increasingly aggressive competition from IBM, in 1981 it acceded to a $253 million buyout offer from Motorola.[31]

Intel, the Search for a Standard LSI Logic Product, and the Microcomputer

The events surrounding Ted Hoff's conception of what is now known as the microprocessor are some of the best known in computer or semiconductor history. But as the story has been told, it has largely focused on Hoff's genius; the larger context of Intel's position in the semiconductor industry has been ignored. The work that led to what is now called the microprocessor represented a convergence of corporate strategy, a particular semiconductor technology, and the opportunities presented by a specific business transaction.

In the late 1960s there were three basic approaches to integrated circuits, particularly LSI integrated circuits. First were standard products, where the semiconductor maker designed and produced a part and sold it to any interested customers. Second were custom products, where the semiconductor producer designed a part specifically and often exclusively for one particular customer. Third was the computer-aided-design approach, where the semiconductor producer would develop a set of tools that customers would use to design much of their own integrated circuits.

The three approaches put an emphasis on different skills and had different advantages. The custom approach required a significant number of circuit designers who could quickly implement a given design. It offered to lock in a particular customer and avoid competitive pressures in pricing. The computer-aided-design approach required the development of a significant infrastructure, but it would then move much of the work to the customer. The promise of the computer-aided-design approach was to make low-volume parts profitable. Standard parts had potentially large volumes and required the fewest number of designers. But standard parts could expose a semiconductor producer to competition from others in the industry.

Moore and Noyce's decision to concentrate on standard parts when they founded Intel had deep roots. Much of Fairchild's success in integrated circuits had come through offering a standard line of products. At the 1966 Fall Joint Computer Conference, Robert Noyce, in considering the impact of LSI, spoke of the cost advantages of standard parts. "How far the industry moves" towards standard parts, Noyce stated, "may very well determine whether or not large arrays [i.e., LSI] are used in significant quantity in the future." While Noyce acknowledged that there were some standard arrays in existence, such as memories and shift registers, he claimed: "The appearance of more standard arrays seems inevitable."[32]

In an intriguing prophecy, Noyce suggested that the industry's reaction to standard parts would "be similar to the reaction that greeted the suggestion that integrated circuits be used."[33] Noyce was implying that just as computer firms objected to semiconductor firms' taking over the design of flip-flops and other circuit elements, they would object to semiconductor firms' taking over the design of computer subsystems. But these objections would disappear if the semiconductor company could offer substantial price advantages.

Intel had started out in semiconductor memory, for that was where the first standard LSI parts were, but Moore and Noyce clearly saw themselves in the semiconductor industry, not the semiconductor memory industry. Intel was a producer of standard LSI products. Intel's approach to standard parts might be termed "use custom to create standard." This was a particularly apt strategy at a time when very few standard LSI parts had been established, in either memory or logic. Intel might design a part in response to the requirements of a particular customer, but it would then try to give the part broad enough applicability that other customers would find it useful. This increased the chances that its parts would draw customers and avoided the biggest dangers of the custom business. A semiconductor company doing custom work would typically design a part based on the prospect of future orders without a firm commitment that such orders would come. If the customer's plans changed, as frequently happened in the course of development, the semiconductor firm might find that all its work had gone for nothing (although typically the customer might pay for the costs of designing and building the initial chip). This was a particularly critical issue for Intel, a start-up in the standard business with few circuit designers. If it spent all its time on projects that then vanished (something Intel had no control over), Intel would have nothing to make. Moore and Noyce used the wide range of customer contacts they had from their days at Fairchild to good advantage at Intel. Intel's first part, the 3101, had been developed in response to a request from Honeywell, as had its 1,024-bit RAM. Noyce described Intel's strategy in a March 1970 article: "Intel is actively soliciting custom business. We're doing this because we want to learn by working very closely with customers what they need to do their job. Hopefully by working with several customers in the same area, we can find the commonality that everybody seems to need, and then we can build that as a standard part. And once it exists as a standard part, the cheapest way for a guy to go will be to use it, because he will have all the advantages of a production-line flow that is already established."[34]

After Intel had launched its first memory chips, it faced the challenge of finding standard parts in logic. In the March 1970 interview published in *Electronic Design,* Noyce gave his views on standard logic parts, suggesting that a computer could be built from three basic LSI components, "random-access-memory, data manipulation registers and read-only memory." Noyce described the data-manipulation registers as basic arithmetic units capable of performing such functions as adding, subtracting, and shifting.[35]

Lee Boysel, another participant in this simulated roundtable discussion, then asked Noyce: "If you're going to standardize, why stop with a universal register? It won't be long before someone will put the whole computer on one chip, stockpile batches of them and microprogram them especially for every customer's needs."

Noyce responded: "I've no doubt that's true. But we see this as an evolutionary process. At the moment our yields generally are good enough to let us integrate a universal register of the type we've been talking about."[36] Noyce claimed that the feasibility of a computer on a chip would be determined by the economics of chip yields. Noyce basically repeated the argument that Gordon Moore had made in his 1965 *Electronics* article, asserting that as one plotted cost per logic gate as a function of the number of logic gates on a chip at first, as one put more circuits on a chip, the cost per circuit went down because the assembly and packaging costs were being amortized over a greater number of circuits. But as one put more and more circuits on a chip, resulting in a larger chip, one reached a point where the yields became prohibitively low, driving costs up dramatically. Noyce claimed that, at that time, a minimum in cost was reached at two hundred logic gates, which was a number far fewer than would be required to put a significant part of a computer on a chip.[37]

Noyce's discussion assumed the use of bipolar circuits; it was possible to put more MOS circuits on a chip, but at this point all computers used exclusively bipolar transistors. (Boysel's computer would shortly change this.) MOS transistors were significantly slower than bipolar, and to the extent that one was planning to use the universal register in existing computers, it had to have at least the speed of existing bipolar parts.

Intel made an effort to implement a computer from the universal registers Noyce had described. At a 1970 conference, Ted Hoff and Stan Mazor presented a paper entitled "Standard LSI for a Microprogrammed Processor." After discussing the advantages of using standard LSI parts for the construction of a computer, they reported on building an experimental computer, which they called a microprocessor. Hoff and Mazor used the standard definition of the term "microprocessor"—a microprogrammed processor—and thus their microprocessor consisted of a number of integrated circuits. The computer was a twelve-bit machine designed to emulate the very popular Digital Equipment Corporation PDP-8. Hoff and Mazor reported that their implementation was four times slower in doing multiplications than the PDP-8. They asserted that Intel was doing further work implementing computers from standard LSI parts.

Intel attempted to build a universal register out of bipolar technology but could not make it work.[38]

The universal register never saw the light of day, primarily because of the parallel work that led to what is now called the microprocessor. To properly understand the origins of this work, one must go back to the founding of Intel. As Moore and Noyce staffed Intel, they were primarily interested in hiring people skilled in the processes of fabricating integrated circuits. Because they were planning to make standard parts, they needed circuit designers, but not a lot of them. From the early 1960s at Fairchild, Noyce and Moore realized that the semiconductor business was increasingly merging with the computer business. For this reason they needed someone who understood computer systems, and so they hired Ted Hoff. Hoff had received his Ph.D. in electrical engineering at Stanford, working on adaptive systems, and in 1968 he was working at the Stanford Research Institute. Hoff had been suggested to Noyce by Jim Angell, a professor of electrical engineering at Stanford who also served as a consultant to Fairchild.[39]

The constraints Intel faced as a start-up forced Moore and Noyce to take a different tack on this matter than they had at Fairchild. Fairchild R&D had a Digital Systems Research Department, which had computer scientists and electrical engineers working on developing a computer built around a new architecture. The computer, called SYMBOL, was a large computer that attempted to put as much function as possible into hardware. Twenty people worked on SYMBOL. While SYMBOL was fairly far afield from Fairchild's main area of business, it had potential linkages and was a project that a large R&D lab could afford to support. Since Moore did not have a computer background (and Fairchild had a flat structure where every department reported to Moore), this was the kind of program that could easily spin out of control.[40]

Intel could not afford to support such an expansive program, but it could afford one person with a computer background. Hoff, as the one person in the company with serious computer skills, would have to work with and relate to people in the rest of the company. His title was manager of applications research, and he reported directly to Moore. He was Intel's technical frontman, being one of the first to deal with potential customers and writing articles in trade publications extolling the virtues of Intel products. Hoff proved to be a valuable commodity at Intel, for not only were semiconductors and computers increasingly related, but also the work of semiconductor engineering increasingly required computer skills. He designed an early memory test box that interfaced with a PDP-8 computer Intel was using as a memory tester, Intel's only in-house computer at the time.[41]

During Intel's first year of existence one of the major applications of MOS technology had come in calculators. While Victor had failed because it had tried to do too much too soon, others coming later had more success. MOS technology was well suited for this application—slowness was no problem, and its density advantage over bipolar was important in a cost-sensitive market. Japanese electronics firms were the leaders in developing calculators using MOS technology. In 1969 Sharp signed an agreement with North American Rockwell under which North American Rockwell would provide both MOS calculator chips and manufacturing know-how. While the Victor calculator had used twenty-nine MOS chips, the Sharp calculator was done in five. The $30 million contract immediately made North American Rockwell one of the leaders in MOS sales. Canon chose Texas Instruments as its supplier of MOS chips, while Ricoh selected American Microsystems Inc.[42]

Then, in an oft-told story, in the summer of 1969, Busicom, a small Japanese calculator company, approached Intel about designing MOS chips for a new family of calculators. Busicom was looking for seven chips, but because Intel had been established to make standard LSI parts, it did not have the circuit designers necessary to design seven parts. (Intel had two MOS circuit designers at the time, and they were occupied designing memories.) Hoff initially worked with Busicom.[43]

Given Intel's initial concentration on memory and its commitment to standard parts, there is the question of why Intel was interested in talking to Busicom. Even the most cursory discussion would have revealed the mismatch between Busicom's requirements and Intel's capabilities. From the beginning Intel's management had been opportunistic and willing to adapt to whatever potential for making money existed, which was a particularly important quality for a still-struggling start-up. It also seems reasonable to believe that Intel had some hope it could use custom to create standard and perhaps lower the number of chips required.

Here Hoff's familiarity with computers came into play. As he examined the complexity of the Busicom approach, he thought the simplicity of the Digital

PDP-8 architecture might provide an alternate approach. The PDP-8 had a simple instruction set but could perform complex operations because of its program memory; it substituted memory for logic. A small general-purpose computer might meet the Busicom requirements and obviate the need for so many different chips. With the support of Intel management, Hoff pursued his approach, while at the same time Busicom engineers worked on their original proposal. Hoff developed an architecture, aided by Stan Mazor, who had recently joined Intel from Fairchild's Digital System Research Department. Hoff and Mazor came up with a three-chip design, a processor, and two memory chips. (This was later increased to four chips.) An initial estimate suggested that it would take nineteen hundred MOS transistors to implement Hoff's processor, resulting in a chip that Intel believed it could make using the silicon gate technology. In the fall of 1969, Busicom managers chose Hoff's approach over that of their own engineers.

And there the project sat for half a year. Neither Hoff nor Mazor were chip designers, and both were pulled off onto other projects. Intel's small MOS design group had more pressing projects, most notably Intel's 1,024-bit memory chip. In April 1970, Intel hired Federico Faggin, who had continued development work on the silicon gate process at Fairchild R&D after the departure of Moore and Noyce, to implement the calculator chip set. Shortly after Faggin joined Intel, Masatoshi Shima, a Busicom engineer, came to Intel to check the progress of the project. He found there had been none. Faggin had to work at a heroic pace, often twelve to sixteen hours a day, to get the project back on track.

Faggin had both a strong commitment to MOS technology and the interdisciplinary skills needed for the task. After graduating from a technical high school in Italy, Faggin worked as a technician for Olivetti, where he designed a small computer. After receiving a doctorate in physics from Padua, he went to work for a small Italian electronics firm that was the Italian licensee of General Microelectronics, the first American MOS firm. Faggin took a strong interest in MOS technology when he was introduced to it. Faggin moved to SGS, another Italian semiconductor firm, where he worked on developing MOS processes. In 1968 he came to Fairchild for a year, where he worked on the silicon gate process.[44]

In designing a chip such as Hoff had proposed, at least four levels of design were required. First was the architecture of the chip, a block diagram defining the main elements and which elements would communicate with each other. The architecture also involved establishing the instruction set. The next level was the logic design, translating the block diagram into discrete logic gates. Circuit design meant translating the logic gates into individual MOS circuits (and designing those circuits). Finally, layout meant translating the circuits into the multilevel masks necessary to actually fabricate a chip. Hoff and Mazor did most of the architecture; Faggin and Shima did most of the rest, under Faggin's leadership.

While Faggin had extensive experience with the silicon gate process, he was not a chip designer and had not designed chips with the complexity of the 4004 before. That he took the job showed his willingness to move beyond his experience. While the design of a memory chip was relatively straightforward, with a basic cell replicated many times, a logic chip was much more complicated, with random logic circuits and irregular structures. Faggin's design of the 4004 was a major engineering feat, involving significant innovations in circuit design and layout. By March 1971, Faggin, working with Shima on the implementation, had operating hardware for all four chips.

In December 1969, Victor Poor, the vice president of CTC, a San Antonio–based maker of computer terminals and a large customer for Intel shift registers, inquired if Intel could modify one of its existing bipolar RAMs for use as a memory stack. Questioning by Intel's engineers revealed that the intended application was an eight-bit serial processor to be used in an intelligent terminal. While Poor had envisioned that the processor would be made up of many chips, Mazor and Hoff proposed putting the entire processor on a single chip. The eight-bit processor, initially called the 1201, was designed by Hal Feeney under Faggin's direction. Feeney had come to Intel from General Instruments, where he had become an enthusiast for MOS technology and the possibilities it presented for putting an entire system on a chip. CTC eventually backed out of the project, but Intel continued on after Seiko expressed interest in using the chip for a calculator. By March 1972 Intel had produced functioning 1201 chips. (The 1201 was later called the 8008.)[45]

A crucial factor in the successful realization of the microcomputer was Intel's use of silicon gate MOS technology, which made a large, complex chip like the 4004 or the 8008 manufacturable. The silicon gate process was denser than the standard MOS processes in use at many other firms, so Intel could build chips that had more components on them than other firms could. When it was still interested in using the 1201, CTC approached Texas Instruments about building a similar part as a second source. Using conventional metal-gate MOS technology, Texas Instruments designed a part that was 2.2 times the area of the Intel chip. Texas Instruments was unable to make the part, and it was never introduced as a product. One suspects the difficulties in making such a chip may have deterred other companies from seriously thinking about it.[46]

After the four-bit processor had been designed for Busicom and had resulted in working calculators, Hoff and Faggin lobbied Intel management to have the part introduced for general use. (The terms of the initial contract gave Busicom the exclusive rights to the parts.) Faggin designed a controller for a tester using the 4004. Although Intel's marketing group appears to have put up some resistance to selling the part, Noyce won back the rights from Busicom to market the chip set for noncalculator uses in exchange for price concessions.[47]

In November 1971, Intel introduced its chip set, which it called the MCS-4, with an advertisement in *Electronic News*. The ad announced a "new era of integrated electronics" and a "microprogrammable computer on a chip." While the former claim has proven accurate, the latter represented hyperbole, for any common-sense definition of "computer" would have included the processor, ROM, RAM, and input/output (I/O). Nowhere was the term "microprocessor" used, for that term as used by Viatron (and Intel) had a different meaning. Because the term "computer on a chip" had been so ambiguously defined, there was no way to challenge Intel when it claimed to have reached the goal. Intel was allowed to define what a "computer on a chip" was. For many years, what are now called unambiguously "microprocessors" were called "microcomputers."[48]

Intel's work on what is now called the microprocessor shows the inadequacy of simple notions of invention in a complex technological system. Intel's microprocessor work received so much attention because Intel was able to make it a successful product (and because Intel subsequently gained dominance of the microprocessor market). The sole credit that Hoff typically receives presumes a disembodied notion of technology, with the idea being the most important thing, and a nonproblematic straight line existing from conception to commercialization. Had Faggin been unable to design the chip, manufacturing unable to make it, upper-level management unwilling to support it, or marketing unable to devise a plan to sell it, Intel's microprocessor work would be seen in a far different way, most likely as a footnote to whatever became the first commercially successful microprocessor.[49]

Finding a Role for the Microcomputer

At the time Intel was proclaiming the beginning of a new era, one person who had thought about the possibilities of that era was Carver Mead, a professor of electrical engineering at Caltech. Mead had started his career as a traditional semiconductor researcher, conceiving and building new semiconductor devices, but he had become increasingly interested in the application of semiconductors as integration made it possible to do more and more on a single chip.[50]

Mead had close relationships with Intel—Moore in particular. Moore was a Caltech Ph.D. and had used connections with Mead to recruit Caltech students for Fairchild and then Intel. When Intel had begun, it had a small operation in Pasadena consisting of one person working to implement an idea of Mead and another Caltech professor regarding light-emitting diodes. By 1971 Intel was processing silicon wafers for students in Mead's classes at Caltech.[51]

In an article published in Caltech's magazine *Engineering and Science* in 1972, but clearly based on years of work, Mead presented his vision of the future of electronics and computers. He compared the way electronics was used at that time to the use of electric power in the early days of the electric motor. Instead of small motors being put on each machine, the investment in belting, shafts, and pulleys led to a large steam engine simply being replaced with a big electric motor. Mead claimed that although microelectronics was clearly best suited to decentralized applications, "the enormous investment in big machines and big software" led to it being used as it fit "into the existing system."[52]

But Mead foresaw this changing. He predicted that within a few years one would be able to put an entire computer on a chip costing twenty-five dollars, which would make it possible to have computers in telephones, washing machines, and cars. Mead himself had been involved with the development of what would now be called a personal word processing system, which had a televisionlike screen for display and entry coupled with a printer. Mead advocated the design of special-purpose user-oriented machines that, while having computerlike power, would be seen in terms of their function, such as a "nurse-medicine machine, a chromosome-microscope machine, an oven-cooking machine." Such systems would not come from "the big companies," Mead asserted, but from "little guys who are willing to change things."[53]

Intel was a "little guy" willing to change things. If Intel followed a direction that was consistent with the path that Mead laid out, it did so for its own reasons. The microcomputer had no market among existing computer makers. While the universal register was designed to be used in minicomputers, the microcomputer was too slow to be used there. In contrast to computer memory, where MOS technology could be used as a direct replacement for magnetic memories, and computer companies understood its advantages, Intel had to go into the wilderness and create customers for its microcomputers.

It was worth creating customers because before the microcomputer was an answer to a customer's problem, it was an answer to Intel's. Moore and Noyce's concentration on standard LSI parts had effectively kept them out of the market for digital logic, the most important market segment in the semiconductor industry. The microcomputer was one of the standard logic parts that Noyce had been expecting to appear in 1966. Intel could now compete in the digital logic business without subjecting itself to the ruthless commodity pricing typical in small-scale integration circuits, without developing the complex infrastructure needed for the computer-aided-design approach, and without submitting itself to the vagaries of the custom business.[54]

Intel's early marketing of the microcomputer focused on replacing random digital logic. Most applications using digital logic required much less than a full general-purpose computer. They would typically use small-scale integration transistor-transistor-logic (TTL) integrated circuits containing a few logic gates each. In an August 1972 ad, Intel claimed that its 4004 could typically replace 90 TTL circuits, while the 8008 replaced up to 125. Intel was asking customers to replace dozens of commodity parts with a few high-priced proprietary parts. In addition to the presumed cost advantages, Intel asserted that the microcomputer approach gave extra flexibility, as design changes could be made in ROM.[55]

Because the microcomputer was so much more complex than other parts and potential customers were less sophisticated than Intel's memory customers, Intel had to provide much more support than it had provided on previous parts. Education was key. Intel coupled its introduction of the microcomputer with a series of seminars explaining what it was and what it could do. While Intel's marketing department had come up with possible applications of the microcomputer, at these seminars two hundred to four hundred people at a time would be looking at it in terms of their particular problems and devising uses for it that Intel had never thought of. Intel became the central node in a network of hundreds of different users and was itself educated about what the microcomputer could do.[56]

Intel's typical early microcomputer customers were small to medium-sized companies that were not building stand-alone computers but needed to put some electronic function into their systems. In an August 1973 advertisement in *Electronic News*, Intel revealed some of its customers and their typical applications. They included a controller of automated typesetting machines made by Datatype, an accounting machine made by Iomec, a gas flow meter made by Daniels Industries, and an intelligent terminal made by Sycor. These customers stood in contrast to Intel's memory customers, which included some of the best-known names in the computer industry. During this period, a member of Intel's board of directors asked when Intel would get a microcomputer customer it could be proud of.[57]

The microcomputer had strong economies of scope with Intel's existing products. Intel was primarily in the memory business, and each CPU chip required memory to operate. Although the 4004 required specially designed ROM and RAM chips, the 8008 was designed to use standard Intel RAMs and ROMs. A buyer of Intel microcomputers would often also be a customer for Intel's erasable programmable ROMs, at least for prototyping. In 1973 *Electronics* claimed that for every kit (the basic microcomputer chip set) sold, ten times its value was sold in memory and peripheral circuits. It was the perfect match for the leading vendor of semiconductor memories.[58]

The Microcomputer and the Shaping of Intel

Those economies of scope did not come automatically. Intel had to reshape itself to be able to take advantage of the possibilities of the microcomputer. In doing so, it had to make difficult decisions about what kind of company it was. Was it in the semiconductor business or the computer business? Intel maintained a very fine balance and continually reinvented itself in response to market needs and opportunities.

In April 1972, under the instigation of Ed Gelbach, Intel's marketing manager, Intel formed a separate Microcomputer Systems Group, reporting to marketing. This group consisted of a dozen people and had responsibilities in manufacturing, engineering, and sales. By August 1972 Intel was advertising that it offered "unprecedented design support" for its microcomputers, including prototyping boards, FORTRAN assemblers, a FORTRAN simulator, and a user-contributed library of programs.[59]

In August 1973 Intel hired William Davidow to manage its microcomputer group. Davidow was the first bona fide computer person in a senior management position. He had received a Ph.D. in electrical engineering at Stanford (writing a dissertation that would now be called computer science). He worked as a computer researcher at General Electric before moving to Hewlett-Packard, where he became marketing manager for its computer group. He then worked for a San Francisco Peninsula start-up firm attempting to make high-speed computer systems using standard bipolar parts. Davidow became Intel's leading strategist and evangelist for its microcomputer work.[60]

At the same time, Intel established boundaries, generally for practical reasons, as to how far it would go in reshaping itself. These boundaries were always open to renegotiation, however. In its early days, Intel's software was written by contractors rather than employees, for qualified software people were reluctant to join a semiconductor company, and Intel had doubts as to whether it could offer them a satisfying career path. The most famous Intel contractor was Gary Kildall, a computer scientist teaching at the Naval Postgraduate School who wrote PL/M, a high-level language for the 8008 based on IBM's PL/1. Previously, another contractor had written a version of FORTRAN for the 8008.[61]

The possibilities of the microcomputer offered temptations to Intel on how far it should integrate forward into systems. Very early on Intel developed its Intellec family of development systems. These were full-blown microcomputer systems designed to help customers develop their own software for microcomputer-based systems in design. Ed Gelbach was preparing a marketing campaign promoting what became the Intellec development system as a computer on a board. When Gordon Moore found out, he was furious and made Gelbach reposition the Intellec as a development system only. Moore believed it would be a serious mistake for Intel to compete with its customers. Whatever gain might come from integrating into the higher-profit systems area would be more than lost from narrowing its base as systems companies viewed Intel as a potential competitor and took their business elsewhere. Intel later went into the computer-on-a-board business as a way to allow customers who needed only a small number of microcomputers to avoid having to build an entire system themselves.[62]

The 8080 and the Personal Computer

The 4004 and 8008 were pioneering products, which showed through their many flaws. They had both been designed for a particular application and then used generally. At the time they were introduced, it was unclear what their exact market would be, or even if they would have a market.

The most glaring flaw was in the packages of the 4004 and the 8008, which clearly show that they were made by a company whose first priority was not maximizing overall performance. The 4004 was packaged in a sixteen-pin DIP, the 8008 in an eighteen-pin DIP. Intel had made a decision early on that it would use only sixteen-pin DIPs, standard packages that were extremely cheap. (By the time the 8008 was made, Intel was using an eighteen-pin package for its 1,024-bit dynamic RAM.) Memories were well suited for low-pincount packages, but a CPU chip had many more lines, requiring some to be multiplexed or shared. Using the same package for both meant that Intel did not have to specially qualify a new and expensive package, which allowed Intel to introduce the 8008 quickly and with minimal cost, but it compromised the 8008's performance.[63]

The 8008 had other problems as well. Computer designers who tried to work with the chip were dissatisfied with some aspects of its architecture. It had essentially no interrupt structure. And in spite of the talk about it being a computer on a chip, roughly twenty additional chips were required to interface the 8008 with memory and I/O.[64]

In the meantime, Intel had developed a new silicon gate MOS process for use in its memory chips. Its outstanding feature was that it used n-channel transistors instead of p-channel transistors, which more than doubled the transistors' speed. The new process had other improvements resulting in the transistors' speed being three to four times that of the original MOS process used in the 4004 and 8008.[65]

In the summer of 1972, Faggin received approval to begin the design of a new chip using the *n*-channel MOS process developed for memory. Faggin, Hoff, and Mazor also worked to fix some of the other problems of the 8008. They improved the interrupt structure and minimized the number of peripheral chips needed for interfacing the chip with memory and I/O. This time they put the chip in a forty-pin package. Thirty new instructions were added to the new chip, but backward compatibility was kept with the 8008, so that any program that ran on the 8008 would also run on the new chip. In April 1974 Intel introduced the 8080 as a product.[66]

Intel called the 8080 a microprocessor from the beginning, although the term "microcomputer" continued to be used, with some attendant confusion. At the same time, Intel continued to show that it knew that the processor chip was not enough; Intel had to help the user build a system with it. Intel introduced fifteen other chips as part of the 8080 product family. These included RAMs, ROMs, and programmable ROMs, as well as peripheral circuits.[67]

Almost simultaneously with the introduction of the 8080 came an important product introduction from Digital Equipment. In March 1974 Digital announced a small "computer-on-a-card" version of its very popular PDP-8, using the 8008. (The new version had much lower performance than its other PDP-8 models.) At the same time, Digital announced that it was working with Western Digital to come up with an MOS microprocessor version of its new PDP-11. Microprocessors now made up full-fledged computers sold by major computer manufacturers.[68]

The general acceptance the microprocessor had received can by gauged by looking at the 1974 Wescon. In contrast to the elite conferences, such as the IEEE International Electron Device Meeting or the International Solid-State Circuits Conference, Wescon was more a trade show, aimed at reaching working engineers and informing them about the latest products available for sale. *Electronic News* reported that the major "buzzword" at the sessions was "microprocessors." Four sessions were devoted to microprocessors, with two of the sessions titled "The Microprocessor Revolution." Six papers were from semiconductor companies describing their newest microprocessors, three were by semiconductor companies describing microprocessor applications, and four were by customers describing how they used microprocessors.[69]

While Intel nurtured the small but growing microprocessor market, the widespread proliferation of the MOS-based calculator had simultaneously given ordinary Americans unprecedented access to computing power and wreaked havoc with much of the semiconductor industry. Here the semiconductor industry felt the dark side of the effects Moore noted in his 1965 *Electronics* article, as it was stuck making a component that got cheaper and cheaper. Lacking a means of expansion into integrated circuits of higher and higher density, firms soon reached a point where they were bitterly fighting one another to maintain a share of a market that was shrinking in dollar terms. The contrast between the calculator and the microprocessor shows the great potential the microprocessor had as a product, for while by 1975 the calculator had essentially gone through its complete life cycle as a viable semiconductor product, the microprocessor was still in its infancy.

After the Japanese firms opened up the calculator market, two ultimately contradictory dynamics were at work among semiconductor firms. The allure of much higher profit margins selling calculators rather than chips proved beyond the industry's ability to resist. Texas Instruments, Rockwell, American Microsystems, National Semiconductor, Mostek, and a host of smaller companies began making calculators. At the same time the industry's ability to build more and more complex MOS integrated circuits continued its relentless pace. While General Microelectronics had found the twenty-nine chips needed for the Victor calculator too difficult to build, by 1969 SCM was introducing an $895 calculator using eight MOS chips. By 1971 Texas Instruments and Mostek built a single MOS chip that contained all the electronics necessary for a calculator. This, coupled with the large number of companies in the field, led to dramatically falling prices. By 1972, calculators had fallen to the $150 to $100 level and manufacturers were increasingly pursuing the consumer, as opposed to business, market. Prices continued to fall and sales continued to rise, reaching 11 million units in the United States in 1974. But the increasing sales did not mean increasing profits. As the prices of both calculators and calculator chips reached what appeared to be rock bottom (they would go lower still) in 1975, at roughly fifteen dollars and three dollars, respectively, for basic models, it became almost impossible for a semiconductor company to make money. Texas Instruments lost $16 million in its calculator business in 1975, and other companies faced similar problems in the light of rapidly declining prices. The price for calculator chips fell faster than the volume of calculator sales rose, with Integrated Circuit Engineering, a consulting firm, estimating that in 1975 about 42 million calculators had been made worldwide, with an average MOS integrated circuit content worth $2.75; two years later, with the average value of the integrated circuitry down to $1.30, sales had risen to only 52 million. The total annual income that the semiconductor industry made from selling calculator chips had shrunk from $115 million to $68 million.[70]

The calculator market had become saturated, and even if the integrated circuit could have been made for free, it would have had only a marginal effect on the price of a calculator and its sales. People had only a limited number of ways they could use calculators, and one or two in most households was enough to meet this need. In the calculator business, no growth path developed for more and more complex integrated circuits. While a few firms, such as Wang Laboratories, Hewlett-Packard, and Texas Instruments, developed sophisticated scientific calculators, this remained a small portion of the market. The losses semiconductor firms suffered in the calculator business forced several small companies out of business, and even larger companies took years to fully recover.[71]

In contrast to calculator chips, which could be used in only a specific product, the microprocessor could be used in a wide range of applications. Thousands of people were looking at microprocessors and thinking of ways they could be used. Potential users did not have to file a business plan for approval with Intel or any other microprocessor maker. They and their companies would take the risks of launching new products. One loser in the calculator wars played an important role in developing a new application for the microprocessor. Micro Instrumentation Telemetry Systems (MITS), a hobbyist electronics firm based in Albuquerque, New Mexico, had made calculator kits for the intrepid would-be owner to assemble, but while MITS was selling kits for $99, Texas Instruments was selling comparable fully assembled units for half that price. In 1974, Ed Roberts, the head of MITS, thought that Intel's new 8080 could serve as the basis for a computer.[72]

The introduction of the Altair 8800 in the January 1975 issue of *Popular*

Electronics stands as one of the signal events in the history of computing. As computer historian Paul Ceruzzi has noted, the Altair represented a remarkable convergence of social and technical forces. MOS memory and the MOS microprocessor were the key components that allowed such a system to be built inexpensively. Work done for time-sharing large computer systems had led to the development of the BASIC programming language, which Bill Gates and Paul Allen adapted to the Altair. Furthermore, the proliferation of time-sharing computers had led to a substantial group of technically savvy people who had experienced the thrill of having a computer at their command and were eager to have a computer of their own.[73]

The Altair was a ridiculous machine by the standards of IBM—or any other computer company producing machines for business applications. Gordon Moore later admitted that he did not think that the Altair or other early personal computers were promising. But it was out of his control. And the Altair set off a torrent of activity as individuals and organizations developed improvements and extensions. While they were not the kind of companies that a respectable board of directors could be proud of in 1975 or 1976, companies like Processor Technology, Cromenco, IMSAI, Digital Research, Apple Computer, and a small Albuquerque firm named Micro-soft transformed that hobbyists' toy into a personal computer that had more and more capabilities and was easier and easier to use. And in doing so they greatly increased the market for microprocessors made by Intel and other companies. As this market grew, Intel and other semiconductor manufacturers were encouraged to build more and more powerful microprocessors that could enable people to make computers that offered still more capabilities and that were still easier to use. While it was not yet fully manifest, here was a growth path for the semiconductor industry into more and more complex MOS integrated circuits that had not existed in the calculator industry.[74]

Conclusion

While it is extremely difficult to quantify the market for microprocessors in the early years, it was small but growing rapidly.[75] By 1975 the microprocessor was seen as a potentially large force in American business, as demonstrated by a *Fortune* article titled "Here Comes the Second Computer Revolution." (The first came with ENIAC.) "Mass applications" of the microprocessor, *Fortune* claimed, "are just beginning to explode, setting off reverberations that will affect work and play, the profitability and productivity of corporations and the nature of the computer industry itself."[76] Many applications were listed, and while most of the customers were small companies, *Fortune* claimed that Ford had found that the use of microcomputers had cut fuel consumption by 20 percent and planned to use them in future models.[77]

In a June 1976 *Proceedings of the IEEE* special issue dedicated to microprocessors, Gordon Moore made the case that the microprocessor was a revolutionary technology. Moore asserted that semiconductor manufacturers had usurped some of the previous responsibilities of equipment manufacturers by taking over a significant portion of the design of their systems. While he predicted that the microprocessor would have a significant effect on society, he said: "The first impact of the revolution is upon the structure of the electronics industry itself."[78] Not everyone saw things that way. In this special issue of the flagship technical journal of the IEEE one can see the maturing of the microprocessor as a technology, with a general consensus being achieved about what a microprocessor was and how it might be used. The old definition of a microprocessor as a microprogrammed processor, no matter how many chips it was implemented in, had been replaced by a more modern definition: a microprocessor now was a CPU on a chip. But there was one notable exception. An engineer from IBM, a company that would logically qualify as one of the equipment manufacturers in danger of having a portion of its function usurped, described "a microprocessor-based portable computer," but his definition of microprocessor was not something that anyone else writing in the issue would have recognized as such. IBM's "microprocessor" took up an entire circuit board and consisted of at least fifteen different integrated circuits. It was as if in a 1950 conference on computers, someone had described their computers as women doing calculations with the help of an adding machine—the pre-1945 definition of computer. It suggested that IBM was oblivious to the way the microprocessor was changing the electronics industry.[79]

Lee Boysel's comment in 1970 that the "computer on a chip was no big deal" was only partly right. Many people had seen it coming, and for a computer company that made its own semiconductors (whether it was Four-Phase or IBM), it was of little moment whether the CPU was on one chip or two or three.

But for a semiconductor company such as Intel, the computer on a chip had great import. It offered a way for Intel to get into markets previously denied it and to bring electronics into a wide new range of areas. Gordon Moore stated that it allowed Intel to "make a single microprocessor chip and sell it for several thousand applications."[80] In 1975 Robert Noyce mischievously called Intel "the world's largest computer manufacturer."[81] It would take years before that was manifest to the rest of the world.

Conclusion/Epilogue

Although the semiconductor industry had clearly established MOS technology by 1975, one could assess its impact in two different ways. If one were to focus on its present accomplishments, one would see a niche technology that had made some new things possible but that had had little effect on either the semiconductor or the computer industries. On the other hand, if one were to strain to look out into the future, one might see a revolution that was yet only dimly manifest.

By 1975 the MOS transistor had not led to any fundamental reorientation in the semiconductor or computer industries. Although it had sent magnetic core memory down the road to obsolescence, this transition had been relatively easy for computer companies to make. IBM had incorporated MOS memories without fundamentally changing the way it developed its computer systems. Although the decreasing costs and increasing densities of MOS memories were responsible for the premature demise of Digital's PDP-11 computer, Digital had been easily able to comprehend and rectify the problem. If one were to assess who had lost with the coming of MOS technology, ironically the biggest losers would be the MOS firms themselves. First there was General Microelectronics, which had made an early commitment to MOS technology but had not been successful. Then came Viatron and Cogar, followed by the firms that had been run over in the stampede to produce MOS calculators. Even by 1975, it was not clear whether the cumulative profits made in MOS technology offset the losses. Intel, which had taken MOS technology further than any other semiconductor company, had risen to fifth place in the semiconductor industry by 1975, an impressive achievement. But each of the top four semiconductor companies, Texas Instruments, Motorola, Fairchild Semiconductor, and National, still relied on bipolar technology for over two-thirds of their sales.[1]

But MOS technology had opened new markets for semiconductors, including calculators, memories, watches, and an assortment of low-cost, low-performance applications. Through the proliferation of the calculator, the general public received firsthand experience with digital electronics and access to unprecedented computing power. Millions of Americans now wore digital watches. The combination of the MOS microprocessor and MOS memory would make it possible for thousands of Americans to have a computer in their homes. By 1979 the consulting firm Integrated Circuit Engineering made the heady assertion that the potential existed to put at least three microprocessors in every house, and an equal number in every car.[2]

A paper written in 1974 emphasized another aspect of MOS's potential, spelling out unstated but common knowledge in the industry: the MOS transistor could be scaled, and these smaller and smaller transistors would lead to more and more powerful integrated circuits. Ironically, the paper came from IBM, which would be most affected by its implications. Robert Dennard and a group from IBM Research gave a detailed exposition of the principles of MOS transistor scaling, showing that as one scaled an MOS transistor by a constant factor, the delay of the transistor scaled down by that factor, while the power per circuit scaled downward with the square. (This meant that smaller transistors would be faster but consume roughly the same power per area.) Dennard's group accompanied the analysis with experimental work making MOS transistors with dimensions as small as one micron. (At the time, IBM's MOS memory chips in production had dimensions as small as five microns.) IBM Research was able to make such small devices through the use of ion implantation, a technique whereby dopant atoms were accelerated into the silicon rather than thermally diffused, allowing for very precise control. Although one could not yet make devices that small in a manufacturing environment (and much more work was required to make that possible), Dennard's group showed that it could be done. And given the economic incentives to produce larger memories and more capable integrated circuits, MOS technology clearly had a long run ahead of it.[3]

Science and Technology, Research and Development

What does the MOS transistor say about industrial research in post–World War II America? Throughout the 1950s and 1960s managers' thinking about research and development was shaped by Vannevar Bush's linear model that new science would produce new technology. But the MOS transistor hardly followed a linear path. And while many believed that science would produce fabulous new technologies, one of the most fabulous technologies of the last third of the twentieth century took advantage of an existing technological base and developed with a relentless incrementalism.[4]

The post–World War II ideals of scientific research played an ironic role in the establishment of an organizational structure where MOS work could flourish. The dominance of Vannevar Bush's linear model that new science led to new technology led industrial research labs in the 1950s and 1960s to abandon contact with existing technologies, venturing out to find new scientific principles that could prove to be the basis for later technologies. The belief that research labs should focus on scientific work and leave the engineering work for development organizations created a technological space between research and development organizations at IBM. As IBM Research began looking at work on technologies with more near-term possibilities, its distance from the product groups allowed it to adopt a technology that interested no one else in the company. While Research did some important scientific work, its major contribution came in serving as an incubator—a place where applications of MOS technology could be explored on a scale so that the program could be quickly reoriented in response to changing circumstances. While such separation between research and development would be considered dysfunctional (and was in many ways), had research and development had a closer relationship at IBM in the early 1960s, IBM Research would probably not have picked up MOS technology. Given the inherent uncertainties of the research and development process, no single R&D organizational structure is optimal for the development of every technology.[5]

The MOS transistor not only did not follow Vannevar Bush's model that new science leads to new technological developments; it was an example of the converse of Bush's model. In 1965 a group of IBM researchers made observations about the behavior of electrons in MOS transistors, which led to a series of investigations culminating in Klaus von Klitzing's observation of the quantum Hall effect, an accomplishment honored with the Nobel Prize for Physics in 1985. Alan Fowler, a physicist who was a member of the IBM MOS research group, wrote: "The development of the MOSFET is an example of the enhancement of physics by technology." Following IBM's 1965 work, Fowler wrote in 1993: "Innovative semiconductor physics increasingly depended upon semiconductor device technology."[6]

When Gordon Moore and Robert Noyce founded Intel, they were not interested in doing semiconductor physics for its own sake—they were forsaking Nobel prizes. Establishing Intel without an R&D lab was a tacit criticism of the validity of Bush's linear model. Noyce and Moore judged that silicon technology had significant staying power, so that it would not be immediately threatened. This has proven true, and the longevity of the MOS technology has meant the end of research in another way: it represents the triumph of technology over science. Throughout the 1960s and 1970s whenever research managers were looking for the next generation of digital electronics technology, a typical mode of operation was to find an interesting scientific effect or a material with promising scientific properties, then attempt to built a technology around it. Josephson junctions, which take advantage of tunneling in superconductors, and the compound semiconductor gallium arsenide, in which electrons move at very high speeds, are notable examples. In spite of the great hopes these aroused, researchers inevitably ran into problems, which delayed their programs. In the meantime, the relentless advance of the silicon planar technology (both MOS and bipolar) passed the other technologies by and left them uncompetitive. After spending hundreds of millions of dollars on such projects, research managers finally lost heart. Today, they are hesitant to fund further frontal assaults on the MOS technology.[7]

The story of the MOS transistor, although the story of a technology that successfully displaced an established technology, does not offer great solace to technologies that seek to do the same to it. The MOS transistor won its position under a very specific set of conditions that are unlikely to be duplicated. Although it took the MOS transistor over ten years to go from initial conception to commercial establishment, it did so in the adolescent years of the semiconductor industry, when capital costs were low. It had the advantage of using the infrastructure and accumulated knowledge that had developed around the bipolar transistor. A company like Intel could work on both technologies and gradually shift resources over to MOS as the business grew. With only a few million dollars necessary to start a company or an effort in MOS technology, a wide range of ventures were pursued in parallel, with great diversity in their timing and approaches. A number of these ventures could fail (or not succeed) without dooming MOS technology as a whole.

It is difficult to be sanguine about the prospects for a competitor to the MOS transistor in the early years of the twenty-first century. Forty years of growth along a Moore's Law trajectory have made the challenge for a competitor incomparably more difficult than it was for MOS technology in the late 1960s. Semiconductor companies and semiconductor equipment companies have a revenue stream of many billions of dollars they can use to fund the next generation of technologies. While the press periodically gives breathless accounts of the possibilities of quantum computing or molecular computing, the momentum of the existing technological system means that a great deal of resources will be expended in trying to come up with ingenious ways of extending it.[8]

Perhaps the later career of Rolf Landauer provides the best commentary on the relation between science, technology, and electronic devices in the last quarter of the twentieth century. Landauer, the instigator of MOS research at IBM, spent the last years of his career as the senior statesman of the IBM physics community and as one of the chief skeptics of trendy technologies based on new physical phenomena. Landauer expressed doubts about Josephson junctions, on which IBM spent millions, optical computing, and quantum computing. Landauer's criticisms were not those of one who had been left behind by science, or one who sought security in the familiar. He had a lively mind and as an attendee at conferences on quantum computing was willing to credit what he considered good work. But at the same time he had a strong sense of what was necessary to convert an idea that looked good on paper into a profitable product, and he demanded that those factors receive due attention. He suggested that papers on quantum computing be accompanied by a disclaimer stating: "This scheme, like all other schemes for quantum computation, relies on speculative technology, does not in its current form take into account all possible sources of noise, unreliability and manufacturing error, and probably will not work."[9] While Landauer did not have a reflexive opposition to new ideas, he had (to use an electrical engineering term) a high noise immunity.

The MOS Transistor and Silicon Valley

The MOS transistor and Silicon Valley grew up together. And while this book makes no pretensions of being a comprehensive history of Silicon Valley, as a comparative study of MOS technology, it provides a basis for understanding how Silicon Valley has developed and functioned. Fairchild Semiconductor was the crucible for the development of both MOS technology and the semiconductor portion of Silicon Valley. The many start-ups formed from Fairchild have been well documented. But to understand why this start-up phenomenon happened, Fairchild Semiconductor and the technologies it developed must be understood in comparison with other firms and their technologies. Fairchild was a hybrid firm with some features of large, vertically integrated East Coast firms and some features of smaller, more entrepreneurial firms.

A defining characteristic of Fairchild at its founding, one that has often been identified with the West, in contrast to the East, was its short-term horizon. Fairchild sought to make money quickly. While a large, vertically integrated company such as AT&T or IBM could put money into research without being concerned about the near-term payback, the eight founders of Fairchild did not have that luxury. They had been able to squeeze venture capital out of Fairchild Camera and Instrument, but they were expected to produce results quickly. After subtracting out William Shockley's personal pathologies, the time horizon was the central point of contention between Shockley and the eight employees who quit to form Fairchild. Shockley, in pursuing the technically challenging four-layer diode, acted as if he were still at Bell Labs, where unlimited amounts of time and money allowed him to focus on a technology that would be viable in the long term. The eight, with very little experience in large industrial research labs, wanted to pursue the approach that would get a product to the market as quickly as possible, which they finally did at Fairchild. And while Fairchild forged ahead with its integrated circuit using its planar process, researchers from RCA sought to make not integrated circuits that they could sell immediately, but ones that would be competitive five or ten years out.

At its founding in 1957 Fairchild Semiconductor was clearly an entrepreneurial firm quite different from its East Coast counterparts, but as it grew in the 1960s, it showed some of the characteristics of the vertically integrated East Coast firms, particularly in its attitude towards research and development. It supported a large R&D lab, with a greatly expanded time frame and a much wider range of projects than could be made into products. Had Fairchild been looking at only technologies that offered immediate returns, MOS would not have been attractive in 1963. Fairchild could afford to pay Bruce Deal, Ed Snow, and Andrew Grove to study the MOS structure, and it could afford to pay others in R&D to pursue a diversity of approaches that might not produce returns for years.

Fairchild ran into problems of technology transfer common among R&D labs, particularly the differing identities and values of its research and manufacturing organizations. Its pursuit of multiple approaches to MOS technology, while prudent, meant that some projects got lost. Furthermore, as Lee Boysel's case shows, Fairchild had enough money to support programs it could not carefully manage. In all these ways, Fairchild was no different from IBM, Bell Labs, or RCA. The difference was that engineers at Bell Labs or IBM rarely left to start their own firms. Their companies were so capital-intensive and the process of developing, manufacturing, and marketing a technology so highly subdivided it would have been absurd to consider forming a start-up based on the experiences of a few people. Cogar showed just how absurd.

But Fairchild was different. Its very existence was testimony to the fact that a few people with talent and money could start their own company and become wealthy in the process. The start-ups that came after Fairchild only served to reinforce that idea in those that remained. An observant person would have seen it done many times: one needed to find a technology on the verge of production, enlist a few engineers, get a salesperson or two, and then raise capital.

Beginning in 1964 with General Microelectronics, Silicon Valley supported a wide range of MOS start-ups focused on getting products to market very quickly. To be sure, many had limited or no success, but employees from failed or failing firms picked themselves up and took their hard-won MOS experience somewhere else. Silicon Valley start-ups succeeded in getting much of the MOS work that was done at Fairchild out into the marketplace. For example, Frank Wanlass took Fairchild's early MOS work with him when he went to General Microelectronics. Intel continued the silicon gate work started at R&D, and Lee Boysel took his computer designs with him when he left to form Four-Phase Systems. In the early 1980s, LSI Logic commercialized Fairchild's computer-aided-design technology to make application-specific integrated circuits (ASICs) technology. Without these start-ups, one suspects that much of this work would have remained bottled up within Fairchild.

The characteristics of the technology Fairchild pursued played a crucial role in supporting Silicon Valley start-ups. Even as it expanded its time horizon, Fairchild R&D favored much simpler technologies than its East Coast counterparts. While Fairchild invented the elegant but complex complementary MOS (CMOS), RCA pursued it. Physicists at both Fairchild and IBM knew that n-channel MOS transistors would be faster than p-channel transistors, but Fairchild chose the slower, simpler approach, while IBM chose the faster, more complex approach. The abundance of capital and much longer time horizon led IBM, RCA, and Bell Labs to pursue much more expensive and complex technologies, while Fairchild chose simpler approaches that could be commercialized more readily. As engineers took technologies from Fairchild, their simplicity proved to be an advantage, while complex technologies proved to be a curse to East Coast start-ups such as Cogar.

Cogar and Intel provide an example of start-ups' differing attitudes towards complexity. Intel initially pursued three projects, including a package holding multiple memory chips that Intel quickly dropped as too complicated. At the same time, Cogar spent much effort trying to develop an even more technically challenging chip packaging scheme. Discussing Intel's early projects in a 1995 interview, Gordon Moore joked that Intel had just recently gotten the multiple chip package to work with its Pentium Pro microprocessor. The past experience of Cogar and Intel engineers had led them to fundamentally different assessments of how difficult a technology they could hope to master.[10]

Not everything that Fairchild touched turned to gold, or turned to gold right away. A group at Fairchild had been working on a computer-aided-design (CAD) approach to MOS integrated circuits when Moore and Noyce left to found Intel. While Fairchild continued work in this area, it was unable to make a successful business out of its CAD work and closed it down. This project was much more complex than other semiconductor work pursued by Fairchild, and it was only in 1981, with the founding of LSI Logic, that Fairchild engineers were able to commercialize it. Had many of the Fairchild R&D programs had an incubation period of a dozen or so years, the dynamic in Silicon Valley would have been much different.[11]

Paradoxically, the simpler approach that Fairchild and other Silicon Valley firms generally adopted—seeking returns as soon as possible—proved to be a long-term advantage. Economic historian Nathan Rosenberg has noted that the uncertainties inherent in technological development often make a sequential development process of small, incremental steps more efficient than one planned out in great detail in advance. Because Silicon Valley firms, unlike most of their East Coast counterparts, got their products to market quickly, they were then able to take further steps in developing them. A crucial element in Fairchild's decision to develop the planar process was that it was the first firm to produce diffused silicon transistors in large quantities and was therefore facing reliability problems in a way that no other firm was facing them. On the other hand, RCA's concentration on CMOS as the ultimate MOS technology locked it out of many new markets that developed along the way, such as dynamic RAM. In the early 1960s IBM Research's managers, looking ahead five or ten years, thought the future development of large-scale integrated circuits would require the development of discretionary wiring (the programmable interconnect pattern) and spent years working on it before they found it was unnecessary. Fairchild, not looking so far ahead, never found this approach attractive.[12]

Of all the East Coast firms with long-term approaches to semiconductor research, the case of RCA is the most poignant. RCA engineers were consistently right about the advantages of certain approaches, but they were right on the wrong time scale. While they were right to think that the bipolar transistor was not optimum for integrated circuits, had they started out with that rather than looking for alternative approaches, they might have been seen as the inventors of the integrated circuit and achieved an early position in the field. Instead, they concentrated on other transistor types, including MOS, that would not be ready for production for many more years. In their announcement of the MOS transistor in 1963, they claimed quite correctly that it would make possible "portable, battery-operated, high-speed computers."[13] By the time portable computers became common, RCA no longer existed as a corporation.

The advantages of the sequential approach are most clear with the development of the microprocessor. Engineers at both IBM and Intel knew that someday it would be possible to put a computer on a chip, and they both aimed towards that goal. Intel's managers were willing to take a chance on their slow four- and eight-bit microprocessors even though they might be a poor imitation of what most would have wanted in a computer on a chip. The important thing was that it had some potential users. Intel was then able to follow that path to the personal computer and a position of leadership in the computer industry—a place beyond the wildest dreams of Intel's managers. "The day of the 'computer on a slice' is nearly dawning," IBM researchers had confidently stated in 1968. They were right, but that day dawned in the West.[14]

IBM's position as a vertically integrated computer systems producer put it at a further disadvantage in its MOS work compared to Intel and other semiconductor firms. While MOS technology had many potential uses, only a small subset of these—computer applications—were legitimate for IBM engineers to pursue. IBM's semiconductor operations were captive to IBM's computer systems. Intel could develop MOS technology into any product it suspected a reasonable number of customers would buy. As a start-up Intel developed MOS watch chips, MOS memory chips, MOS microprocessors, and MOS shift registers, while the much larger IBM concentrated on memory chips. With the microprocessor, Intel was able to start out with applications in calculators and scales and then move into personal computers. That path was not open to IBM.

After several cycles of semiconductor start-ups in the San Francisco Peninsula, institutions developed that supported future start-ups. Cartesian, the wafer processing firm started to support Lee Boysel's efforts, also offered wafer processing and mask production services. Intel used Cartesian to process its early masks, giving Intel one less thing to worry about. In 1970, after having worked for General Microelectronics and National Semiconductor, Regis McKenna started his own marketing services company, which went on to help Intel develop strategies for marketing its microprocessor. On the other hand, the paucity of East Coast semiconductor start-ups and their lack of success meant that no comparable infrastructure developed in the East, making successful East Coast start-ups less likely.

The fact that there were numerous semiconductor firms and start-ups lessened the trauma when any Silicon Valley firm failed. One might see in the aftermath of Cogar an entirely different dynamic in effect. Cogar's engineers faced a harsh choice: if they wanted to stay in semiconductors they had to leave the Hudson Valley. The only semiconductor producer in the area, IBM, would not hire them back. If they wanted to stay in the area, they had to find a new line of work. It was a difficult experience for many. One imagines that many IBM engineers saw Cogar's fate as evidence of the folly of leaving IBM. Cogar ultimately contributed to the success of Silicon Valley, as a contingent of former Cogar employees joined Fairchild, and a few went from there to start their own firms.

Even as they and many of their early employees came from Fairchild, Gordon Moore and Robert Noyce were able to create at Intel a new company with dramatically different characteristics from those of its parent. Because of their established credentials and the money they had made at Fairchild, they were able to raise money for their venture without ceding control. The maturity of the industry and the work they had done at Fairchild enabled Intel to omit the research laboratory. Intel adopted a very aggressive stance, quickly commercializing its technologies, with a much more narrow focus than Fairchild. While Intel supported other types of life in Silicon Valley—from AMD, which served as a second source of its microprocessors, to Regis McKenna's marketing organization—it did so on the condition that it receive benefits at least equal to those it conferred.

Looking at Silicon Valley as an ecological region demonstrates that just as William Shockley's pathologies encouraged the area's development, so too did Fairchild's difficulties in the late 1960s. Fairchild's most important contribution to the development of Silicon Valley may have come in 1968, just after Intel's formation, when the arrival of Lester Hogan and his management team resulted in widespread disillusionment in Fairchild's staff. Many left to start their own companies or join others in the area. Hogan and his group had little appreciation for the talent then at Fairchild or its intellectual property and pursued no legal remedies to halt the outflow. Fairchild provided not only many of Intel's initial personnel, but also later waves of employees with the experience and management talent to help the firm grow.

Stanford, commonly seen as central in the development of Silicon Valley, played only a peripheral role in the development of the semiconductor industry in the region. While Frederick Emmons Terman, Stanford's engineering dean and later provost, had an interest in using the electrical engineering department and the university as a whole as a vehicle for regional development, Stanford's role in the semiconductor industry was modest at best. Terman encouraged Shockley to locate in the area (although the presence of Shockley's mother in Palo Alto gave him other reasons to locate there), and Fairchild's R&D lab was located in the Stanford Industrial Park. In 1961 Terman described himself as only "an acquaintance" of Robert Noyce.[15]

Stanford's electrical engineering department, while one of the nation's strongest in solid-state electronics, provided no essential content to Fairchild, Four-Phase, or Intel. A large contingent of Stanford's solid-state faculty came from Bell Labs and was much more representative of the large East Coast industrial-lab approach to semiconductors than it was of the work done at Fairchild. Bell Labs was not well represented at Fairchild, nor was it particularly well respected. Nor was Stanford an overwhelmingly important source of talent to the area. In 1965, Fairchild's Digital Systems Research Department reported that on a recruiting visit to Stanford, "A typical student comment was: 'Is Fairchild interested in Ph.D.'s? I didn't know they wanted any.' "[16] It is suggestive that when Frederick Terman's son received his Ph.D. in electrical engineering at Stanford, Fairchild offered him a job, but he chose IBM Research on the East Coast instead. (In fact, the Stanford electrical engineering department was a more important source of talent for IBM than for Fairchild, and several of Fairchild's Stanford Ph.D.'s came through IBM.) Even though Stanford had its Honors Cooperative Program to allow employees in the area to work on advanced degrees, Fairchild conducted its own course in semiconductor physics. For Intel, although Ted Hoff came from Stanford, ultimately Berkeley—based on Andrew Grove's connections—proved to be a more important source of talent. At Intel, Gordon Moore maintained close connections with Caltech, the school where he did his Ph.D. work. At the time of its founding, Intel employed several Caltech faculty members on a small project.[17]

This suggests that there have been several Silicon Valleys and that even studying a local economy, distinctions must be carefully made between time periods and industrial sectors. Stanford played a much more prominent role in some industries, such as the microwave tube industry and the computer industry in the 1980s, than it did in the semiconductor industry. While Fairchild's direct effects on the Silicon Valley economy had largely dissipated by the mid-1970s, Xerox PARC and later the Stanford electrical engineering department took over Fairchild's role as the incubator of new technologies.

MOS Technology, Intel, IBM, and the End of the Mainframe

The story of semiconductor technology in America over the last quarter of the twentieth century has many plot lines and is deserving of its own complete history. This study will conclude by pointing to the continued development of MOS technology, which routed bipolar technology for digital applications, made the mainframe computer obsolete, and dramatically changed the balance of power between the semiconductor and computer industries. In all of this, one might see a continuation of patterns and technological trends that had been established prior to 1975.

One imagines that as IBM executives watched Cogar collapse in 1972, they had a certain satisfaction that an annoying pest had been removed from the scene. But had they looked deeper they might have seen Cogar as IBM's miner's canary and recognized the ominous implications Cogar's failure had for IBM. Cogar suggested that IBM's semiconductor operations were not well suited for an environment that required rapid maneuvering. If a Silicon Valley semiconductor firm could get IBM into a position where it could compete on equal terms, IBM might not be able to respond effectively.

Cogar was long forgotten by the early 1980s, as IBM racked up a series of triumphs that seemed to solidify its position not only as the world's leading computer company, but as the world's leading industrial enterprise. In 1982, the U.S. government withdrew the antitrust case that had placed a cloud over IBM and consumed great amounts of time and energy. In 1980, IBM introduced a new and highly successful line of mainframe computers, the 3081. While these multimillion-dollar machines were IBM's most important product, the next year it displayed a keen sense of timing in an emerging market by introducing its wildly popular personal computer.[18]

In December 1982, as IBM ended a year of record profits, it announced that it would buy a 12 percent stake in Intel, injecting $250 million into the much smaller firm, which was struggling through a recession. While this move suggested IBM was strong and Intel weak, just a decade later, in December 1992, the tables would be turned as IBM, battling for its very survival in the face of a sharply declining market for mainframe computers and its inability to keep pace with technological change, announced a $6 billion restructuring charge against earnings and the end of its long policy against layoffs. Although IBM continued to be a force to be reckoned with, Intel and Microsoft emerged as the leaders of the new order.[19]

A careful examination of both IBM and Intel in 1982 would suggest that the strength of IBM could become its weakness and the weakness of Intel could be its strength. While IBM in the early 1980s gave the outward appearance of strength, it carried two legacies of the 1960s and 1970s that lessened its ability to respond to the changes that lay ahead. The first was the dominance of the mainframe computer, responsible for roughly two-thirds of IBM's profits in the 1980s. IBM's semiconductor operations were very responsive to needs and opportunities in the area of large computing systems but much less responsive to opportunities elsewhere. The IBM manager who in 1965 recommended the termination of the MOS program because it had little application to mainframes was not an aberration but an accurate reflection of IBM's strategic interests. While IBM moved quickly in MOS memory, which had applications in mainframes, MOS logic, which did not, languished for years.[20]

The other legacy was IBM's lack of responsiveness to competitive threats in the semiconductor industry. While Thomas Watson Jr. had lost sleep over whether IBM's slowness in adopting integrated circuits would cost sales, later IBM executives displayed no such anxiety. The characteristics of IBM's semiconductors were hidden under the covers of its machines and did not directly affect whether customers chose IBM equipment. The semiconductor industry imposed a certain discipline on its members, essentially requiring each firm with aspirations of being a major player to have a certain product line that met industry standards. IBM was not at all bound by semiconductor industry standards or practice.[21]

IBM's semiconductor operations existed inside a walled environment isolated from competitors, making it much less prepared for a fight that might require it to maneuver rapidly. Intel was a streetfighter that constantly faced bruising struggles with such rivals as AMD, Motorola, Texas Instruments, and Mostek. While Intel's proprietary technologies gave it some advantage against its competitors, Intel knew that its survival depended on never letting down its guard. In the early 1980s, Intel had just completed a particularly fierce campaign against Motorola, code-named Operation Crush, in which it had mobilized its forces in a matter of days to protect its microprocessor franchise. While IBM qualified as one of John Kenneth Galbraith's industrial giants whose ability to manage the demand for its products ensured that it could not lose money (or so Galbraith could think in 1967), Intel had no such solace. It had grown up a small company in a cyclical industry littered with failures. Immediate responses were imperative both to take advantage of good times and for protection in bad times. In the 1981–82 recession, Intel instituted typically drastic measures: first, a "125% solution" where engineers worked two more hours per day without pay to speed up the development of products, and then, when the recession lingered, 10 percent wage cuts. While Intel would scrap for any advantage it could, developing products on a experimental basis to see if they would give it some competitive advantage in the market, IBM provided no room for that type of economic experimentation—a semiconductor product would be used in large volumes or not at all. Despite the proficiency of IBM's technical staff, this environment hardly kept IBM in fighting form.[22]

Perhaps the best symbol of this languor was IBM's second-generation MOS technology, called SAMOS.[23] With its development IBM made a conscious decision to take a different direction than the rest of the industry, choosing a technology that ostensibly offered very low cost in exchange for very slow speed. The technology was suitable for memory only, not logic, a limitation that made some sense given the importance of the mainframe at IBM.

SAMOS suffered from problems that were apparent to some at IBM as it was being developed. Engineers and managers at IBM Research argued passionately that SAMOS was an inferior technology that would foreclose the use of MOS for computer logic at IBM. Research contended that IBM should follow industry standards and develop a silicon gate technology similar to Intel's, which would provide a base for the development of a wide range of memory and logic products. But Research, still in a marginal position with regards to IBM's semiconductor operations, had limited means with which to influence development decisions. By 1976, after potential alternatives had been shelved, a Research manager asserted that "IBM was on a disastrous course with SAMOS" and that within IBM's semiconductor organization there was a "SAMOS establishment" that would "not be moved by mere technical argument."[24] SAMOS kept IBM's systems designers from familiarizing themselves with MOS technology and skewed IBM's use of bipolar and MOS technologies much more heavily towards bipolar, putting IBM years behind in its use of MOS.

SAMOS was IBM's equivalent of the East German Trabant, a grossly inferior technology whose existence could only be explained by the absence of market competition. A senior IBM technologist evaluating SAMOS in 1975 recognized IBM's penchant for stagnation, saying: "We seem to get stuck into a program and stay with it in spite of the improvements in technology that seem to be possible, and are indeed practiced in the competitive world."[25] As Moore's Law continued its relentless progression, IBM pursued SAMOS down a decade-long cul-de-sac. During the time that IBM did not have a competitive MOS technology, the number of transistors that could be put on a chip increased by a factor of 250. In the semiconductor industry, a company that had developed a technology like SAMOS would have beaten a retreat to industry standards within a matter of weeks or months at the first sign of failure in the marketplace. IBM's semiconductors, surrounded by IBM's prowess in computer design, production, marketing, and service, existed in a very forgiving environment where even a noncompetitive technology could be used without hurting sales. Over five years passed before IBM took decisive action to develop a more competitive technology, which even then it did not with the urgency of someone trying to avoid disaster, but in a gradual, orderly way, squeezing several more generations out of SAMOS and introducing a new technology only years later.

But SAMOS proved that a technology could be a failure even within the IBM environment. Instead of being cheaper, it turned out to be more expensive than others' memory technologies. That SAMOS was an error of monumental proportions became clear when it ultimately proved unusable in the main memories of IBM's mainframe computers, the largest and most important application of MOS within IBM. IBM had to turn to buying memory chips from outside vendors. With IBM's MOS production facility in Burlington, Vermont, geared to produce a product that IBM had little use for, IBM was forced to make an arrangement to license both Intel's silicon gate MOS technology and Intel's sixty-four-kilobit memory chip and then transfer the technology to Burlington to keep the plant and its workers productively employed. The exchange with Intel was crucial to the modernization of IBM's MOS technology.[26]

As IBM and Intel exchanged technologies, engineers from each company found that while they were ostensibly engaged in the same job, they lived in fundamentally different worlds. Intel engineers were amazed at the size of IBM's facility and the level of automation IBM employed. IBM engineers were surprised by the thinness of the oxides that Intel was able to make. Intel engineers saw practices they considered years out of date but also technologies for packaging chips that were more advanced than anything Intel had.[27]

But while IBM transferred Silicon Valley technology, introducing Silicon Valley competitiveness and maneuverability was something else entirely. IBM would need these qualities because the microprocessor would prove to be the ladder that would finally allow outsiders to get over IBM's wall and compete on its own ground. With the microprocessor, RAM, and other key components readily available off the shelf, anyone could make a personal computer equivalent to IBM's. While the failure of Cogar might have implied that IBM's methods might not work well in a competitive environment, IBM's inability to compete in the personal computer demonstrated it unequivocally. Although IBM's original personal computer made no use of IBM semiconductor technology, it could have served as a good opening gambit if IBM had subsequently been able to use its own semiconductor technology to give it a proprietary advantage. During more than a decade, in spite of numerous opportunities, IBM proved completely unable to coordinate its internal operations to bring its own semiconductor technology to bear on the personal computer. One might wonder if a company ever had greater resources and less ability to use them. The years within the walled compound had created an IBM that could not defend itself against predators. IBM researchers had developed the reduced instruction set computer (RISC) architecture, which was capable of producing very fast microprocessors, and some in IBM had proposed developing a personal computer using an internal microprocessor based on that architecture that would provide speed advantages over Intel's x86 architecture. The plan went nowhere. IBM negotiated an agreement with Intel giving IBM the right to produce Intel's microprocessors for IBM's personal computers. With this agreement IBM could have added function onto Intel's microprocessors to give it a proprietary advantage or begun a migration path away from Intel's processors to its own proprietary microprocessor designs, but IBM never took advantage of these possibilities. Ironically, those in IBM's personal computer division considered Intel a more reliable source than IBM itself. A market relationship with Intel guaranteed that IBM's cash would command a certain level of service, responsiveness to its requests, and respect. If the personal computer organization bought components internally, it would always be relegated to second place by its much larger and more profitable older sibling, the mainframe business.[28]

By 1992, when consumers no longer required the security of the IBM brand name on a personal computer, IBM was just one among many producers of personal computers. In spite of its incomparable capabilities in almost all aspects of computer research, development, and manufacture, it had been reduced to the position of a jobber, adding little more value to its personal computer than a college student assembling one in a dorm room. Jack Kuehler, the president of IBM, was forced to admit that the companies that made money in the sale of a personal computer were Intel and Microsoft. It was the ultimate fulfillment of Gordon Moore's 1976 article stating that with the microprocessor, the semiconductor producer had usurped much of the work of the equipment manufacturer. But in a painful twist, Moore's 1976 article had suggested that with the semiconductor manufacturers moving into the design and production of subsystems, equipment manufacturers would be forced to focus on software. With IBM having ceded that field to Microsoft, it had precious little to contribute to the personal computer.[29]

IBM's debacle in the personal computer business was an embarrassing missed opportunity, but the real trauma and financial damage came with the demise of the mainframe computer. These machines, selling for millions of dollars, yielded huge profits for IBM. At the time IBM introduced the personal computer, the mainframe offered so much greater capability that calling them both computers would be like calling both a 747 and an ultralight airplanes; they may have performed the same function, but one hardly considered using them for the same jobs.[30] The personal computer had a standard memory of 64,000 bytes; the model 3081 mainframe had a memory of 64 million bytes. The 3081 sold for around $4 million, while a personal computer might sell for several thousand dollars. While the customers for both the mainframe and the 747 were willing to pay an extraordinary price to guarantee the reliable operation of their machines, those who bought personal computers and ultralights took their chances. While both the ultralight and the personal computer could be built in a garage, both IBM and Boeing had built large and complex organizations to create these enormously sophisticated technologies. With both sets of technologies it would be hard to come up with a meaningful scale to compare their performance. And just as the price of an ultralight had no bearing on how much an airline was willing to pay for a 747, so the price of a personal computer had no bearing on the price of a mainframe. But from 1981 until 1993, as MOS technology scaled down to smaller and smaller dimensions, the MOS microprocessor essentially rolled up the IBM product line, reaching a point where microprocessors offered computing power that was close enough to the mainframe to affect its price. It was as if for the price of an ultralight a decade ago, it had become possible to buy components that could be easily snapped together to produce not a 747, but maybe a 737. The commercial aircraft industry would never be the same.[31]

The bipolar-based mainframe computer as it developed in the thirty-odd years prior to 1992 was an extremely complex technological system. Because bipolar integrated circuits contained relatively few transistors, the computer consisted of thousands of integrated circuits (twenty thousand in a mid-1970s mainframe, five thousand by the early 1990s). Bipolar transistors used in mainframes operated at high speeds and power levels, which required the development of extremely complex packaging and cooling systems to remove the heat and keep the systems' temperatures at reasonable levels. The development time for a new mainframe computer extended five years or more, requiring that most key technical decisions be made years before it was produced. Tens of thousands of people worked on designing these systems, developing and building the chips, designing the necessary packaging and cooling systems, making sure that all the parts interacted electrically in acceptable ways, and tracking the schedules of the thousands of components, to make sure that they would all come in on time. While the overhead was huge, the premium IBM could command from its customers for its ability to design, mass-produce, and guarantee the operation of these complex systems could justify almost any expense. The revenues from mainframes kept IBM among the country's most profitable enterprises. The mainframe's extraordinary profits underwrote IBM's unique culture, with its country clubs and full employment policy, which allowed underperforming or technologically outdated workers to be kept on somewhere in the company.[32]

The centrality of the mainframe at IBM meant that as IBM rejuvenated its MOS technology in the 1980s, its first priority was not developing microprocessors for personal computers, but developing RAM chips, which provided the main memory for large computer systems. Throughout the 1980s IBM not only worked on its own RAM programs, but spent a great deal of energy trying to reinvigorate domestic dynamic RAM (DRAM) production, playing a leading role in the proposed memory cooperative U.S. Memories. Although RAMs were increasingly becoming commodities, they were still an essential component of mainframes, and IBM worried about being dependent on Japanese suppliers. Ironically, just as IBM was redoubling its efforts in DRAM (and investing in Intel), Intel was moving out. As Japanese competition reduced the amount of money Intel made from DRAMs, Intel's middle-level managers reallocated resources into other, more profitable products. In 1984 Gordon Moore and Andrew Grove made the strategic decision to exit the DRAM business entirely. While IBM zigged, Intel zagged.[33]

The MOS microprocessor-based system was still complex, but it had much of its complexity built into the integrated circuits. With the microprocessor and the other components provided on the open market, the computer could no longer be controlled by a single corporation or command such high prices. As the MOS transistor scaled downward, making transistors smaller and faster, the difference in performance between a bipolar-based mainframe and an MOS-based computer narrowed. And as the performance difference narrowed, so did the premium customers were willing to pay for IBM's ability to produce complex mainframe computers.

Figure C.1 gives a retrospective look at the situation IBM's mainframe business faced in the 1980s. As IBM considered how to continue the development of the mainframe along its traditional trajectory, MOS technology would not have been an option—it resulted in a machine with orders of magnitude lower performance. But MOS-based systems increased in performance at a faster rate than bipolar systems, resulting in MOS technology being sufficient in the late 1980s to meet the needs of workstations and mid-range computers that had once used bipolar.

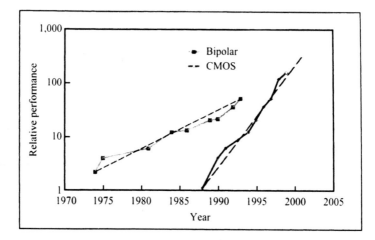

FIGURE C.1 • COMPARISON OF PERFORMANCE TRAJECTORIES OF IBM'S LARGE COMPUTING SYSTEMS USING BIPOLAR AND CMOS (COMPLEMENTARY MOS). *Source:* R. D. Isaac, "The Future of CMOS Technology," *IBM Journal of Research and Development* 44 (2000): 375. Used by permission.

MOS technology advanced in density and speed more quickly than bipolar due to a host of technical and organizational factors. Because the MOS transistor scaled in a way that bipolar did not, each new generation of MOS came more easily than a new bipolar generation. Because MOS had by the mid-1970s captured the bulk of the low-cost, high-density market and bipolar stayed in the high-performance area, Moore's Law became a less relevant guidepost for bipolar technology. While a company making MOS DRAM chips in the late 1970s or 1980s knew that it had to follow a Moore's Law curve and build a new chip with four times greater density every three years if it wanted to stay in business, IBM's bipolar operations had no such unforgiving taskmaster. For the development of IBM's high-end systems, Moore's Law could be ignored. IBM had certain performance objectives for its systems, but it had great freedom about how to reach them. As IBM replaced its 3081 mainframe computer system, introduced in 1980, with its 3090 mainframe system, introduced in 1985, it used essentially the same semiconductor technology and the density of components on an integrated circuit actually went down—it went the wrong way on the Moore's Law curve! For a semiconductor company, maintaining the same basic semiconductor technology for a decade would inexorably lead to bankruptcy court, but IBM was able to achieve the overall system performance goals it had set by advancements in circuit design, packaging, and architecture.[34]

Many have pilloried IBM's management in the 1980s and the early 1990s for its poor performance in responding to the decline of the mainframe computer, but the fundamentally different trajectories of MOS and bipolar technology made a graceful transition difficult. Moreover, IBM was not oblivious to competitive threats, but it expected competition from above, not from below. IBM saw the Japanese as the biggest threat, worrying about the potential of Hitachi and Fujitsu to develop faster versions of bipolar technology, which would enable them to produce faster computers and take away IBM's high-end customers. Even in the late 1980s, a mainframe made entirely from MOS technology without a fundamentally new architecture would lead to much lower performance and would have been an extremely unattractive option for IBM.

In 1989 and 1990, Intel and IBM each made important new announcements that brought their franchise product lines to new levels of development. Intel's 486 microprocessor, introduced in April 1989, showed what 18 years of scaling down MOS transistors and moving out along a Moore's Law curve had done to the MOS microprocessor since 1971. While a 4004 processed four bits of data at a time, the 486 processed thirty-two. While a 4004 had 6.5-micron MOS transistors, the 486 had 1-micron MOS transistors. While the 4004 had 2,300 transistors, the 486 had 1.2 million. While the 4004 operated at 108 Khz, the 486 operated at 25 Mhz. CMOS, invented by Frank Wanlass in 1963 but rarely used in the 1960s and the 1970s, had become the standard MOS technology, and its low power consumption had made it possible to put those 1.2 million transistors on a chip. With 1.2 million transistors on a chip, Intel could sweep up other functions that had previously been put on other chips on a computer and incorporate them into the microprocessor. The cache, floating point unit, and memory management unit were all integrated onto the 486. Claims that the 486 was equivalent in power to a mainframe were an exaggeration, but what was true was that the 486 greatly decreased the difference in power between a mainframe and a personal computer. The 486 was initially priced at $950.[35]

In 1990, IBM introduced its latest bipolar-based mainframe computer, the ES/9000, appropriately code-named Summit, for it would represent the apogee of IBM's mainframe computer line. The system, models of which listed for as much as $22.8 million, had been in development since the early 1980s, when the MOS-based microprocessor would have seemed to pose no threat. In 1992, IBM engineers confidently asserted their belief that Summit's bipolar technology would "provide a sound basis for meeting the needs of high-end computing systems throughout the next decade."[36] But Summit brought IBM to a precipice from which there would be only a steep slide down. The power of MOS-based microprocessors meant that IBM's customers, even those with the most demanding applications, had other options to meet their computing needs. They could use engineering workstations, or specialized computers with many microprocessors working together. While these options were not painless and might involve incompatibilities with existing software or decreased performance, the cost premium the mainframe commanded (sometimes a factor of twenty over alternate approaches) gave companies that were struggling to cut costs in the early 1990s a great deal of incentive to consider these other options. While in the late 1970s, IBM's main problem with mainframes was in producing enough to satisfy demand, by the early 1990s, the price IBM could command for a mainframe was collapsing. In some cases IBM had to offer discounts of 50 percent off its list prices, price cutting that would have been inconceivable in earlier generations of mainframes. The mainframe as a concentrated center of computing power was not extinct, but the system IBM had developed to build and sell that computing power was.[37]

The mainframe had developed as a technology and as a business that generated profits that could support any costs. Mainframes seemed to be able to support a limitless overhead, whether it be the use of extremely complex semiconductor or packaging technologies, forty full-time salespeople assigned to one large account, or less competent employees kept on in almost make-work positions. With the increasing power of the MOS microprocessor, the mainframe could no longer support the top-heavy structure that IBM had placed on it. As the computer market converged, with the same basic MOS technology capable of producing both mainframes and personal computers, the price charged for a given level of performance had to converge across computer systems as well.[38]

In 1993 IBM cut off its line of bipolar-based mainframes, making the decision to build all its future computers using MOS technology. It was a wrenching decision made on the basis of IBM's need to dramatically cut costs to ensure its continued survival. The move to MOS technology for IBM's large systems was one critical part of its efforts to redefine the mainframe computer, now called a

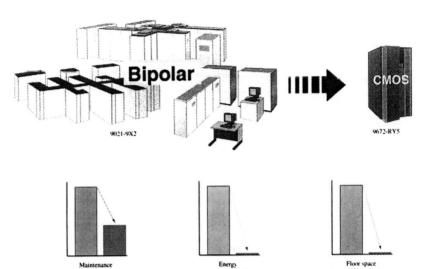

Measurement	9021-9X2	9672-RY5	Change (%)
Floor space (w/clearance) (ft²)	671.6	51.9	93
Power (kVA)	144	5.0	97
Weight (lb)	31857	2067	93

FIGURE C.2 • COMPARISON OF IBM'S LARGE COMPUTER SYSTEMS IN THE 1990S USING BI-POLAR AND CMOS SEMICONDUCTOR TECHNOLOGY. Source: G. S. Rao et al., "IBM S/390 Parallel Enterprise Servers G3 and G4," *IBM Journal of Research and Development* 41 (1997): 400. Used by permission.

server. Because of the much greater density of MOS technology, many fewer total integrated circuits were required for the entire system. The central processor was put on a single MOS chip rather than the 390 required using bipolar technology. Using MOS rather than bipolar technology, expensive and complex water cooling systems were no longer required. One is tempted to say that the bipolar mainframe—with its enormous size, weight, and power consumption—was a metaphor for IBM itself. But the link was more than metaphoric. In looking at the dramatic decreases in the number of integrated circuits used, the number of total parts used, and the physical size and weight of the overall system (all made possible by the enormous density advantage of MOS technology), one can almost see the people who were "surplused" (in IBM's infelicitous phrase) as IBM made the transition between the two machines (figure C.2 and table C.1).[39]

TABLE C.1 COMPARISON OF PHYSICAL
CHARACTERISTICS OF BIPOLAR AND
CMOS-BASED S/390 SYSTEMS

	ES/9000 9X2	S/390 G6
Technology	Bipolar	CMOS
Total number of chips	5,000	31
Total number of parts	6,659	92
Weight (lb)	31,145	2,057
Power requirement (kVA)	153	5.5
Chips per processor	390	1
Maximum memory (GB)	10	32
Space (sq ft)	671.6	51.9

NOTE: The CMOS-based system, the S/390 G6, was first shipped in 1999; the bipolar-based ES/9000 9X2 was first shipped in 1994. The CMOS system had twice the performance of the bipolar system.
SOURCE: R. D. Isaac, "The Future of CMOS Technology," *IBM Journal of Research and Development* 44 (2000): 376. Used by permission.

Although IBM had successfully navigated the transition from punched card data-processing machinery to electronic computers, and from vacuum tube–based computers to transistorized computers, the transition from the bipolar-based mainframe to the MOS server proved to be much more difficult. The differences between the technologies suggest why IBM management resisted making the change. While IBM had an almost unique ability to produce large quantities of the bipolar-based mainframe, with the complexity of the server confined to a relatively few CMOS chips, many more potential competitors existed, making profit margins much lower. Moreover, in the mid-1990s, while MOS offered performance close to that of the bipolar-based mainframe, bipolar-based mainframes still provided greater overall performance, and some customers still wanted these types of machines. IBM's switch from bipolar to MOS, made to cut costs immediately, was based on the long-term calculation that within a few years MOS technology would provide a performance advantage over bipolar for large systems. But by switching its systems over to MOS, IBM did something it would have done under only extreme duress: it essentially handed over its best customers to its archrival Hitachi, which continued to develop bipolar-based systems and took the customers who were looking for the highest level of performance possible.[40]

In the postmainframe era, IBM made fundamental changes in the company and in its semiconductor operations. It brought in new management, led by Louis Gerstner. To reduce costs, it laid off many workers in New York's Mid-

Hudson Valley, the home of IBM's bipolar semiconductor technology effort and its mainframe operations. It completely revamped its semiconductor operations. For the first time ever, IBM began making aggressive efforts to sell its semiconductors on the open market. No longer would its semiconductor operations be a mere servant of IBM's computer business. Now IBM could stand to profit directly from its semiconductor innovations, selling even to its competitors in a way it had not done before. Now it could make judgments about the optimal profits it could earn from each wafer it processed. As a result, IBM moved away from its emphasis on producing memory chips, important for mainframes but increasingly commodities dominated by Korean semiconductor manufacturers. IBM's implementation of copper interconnection technology for its MOS circuits marked a much greater degree of cooperation between Research and development than had existed in the 1960s, 1970s, or 1980s and also represented unprecedented speed in moving a new semiconductor technology into products. The name for IBM's new server microprocessor, the Alliance microprocessor, symbolized not an alliance of two foreign powers but, perhaps more radically for IBM, an alliance between IBM Research and the corporation's development organizations. IBM also greatly picked up the pace of technological change of its new MOS-based server and reduced its development time, planning for the introduction of a new system every year, sustaining a performance improvement of 40 percent to 50 percent per year, in contrast to the old regime, in which IBM introduced a new model every four to five years at an annual performance improvement rate of 20 percent per year.[41]

These fundamental changes in the way IBM operated addressed areas that had hampered the development of IBM's semiconductor and computer technology in the past. Considering the technological talent IBM had in its research, development, and manufacturing organizations, had these changes been implemented in the 1960s and 1970s, they would have made IBM an unstoppable competitor. But in the fiercely competitive environment of the 1990s, these changes could not restore the security IBM had enjoyed earlier. Those days were gone, swept away by forces that had been unleashed at IBM's research laboratory and other research laboratories around the country but that these laboratories and their corporations had been unable to control.

APPENDIX 1: ORGANIZATIONAL CHARTS

TABLE A.1 • IBM RESEARCH ORGANIZATION, 1963

IBM Director of Research

Assistant Director of Research
    Director of Mathematical Sciences
    Director of General Science
    Director of Solid State Sciences
    Director of Experimental Machines
    Director of Engineering Science
    Director of Experimental Systems
    Manager of San Jose Research Lab
    Manager of Watson Lab, Columbia University

NOTE: The LSI program initially included members from both the Experimental Machines group (which dealt with circuits and systems) and the Solid State Sciences group. Research later reorganized so the members of the program were in one group called Solid-State Electronics.

SOURCE: From IBM Organization Directory, 25 January 1963, Box TMB 106, IBM Technical History Project, IBM Archives, Somers, New York.

ORGANIZATION

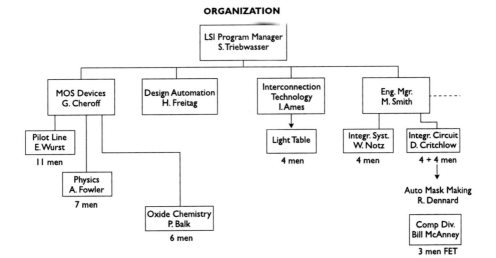

FIGURE A.1 • ORGANIZATION OF RESEARCH LARGE-SCALE INTEGRATION (LSI) PROGRAM, MAY 1966. From C. Kinberg and I. Ringstrom, "Review of LSI Program, Research, Yorktown Heights, Visit on May 5, 1966," Box TAR 243, IBM Technical History Project, IBM Archives, Somers, New York.

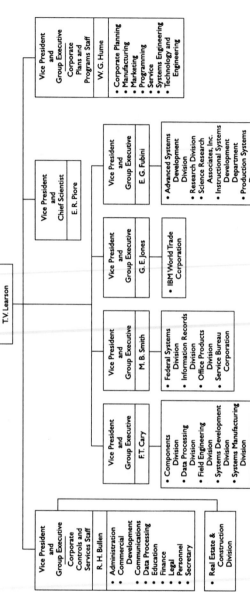

Board of Directors

Corporate Office
- Chairman of the Board — T.J. Watson, Jr.
- Chairman of the Executive Committee — A.L. Williams
- Vice Chairman of the Board — A.K. Watson
- President — T.V. Learson

• Organization

Management Review Committee
T.J. Watson, Jr.
A.L. Williams
A.K. Watson
T.V. Learson

Vice President and Group Executive — Corporate Controls and Services Staff — R. H. Bullen
- Administration
- Commercial Development
- Communications
- Data Processing
- Education
- Finance
- Legal
- Personnel
- Secretary
- Real Estate & Construction Division

Vice President and Group Executive — F.T. Cary
- Components Division
- Data Processing Division
- Field Engineering Division
- Systems Development Division
- Systems Manufacturing Division

Vice President and Group Executive — M. B. Smith
- Federal Systems Division
- Information Records Division
- Office Products Division
- Service Bureau Corporation

Vice President and Group Executive — G. E. Jones
- IBM World Trade Corporation

Vice President and Chief Scientist — E. R. Piore

Vice President and Group Executive — E. G. Fubini
- Advanced Systems Development Division
- Research Division
- Science Research Associates, Inc.
- Instructional Systems Development Department
- Production Systems Department

Vice President and Group Executive — Corporate Plans and Programs Staff — W. G. Hume
- Corporate Planning
- Manufacturing
- Marketing
- Programming
- Service
- Systems Engineering
- Technology and Engineering

312

FIGURE A.2 • IBM ORGANIZATION, APRIL 1966. Research was in the group headed by E. G. Fubini, while the Components Division, which had responsibility for semiconductor development and manufacturing, reported to F. T. Cary, the head of the Data Processing Group. The Data Processing Group also included the divisions responsible for the development, manufacturing, and marketing of IBM's large computer systems. A. R. DiPietro was on the corporate staf (far right). Redrawn from Emerson W. Pugh, Lyle R. Johnson, and John H. Palmer, *IBM's 360 and Early 370 Systems* (Cambridge, Mass.: MIT Press, 1991), 664. Used with permission.

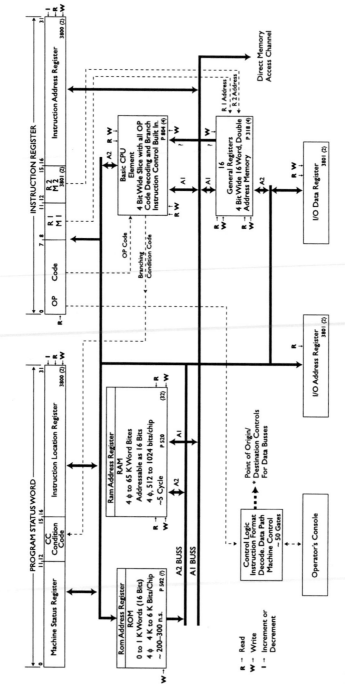

FIGURE A.3 • LEE BOYSEL'S MOS COMPUTER DESIGN MANIFESTO, SEPTEMBER 1967 — BOYSEL'S PLAN FOR A COMPUTER BUILT OUT OF MOS TECHNOLOGY. Each block represents an MOS part. The Fairchild part number is given in the right-hand corner ("P" indicates a part that is planned). Redrawn version of diagram provided by Lee Boysel.

Robert Noyce (Fairchild)
President

Gordon Moore (Fairchild R&D)
Executive Vice President

Ted Hoff (Stanford Research Institute)
Applications Research

Andrew Grove (Fairchild R&D)
Operations

Gene Flath (Fairchild manufacturing)
Manufacturing
Tom Rowe (Fairchild manufacturing)
Gary Hart (Fairchild manufacturing)
George Staudacher (Fairchild manufacturing)
George Chiu (Fairchild manufacturing)
Larry Brown, technician (Fairchild R&D)
Bob Holmstrom (Sprague)
Ted Jenkins (Fairchild R&D)
Gus Skousen (Signetics)

Les Vadasz (Fairchild R&D)
MOS Engineering
Joel Karp (Philco Microelectronics)

Dick Bohn (Sylvania)
Bipolar Engineering
H. T. Chua (Fairchild)
Tom Innes (Sylvania)

Desmond Fitzgerald (Fairchild R&D)
Quality Assurance

FIGURE A.4 • INTEL ENGINEERING PERSONNEL AND PREVIOUS EMPLOYMENT, CIRCA SPRING 1969. While Fairchild was the source of a large number of Intel's early employees, they came from both R&D and manufacturing, and a significant number of employees came from other firms. *Sources:* List of first fifty Intel employees provided by Intel Museum, Santa Clara, California; interviews with various Intel employees.

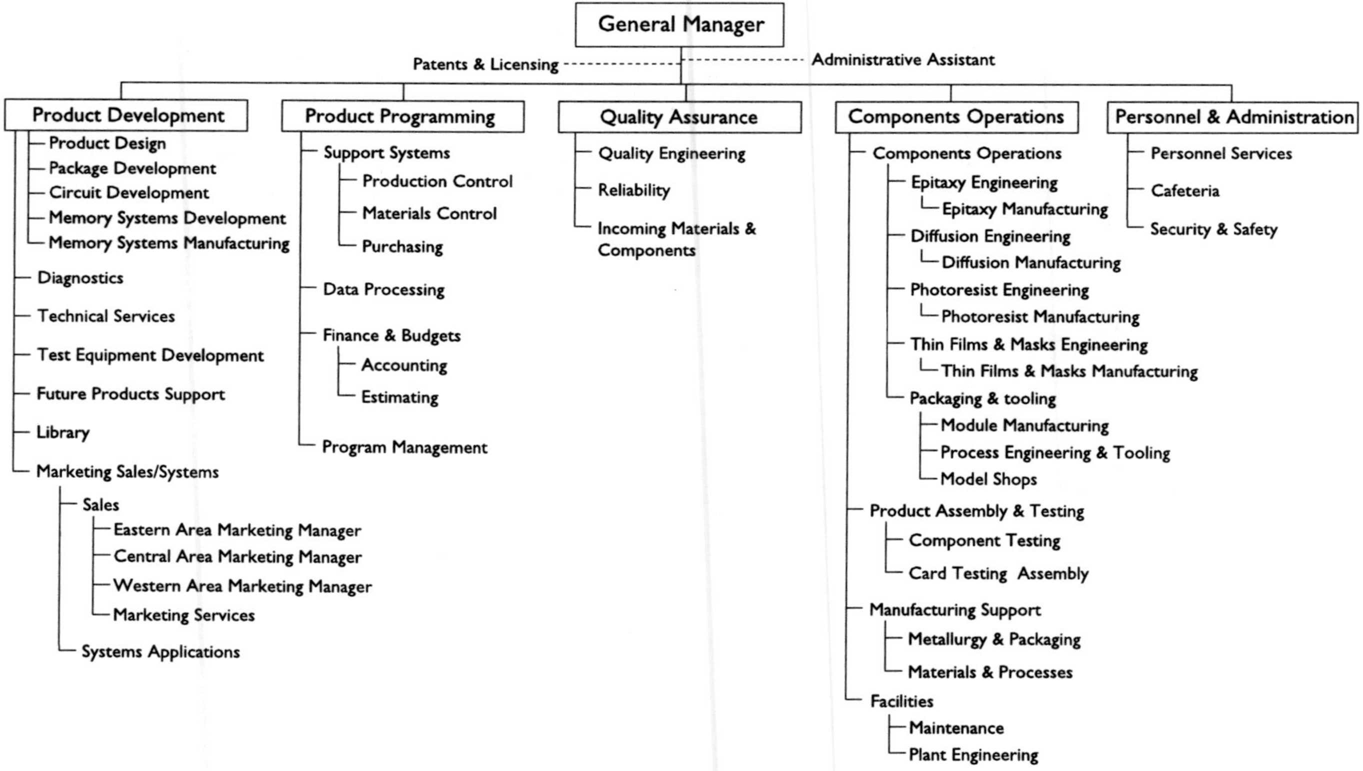

FIGURE A.5 · COGAR ORGANIZATION CHART, CIRCA SPRING 1972. Although Cogar was still a small start-up company making very little product, it had formed a very large and complex organization. In the Components Operations section there are pairs of similarly named departments (epitaxy engineering/epitaxy manufacturing, photoresist engineering/photoresist manufacturing, etc.). Each pair had one manager, with the idea that they would become two separate departments as the company grew. One can see in this chart the many functions that Cogar internalized, from the technically challenging, such as mask manufacturing or test equipment development, to the more trivial, such as library or cafeteria. The department marked "manufacturing support" had formerly been a research group, but as Cogar began to run short on funds and needed help in production, it was converted. Redrawn version of chart provided by Robert Meade.

APPENDIX 2: SOURCES FOR TABLES

Sources for tables 1.1 and 1.2 are below, grouped by the column to which they apply.

Position at Bell Labs: Bell Telephone Laboratories, *Organizational Directory, January 1960*, AT&T Archives, Warren, New Jersey.

IRE status: *IRE Directory, 1961*, IEEE Headquarters Library, Piscataway, New Jersey.

Year joined Bell Telephone Laboratories semiconductor effort: (biographical information in published articles) *Bell Telephone Laboratories Record,* September 1955, 329 [Theuerer]; *Proceedings of the IRE* 40 (November 1952): 1606 [Early], *Bell System Technical Journal* 41 (May 1962): 1172 [Lindner]; *Bell System Technical Journal* 40 (January 1961): 349 [Gummel]; *IRE Transactions on Electron Devices* (November 1962): 513 [Kleimack]; *IEEE Transactions on Electron Devices* 23 (July 1976): 787 [Kahng]; *Bell System Technical Journal* 34 (November 1955): 1330 [Ross]; *Bell System Technical Journal* 39 (July 1960): 1121 [Atalla]; *IEEE Transactions on Electron Devices* 24 (May 1977): 628 [Labate]. (For those I could not date this way, I used organizational and telephone directories for Bell Labs and established the date by the first year they were located in the transistor development or semiconductor research organization.)

Contributor, *Transistor Technology:* H. E. Bridgers, J. H. Scaff, and J. N. Shive, eds., *Transistor Technology* (Princeton, N.J.: D. Van Nostrand, 1958).

Participant, Bell Telephone Laboratories diffusion symposium: "Tentative Program for Symposium on Diffused Semiconductor Devices, January 16–17, 1956, Murray Hill, New Jersey," Location 85 10 01 02, AT&T Archives, Warren, New Jersey.

Participant, Solid-State Device Research Conference: J. Kurshan, ed., "Report on Semiconductor Device Research Conference, University of Illinois, June 19–20, 1952," RCA Princeton Engineering Memorandum 231, 9 July 1952, 1-3, RCA Collection, Hagley Museum and Library, Wilmington, Delaware (hereafter RCA-Hagley); J. Kurshan, ed., "Report on the Conference on Transistor Research, Pennsylvania State College, July 6–8, 1953," RCA Princeton Engineering Memorandum 325, 27 July 1953, 4–6, RCA-Hagley; J. Kurshan et al., "Report on the Conference on Semiconductor Device Research, Minneapolis, Minn., June 28–30, 1954," RCA Princeton Engineering Memorandum 452, 17 August 1954, RCA-Hagley; J. Kurshan, ed., "Report on the Conference on Semiconductor Device Research, Philadelphia, Pennsylvania, June 20–22, 1955," RCA Princeton Engineering Memorandum 565, 15 July 1955, RCA-Hagley; E. Johnson, ed., "Report on the Conference on Semiconductor Device Research, Purdue University, Lafayette, Indiana, June 25–27, 1956," RCA Princeton Engineering Memorandum 748, 17 August 1956, RCA-Hagley; E. O. Johnson, ed., "Report on the Conference on Semiconductor Device Research, University of Colorado, Boulder, Colorado, July 15–17, 1957," RCA Princeton Engineering Memorandum 925, 12 August 1957, RCA-Hagley; H. S. Sommers Jr. et al., "Report on Conference on Semiconductor Device Research, Ohio State University, June 18–20, 1958," RCA Princeton Engineering Memorandum 1186, 3 July 1958, RCA-Hagley; C. W. Mueller, ed., "Report on 1959 IRE-AIEE Solid State Device Research Conference, Cornell University, June 17–19, 1959," RCA Princeton Engineering Memorandum 1449, 8 July 1959, 3–8, RCA-Hagley.

NOTES

INTRODUCTION

1. Doug Andrey, Semiconductor Industry Association, conversation with the author, 19 June 2000. Microcontrollers are typically less powerful microprocessors with more function integrated on the chip, enabling them to function as single-chip computers. They are used in such applications as automobiles, cameras, and consumer electronics.

2. Bruce Deal and Jan Talbot, "Principia Moore," *Electrochemical Society Interface,* spring 1997, 19. In 1999 the Semiconductor Industry Association reported digital integrated electronics sales of $130 billion, with bipolar sales making up roughly $1 billion, the rest being MOS. Andrey, conversation. In 1965 Gordon Moore gave a paper on the MOS transistor at the IEEE Annual Convention. He called it an "interesting gadget," noting several possible applications for it, but concluded it would "not cut heavily into the general area presently occupied by transistors." Gordon E. Moore, "The MOS Transistor as an Individual Device and in Integrated Arrays," *1965 IEEE International Convention Record,* Part 5, 48. The triumph of the MOS transistor has been most complete in digital electronics, the largest market for semiconductors. The bipolar transistor is still heavily used for analog applications.

3. In asserting the simplicity and scalability of the MOS transistor, two important qualifications have to be made. The MOS transistor was simpler than the bipolar transistor, but as it developed over time, it became an extraordinarily complex technology in its own right. While the MOS transistor scaled, that scaling was far from simple; almost every new generation of MOS technology required a host of innovations to make this scaling possible. The complexity of an MOS process from the mid-1990s can be see in C. W. Koburger III et al., "A Half-Micron CMOS Logic Generation," *IBM Journal of Research and Development* 39 (January/March 1995): 215–27.

4. Intel's most recent MOS process is described in Intel Corporation, *Annual Report, 2000,* 2. The gate dimension from the RCA transistor is from C. W. Mueller and K. H. Zaininger, "MOS-Unipolar," 14 December 1960 (in author's possession). I thank Charles Mueller for sending me this document.

5. This fundamental continuity can be seen in Intel's four chief executive officers, all Ph.D.-level semiconductor technologists.

6. When I queried Gordon Moore about his (and Intel's) response to technologies such

as Josephson junctions, gallium arsenide for digital logic, high electron mobility transistors, and ballistic transistors, Moore, in his typically understated way, said that he "had developed fairly strong ideas about what made sense and what didn't." He considered that none of these technologies had advantages for making the complex digital circuits that Intel concentrated on. He said "it was easy to focus." Gordon Moore, note to author, 21 April 2001.

7. The consulting firm Integrated Circuit Engineering estimated Fairchild's 1985 MOS sales at $25 million, accounting for roughly 5 percent of Fairchild's total integrated circuit sales. This placed it in twenty-seventh place in MOS sales among the fifty companies tracked. Integrated Circuit Engineering, *Status 1987: A Report on the Integrated Circuit Industry* (Scottsdale, Ariz.: Integrated Circuit Engineering: 1987), 2-11–2-12. Fairchild Semiconductor was acquired by National Semiconductor in 1987, and in 1997 National spun off a group of assets, including some associated with the original Fairchild, under the name Fairchild Semiconductor. Although there are some continuities between the old Fairchild and the new Fairchild (the new Fairchild is based in Portland, Maine, at one of the old Fairchild sites), this is more a case of using an existing brand name to jump-start a new enterprise—one cannot consider them the same company with a continuous history. Richard B. Schmitt, "Schlumberger Reaches Accord to Sell Fairchild—Agreement with National Semiconductor Corp. Valued at $122 Million," *Wall Street Journal*, 1 September 1987, 1; Crista Hardie, "National to Spin Off Logic," *Electronic News*, 24 June 1996, 1; Crista Hardie, "Fairchild Banner Flies Once Again in Maine," *Electronic News*, 17 March 1997, 14.

8. John Hoerr, "System Crash," *American Prospect*, 1 December 1994, 68–77; Gene Bylinsky, "America's Best-Managed Factories," *Fortune*, 28 May 1984, 16–24. The story of the ordeal of IBM, culminating in the disastrous early 1990s, has not been fully told. The best outline of it so far has been given by Charles H. Ferguson and Charles R. Morris in *Computer Wars: The Fall of IBM and the Future of Global Technology* (New York: Times Books, 1994). For a statement of IBM's cessation of bipolar work, see John Armstrong, "Reinventing Research at IBM," in *Engines of Innovation: U.S. Industrial Research at the End of an Era*, ed. Richard S. Rosenbloom and William J. Spencer (Boston: Harvard Business School Press, 1996), 152. In 2000 IBM announced plans to restart large-scale semiconductor production at East Fishkill, although it would employ far less people than it had in its heyday. Barnaby J. Feder, "Courted by State, I.B.M. Plans Huge Chip Factory," *New York Times*, 11 October 2000, B1. In light of its subsequent fate, one might question the validity of *Fortune*'s 1984 assessment. Bylinsky did not recognize that what he saw (and praised) was the reconfigured remnants of the failed effort to fully automate the process of semiconductor production. The FMS (Future Manufacturing System) and its demise are described in William MacGeorge, interview by author, 30 January 1996.

9. Clayton M. Christensen, *The Innovator's Dilemma: When New Technologies Cause Great Firms to Fail* (Boston: Harvard Business School Press, 1997). For further discussion of technological trajectories, see Giovanni Dosi, "Technological Paradigms and Technological Trajectories: A Suggested Interpretation of the Determinants and Directions of Technical Change," *Research Policy* 11 (1982): 147–62; Richard R. Nelson and Sidney G. Winter, *An Evolutionary Theory of Economic Change* (Cambridge, Mass.: Harvard University Press, 1982), 255–62; and Donald MacKenzie, *Knowing Machines: Essays on Technical Change* (Cambridge,

Mass.: MIT Press, 1996), 54–59. Gordon Moore's article suggesting the trajectory that became known as Moore's Law was published as "Cramming More Components onto Integrated Circuits," *Electronics*, 19 April 1965, 114–17. Christensen's work is consistent with a large body of writing among historians of technology over the last two decades which insists that technology develops through the interactions of a variety of social groups. The best introduction to this work is Wiebe E. Bijker, Thomas P. Hughes, and Trevor Pinch, eds., *The Social Construction of Technological Systems: New Directions in the Sociology and History of Technology* (Cambridge, Mass.: MIT Press, 1987). See also Donald MacKenzie and Judy Wajcman, eds., *The Social Shaping of Technology*, 2d ed. (Buckingham, England: Open University Press, 1999), 1–27.

10. Christensen, *Innovator's Dilemma*, xv.

11. Quoted in R. M. Warner Jr. and B. L. Grung, *Transistors: Fundamentals for the Integrated Circuit Engineer* (New York: Wiley, 1983), 61.

12. Integrated Circuit Engineering Corporation, *Status 1979: A Report on the Integrated Circuit Industry* (Scottsdale, Ariz.: Integrated Circuit Engineering, 1979), 5–2. This book focuses on American MOS work, although by the late 1970s, Japanese semiconductor manufacturers had become a major force in MOS technology.

13. The author worked for IBM in the 1980s as an electrical engineer.

14. For an introduction to the industrial system of Silicon Valley, see AnnaLee Saxenian, *Regional Advantage: Culture and Competition in Silicon Valley and Route 128* (Cambridge, Mass.: Harvard University Press, 1994). Silicon Valley has been the subject of a number of recent works, including Martin Kenney, ed., *Understanding Silicon Valley: The Anatomy of an Entrepreneurial Region* (Stanford, Calif.: Stanford University Press, 2000); and Chong-Moon Lee, William F. Miller, Marguerite Gong Hancock, and Henry S. Rowen, eds., *The Silicon Valley Edge: A Habitat for Innovation and Entrepreneurship* (Stanford, Calif.: Stanford University Press, 2000).

ONE · HOW A BAD IDEA BECAME GOOD (TO SOME)

1. Michael Riordan and Lillian Hoddeson, *Crystal Fire: The Birth of the Information Age* (New York: W.W. Norton, 1997), 70–141; Lillian Hartmann Hoddeson, "The Entry of the Quantum Theory of Solids into the Bell Telephone Laboratories, 1925–40: A Case-Study of the Industrial Application of Fundamental Science," *Minerva* 18 (1980): 422–47; Lillian Hoddeson, "Research on Crystal Rectifiers during World War II and the Invention of the Transistor," *History and Technology* 11 (1994): 121–30; Lillian Hoddeson, "The Discovery of the Point-Contact Transistor," *Historical Studies in the Physical Sciences* 12 (1981): 41–76.

2. Riordan and Hoddeson, *Crystal Fire*, 110–28; William Shockley, "The Path to the Conception of the Junction Transistor," *IEEE Transactions on Electron Devices* 23 (July 1976): 604–6; Hoddeson, "Discovery of the Point-Contact Transistor," 62–65. The latter article, based in part on the laboratory notebooks of all the key participants, provides the greatest technical detail on the work leading to the invention of the point-contact transistor. Bardeen's work on surface states was published as John Bardeen, "Surface States and Rectification at a Metal Semiconductor Contact," *Physical Review* 71 (15 May 1947): 717–27.

3. In early December, at Bardeen's suggestion, Brattain switched from using silicon to

using germanium, which, unlike silicon, has a water-soluble oxide. Riordan and Hoddeson, *Crystal Fire*, 128–41; Hoddeson, "Discovery of the Point-Contact Transistor," 64–76; Shockley, "Path to the Conception," 605–12. For Bell's original publication on the transistor, see J. Bardeen and W. H. Brattain, "The Transistor, A Semi-Conductor Triode," *Physical Review* 74 (15 July 1948): 230–1.

4. Cat's whisker crystal detectors, formed originally by probing a piece of the semiconductor galena with a sharp piece of metal, had been widely used by radio hams but had been largely displaced by vacuum tubes in the 1930s. During World War II crystal rectifiers were used to detect high-frequency signals used in radar. Riordan and Hoddeson, *Crystal Fire*, 19–21, 99–105. On the construction of cat's whisker crystal rectifiers during World War II, see Henry C. Torrey and Charles A. Whitmer, *Crystal Rectifiers*, MIT Radiation Laboratory Series (New York: McGraw-Hill, 1948), 15: 301–22. For further details on the problems of the point-contact transistor, see Riordan and Hoddeson, *Crystal Fire*, 168–70, 180–81. For one example of users' struggles with the point-contact transistor, see Charles J. Bashe, Lyle R. Johnson, John H. Palmer, and Emerson W. Pugh, *IBM's Early Computers* (Cambridge, Mass.: MIT Press, 1986), 374–83.

5. William Shockley, "The Theory of p-n Junctions in Semiconductors and p-n Junction Transistors," *Bell System Technical Journal* 28 (July 1949): 474. Shockley's work leading to his idea of a junction transistor is described in Shockley, "Path to the Conception," 612–16, and Riordan and Hoddeson, *Crystal Fire*, 143–55. The elemental semiconductors germanium and silicon are from column IV of the periodic table—they have four valence electrons. Adding a small amount of material from column V, such as phosphorus, results in a crystalline structure where the phosphorus atoms contribute an electron that is very loosely bound and relatively free to move about the crystal. Such material is called n type. Adding a small amount of material from column III, such as boron, to a crystal of germanium or silicon results in a crystal structure where the boron contributes only three electrons to the crystal. The electron vacancy at the boron site may be supplied by a neighboring atom. This mobile vacancy, a hole, is the virtual counterpart to the electron. Semiconductor material that is doped with material from column III is called p type. These concepts are nicely explained in William Shockley's classic text *Electrons and Holes in Semiconductors* (Princeton, N.J.: D. Van Nostrand, 1950), 13–14. Thomas Misa has aptly described the junction transistor as a semiconductor "sandwich" consisting of three layers: the emitter; the base, on the inside; and the collector. The emitter and the collector are made from the same type of doped semiconductor material (n or p), while the base is made from the opposite type (p or n). Thomas J. Misa, "Military Needs, Commercial Realities and the Development of the Transistor," in *Military Enterprise and Technological Change*, ed. Merritt Roe Smith (Cambridge, Mass.: MIT Press, 1985), 269–70.

6. Riordan and Hoddeson, *Crystal Fire*, 177–90; Gordon K. Teal, "Single Crystals of Germanium and Silicon—Basic to the Transistor and Integrated Circuit," *IEEE Transactions on Electron Devices* 23 (July 1976): 621–34; Andrew Goldstein, "Finding the Right Material: Gordon Teal as Inventor and Manager," in *Sparks of Genius: Portraits of Engineering Excellence*, ed. Frederik Nebeker (Piscataway, N.J.: IEEE Press, 1993), 93–126.

7. J. A. Morton, untitled document, 5, Location 85 10 01 01, AT&T Archives, Warren,

New Jersey (hereafter AT&T Archives). From internal evidence, this document was written before October 1948.

8. Mervin J. Kelly, "The Bell Telephone Laboratories—An Example of an Institute of Creative Technology," *Proceedings of the Royal Society* 203A (10 October 1950): 292. More details on Bell Labs' transistor development effort are provided in Riordan and Hoddeson, *Crystal Fire*, 168–224; Misa, "Military Needs," 254–87.

9. A 1965 sketch of Morton said: "He is a strong personality—outspoken, articulate and persuasive. He is never neutral, and can never be ignored." "Jack Morton," 25 August 1965, Location 11 08 03 01, AT&T Archives. James Early, who joined Bell Labs in 1951, described Morton's microwave triode as "an astounding accomplishment." James Early, interview by author, 13 January 1995. Morton was murdered. "Find Morton Dead in Burning Car," *Electronic News*, 20 December 1971, 10. Further biographical details on Morton are provided in Riordan and Hoddeson, *Crystal Fire*, 169, and his obituary, "Jack A. Morton, Bell Labs Vice President, (1913–1971)," *Bell Laboratories Record* 50 (January 1972): 29.

10. Morton, untitled document, Location 85 10 01 01, AT&T Archives; J. A. Morton, "Principles of Device Development," in *Transistor Technology*, vol. I, ed. H. E. Bridgers, J. H. Scaff, and J. N. Shive (Princeton, N.J.: D. Van Nostrand, 1958), 175. The original *Transistor Technology* was published by Bell Labs in 1952 at the request of the military to spread Bell Labs' knowledge of the transistor throughout the industry. It was republished commercially and expanded in 1958. *Transistor Technology*, I, v. Morton's longtime interest in the organization of the R&D process can be seen in Jack A. Morton, "From Research to Technology," *International Science and Technology*, May 1964, 82–92.

11. W. Shockley, M. Sparks, and G. K. Teal, "p-n Junction Transistors," *Physical Review* 83 (1 July 1951): 151–62; J. M. Early, "Effect of Space-Charge Layer Widening in Junction Transistors," *Proceedings of the Institute of Radio Engineers* 40 (November 1952): 1401–6; J. M. Early, "Design Theory of Junction Transistors," *Bell System Technical Journal* 32 (November 1953): 1271–1312; J. J. Ebers and J. L. Moll, "Large-Signal Behavior of Junction Transistors," *Proceedings of the Institute of Radio Engineers* 42 (December 1954): 1761–72. For a statement of the incompleteness of the point-contact transistor design theory, see R. M. Ryder, "A Descriptive Summary of the Design Theory of Transistors," in Bridgers, Scaff, and Shive, *Transistor Technology*, I, 225–34.

12. A. E. Anderson and R. M. Ryder, "Development History of the Transistor in Bell Laboratories and Western Electric, 1947–1975," vol. I, General Narrative, 47–48, 62–65, Location 85 10 02 02, AT&T Archives; Riordan and Hoddeson, *Crystal Fire*, 218–22.

13. In 1958 Bell Labs executive vice president James B. Fisk told *Fortune:* "We can't keep the Bell System in a constant uproar for gains of a few percent." Francis Bello, "Tomorrow's Telephone System," *Fortune*, December 1958, 118–19. For a general perspective on the economic problems of new technology in the telecommunications environment, see Nathan Rosenberg, "Telecommunications: Complex, Uncertain, and Path-dependent," in *Exploring the Black Box: Technology, Economics, and History* (Cambridge: Cambridge University Press, 1994), 203–31.

14. J. A. Morton, "Bell System Transistor Program," Hershey Conference, 23–26 September 1958, 1–5, Location 58 01 01 02, AT&T Archives; A. Eugene Anderson, "Transistor

Technology Evolution I: Point-Contact to Diffused Transistors," *Western Electric Engineer*, July 1959, 3–12.

15. John Brooks, in his authorized history of the Bell System, called ESS No. 1 "the greatest development effort in Bell System's history," requiring four thousand man-years of effort and development costs of $500 million. John Brooks, *Telephone: The First Hundred Years* (New York: Harper & Row, 1976), 278–79. A complete history of ESS No. 1 is provided by A. E. Joel et al., *A History of Engineering and Science in the Bell System: Switching Technology (1925–1975)* (n.p.: Bell Telephone Laboratories, 1982), 199–261. P. A. Gorman, Bell System Presidents' Conference, 11–14 May 1964, 3, Location 97 05 01 02, AT&T Archives. Gorman's exact statement shows the temporal and financial scale that the Bell System operated on: "By 2000 A.D., when it's anticipated our whole network will have been converted to electronic switching, our gross investment in ESS will amount to more than $12.5 billion."

16. Anderson and Ryder, "Development History of the Transistor," I, 47–51; Riordan and Hoddeson, *Crystal Fire*, 220–23; John Moll, interview by author, 11 January 1995. Anderson and Ryder note that although germanium was better understood and more tractable, the high temperature advantages of silicon were understood, and some suspected that the silicon surface would be more stable than germanium.

17. Morton, "Bell System Transistor Program," 5. Anderson and Ryder, "Development History of the Transistor," I, 47–51. The three brought from research were Ian Ross, George Dacey, and John Hornbeck. The organization chart of Bell's transistor development effort as of July 1, 1955, is given in an unpaginated appendix in Anderson and Ryder, "Development History of the Transistor." Their earlier positions in research are shown in Bell Telephone Laboratories, Incorporated, *Telephone Directory*, January 1954, 64, AT&T Archives.

18. Transistor technology's evolution, A. Eugene Anderson claimed in 1959, "has entered a plateau wherein widespread applications can now be made and large-scale, low-cost manufacture predicted with confidence." In a later article, Anderson analyzed the costs of various transistor technologies, showing that silicon diffused transistors had significant cost advantages over other types. He called these results "comforting," "particularly to an organization so committed [to silicon diffused technology]." Anderson, "Transistor Technology Evolution—I," 4; A. Eugene Anderson, "Transistor Technology Evolution—III: The Future in Terms of Costs," *Western Electric Engineer*, January 1960, 15. Riordan and Hoddeson quote Morton on the maturing of semiconductor technology in *Crystal Fire*, 254.

19. Anderson and Ryder, "Development History of the Transistor," I, 49. Anderson and Ryder note that Morton also dismissed work on the transitional alloy junction technique as "diddling." Texas Instruments' work, led by Gordon Teal, is described in Teal, "Single Crystals of Germanium and Silicon," 634–35, and Riordan and Hoddeson, *Crystal Fire*, 206–9.

20. Bridgers, Scaff, and Shive, *Transistor Technology*, I, 382.

21. Richard Petritz, "Contributions of Materials Technology to Semiconductor Devices," *Proceedings of the IRE* 50 (May 1962): 1030; James Early, "Semiconductor Devices," *Proceedings of the IRE* 50 (May 1962): 1007. Early stated: "Control and understanding of the structures and behavior of the junction surfaces lagged far behind the advances in bulk properties. Even today most changes in the behavior of diodes and transistors during normal use result from changes at the junction surfaces."

22. M. M. Atalla, interview by author, 17 October 1994. Atalla's background is described

in the biographical sketches in *Bell System Technical Journal* 32 (September 1953): 1267 and *Bell System Technical Journal* 38 (May 1959): 905.

23. Frosch and Derick's work is described in Ian M. Ross, "Invention of the Transistor," *Proceedings of the IEEE* 86 (January 1998): 19–20; Anderson and Ryder, "Development History of the Transistor," I, 75–77; Riordan and Hoddeson, *Crystal Fire*, 221–22; C. J. Frosch and L. Derick, "Surface Protection and Selective Masking during Diffusion in Silicon," *Journal of the Electrochemical Society* 104 (September 1957): 547–52. M. M. Atalla, E. Tannenbaum, and E. J. Scheibner, "Stabilization of Silicon Surfaces by Thermally Grown Oxides," *Bell System Technical Journal* 38 (May 1959): 749–83. Atalla's work was partially funded by the Army under a general program to improve transistor reliability.

24. Morton, "Bell System Transistor Program," 19–20. To say that Bell saw potential in Atalla's work is not to say that the work was implemented on a production basis or that plans had been made to do this. Bell Labs engineers believed that an oxide that had been exposed to a diffusion process was contaminated and had to be removed. Also, covering the surface with oxide raised compatibility problems with Bell Labs' existing process for making transistors. Anderson and Ryder, "Development History of the Transistor," I, 100–101; J. Earl Thomas, interview by author, 5 October 1995; Charles A. Bittman, interview by author, 14 October 1994.

25. Morton, "Bell System Transistor Program," 19; H. S. Sommers Jr. et al., "Report on Conference on Semiconductor Device Research Ohio State University, June 18–20, 1958," Princeton Engineering Memorandum 1186, 11, 3 July 1958, RCA Collection, Hagley Museum and Library, Wilmington, Delaware (hereafter RCA-Hagley).

26. D. Kahng, "A Historical Perspective on the Development of MOS Transistors and Related Devices," *IEEE Transactions on Electron Devices* 23 (July 1976): 656, credits Atalla with the idea for the basic structure. Kahng's laboratory notebook credits Atalla and Richard Lindner, an electrical engineer in the group, with the idea. Dawon Kahng, Laboratory Notebook 32560, 3, AT&T Archives. I have not found anything particularly revealing in Atalla's notebooks in the AT&T Archives. Ernest Labate, a technician in the group, in 1957 conceived of a device with striking similarities to what Atalla later proposed, although Labate's device did not explicitly use silicon dioxide as an insulator. Ernest E. Labate, "Proposal," 8 March 1957, laboratory notebook page (in author's possession).

27. Atalla says he assigned Kahng this task. Atalla, interview. Dawon Kahng, Laboratory Notebook 32560, 3–13, AT&T Archives; D. Kahng, "Silicon-Silicon Dioxide Surface Device," Technical Memorandum MM61-2821-2, 22, 16 January 1961, AT&T Archives.

28. These three conferences have had some minor name variations over the years, but they continue as the Device Research Conference, the International Electron Device Meeting (IEDM), and the International Solid-State Circuits Conference (ISSCC), three of the leading solid-state device conferences in the United States. Misa, "Military Needs," 267–68, discusses Bell Labs' efforts, with the help of the military, to diffuse transistor technology throughout the industry. A description of Bell Labs' role in the diffusion of semiconductor technology by a Bell Labs manager is I. M. Ross, "BTL Role in the Growth of the Semiconductor Industry," Location 85 10 01 02, AT&T Archives.

29. J. Kurshan, ed., "Report on the Conference on Transistor Research, Pennsylvania State College, July 6–8, 1953," RCA Princeton Engineering Memorandum 325, 2, 27 July

1953, RCA-Hagley. The history of the Electron Tube Conference and its successors is described in Charles Susskind, "Conference on Electron Device Research: A Pattern To Be Copied," *IEEE Spectrum,* December 1967, 100–102; Leon S. Nergaard to H. C. Casey, 7 June 1982, IEEE Electron Device Society Records, IEEE Headquarters, Piscataway, New Jersey.

30. Ian M. Ross discusses the SSDRC in "Invention of the Transistor," 26. In his report on the 1960 conference, the chair stated that representatives of the press asked to attend. They were permitted but accepted restrictions on what they could report. A. L. McWhorter to R. L. Pritchard, 9 October 1960, IEEE Electron Device Society Records, IEEE Headquarters, Piscataway, New Jersey. The press reported on the conference as if they had not attended, calling it a "closed-door" meeting. Carl J. Kovac, "Many New-Type Semicons Studied," *Electronic News,* 20 June 1960, 1, 32.

31. The operation of the conference is described in Kurshan, "Report on the Conference on Transistor Research," 2–3; E. Johnson, ed., "Report on the Conference on Semiconductor Device Research, Purdue University, Lafayette, Indiana, June 25–27, 1956," Princeton Engineering Memorandum 748, 17 August 1956, RCA-Hagley; K. H. Zaininger, "What the Competition Knows about the MOS-Structure," Princeton Engineering Memorandum 2483, 1–2, 12 August 1964, RCA-Hagley.

32. Early, "Design Theory," 1271–1312; J. M. Early, "P-N-I-P and N-P-I-N Junction Transistor Triodes," *Bell System Technical Journal* 33 (May 1954): 517–33; Early, interview. While the Army Signal Corps supported further work on the device, it did not represent one of Bell Labs' prime focuses.

33. Early, interview; H. Christensen and Gordon Teal, "Method of Fabricating Germanium Bodies," U.S. Patent 2,692,839, 26 October 1954.

34. Ross, "Invention of the Transistor," 18–19; Ian Ross, interview by author, 17 September 1996; Early, interview; H. C. Theuerer, J. J. Kleimack, H. H. Loar, and H. Christensen, "Epitaxial Diffused Transistors," *Proceedings of the IRE* 48 (September 1960): 1642–43.

35. N. J. Herbert to J. F. Paulson, 16 May 1960, Filecase 38589–51, AT&T Archives.

36. N. J. Herbert, "Introduction of Epitaxial Silicon Transistors in Western Production, Meeting No. 1," 1 November 1960, Filecase 38589–51, AT&T Archives.

37. "Epitaxial Transistor Cuts Transistor Switching Time," *Electronics,* 24 June 1960, 11; Kovac, "Many New-Type Semicons Studied," 1, 32; "Program of Meeting" [1960 Electron Device], *IRE Transactions on Electron Devices* 8 (March 1961): 174–83; Walter Johnson, "New Microwave Devices, Methods Cited at Parley," *Electronic News,* 31 October 1960, 8; "Heavy Orders Delay Deliveries of Motorola Ge, Si Epitaxials," *Electronic News,* 2 January 1961, 20; Jack Robertson, "TI Confirms Introduction of 2 Si Epitaxials," *Electronic News,* 6 February 1961, 31; George M. Drake, "Fairchild Sampling Industry with Si Epitaxial Transistor," *Electronic News,* 13 February 1961, 15; "Parley to Stress Epitaxials," *Electronic News,* 13 February 1961, 1, 16. Fairchild Semiconductor had conceived of an epitaxial transistor prior to Bell Labs' announcement. Gordon E. Moore, "The Role of Fairchild in Silicon Technology in the Early Days of 'Silicon Valley,'" *Proceedings of the IEEE* 86 (January 1998): 60; Gordon Moore, interview by author, 15 May 1996.

38. R. W. Ketchledge, "ESS No. 1—Results of a Study of Epitaxial Diffused Transistors," 22 July 1960, Filecase 36279–1, AT&T Archives.

39. The speed comparison is from Anderson and Ryder, "Development History of the Transistor," I, 170.

40. Ian M. Ross and Eugene D. Reed, "Functional Devices," in *Microelectronics: Theory, Design, and Fabrication,* ed. Edward Keonjian (New York: McGraw-Hill, 1963), 374. Further details on Bell's response to integrated circuits can be found in Ross, "Invention of the Transistor," 23; Ross, interview; R. M. Warner Jr. and B. L. Grung, *Transistors: Fundamentals for the Integrated Circuit Engineer* (New York: Wiley, 1983), 58–59. For a statement of Bell's lagging position in integrated circuits in 1962, see J. M. Goldey and J. H. Forster, "IRE 4.10 Meeting, August 23 and 24, 1962, Seattle, Washington," 5 October 1962, Casefile 38589–118, AT&T Archives.

41. Ross, "Invention of the Transistor," 21, and Anderson and Ryder, "Development History of the Transistor," IV, 31–74, discuss the embarrassment of transistor reliability problems. A 1965 evaluation by engineers at Motorola stated: "The IGFET [MOS transistor] seeks to exploit surface effects which have plagued the transistor since its inception and which, for the most part, have been 'solved' through circumvention." (The Motorola engineers referred to the MOS transistor as an isolated-gate field-effect transistor.) Raymond M. Warner Jr., ed., *Integrated Circuits: Design Principles and Fabrication* (New York: McGraw-Hill, 1965), 229.

42. Anderson and Ryder, "Development History of the Transistor," I, 171–72. Ian Ross, in an article written to mark the transistor's fiftieth anniversary, gives an appraisal of Kahng and Atalla's work that sounds like it reflects the attitudes of experienced Bell semiconductor engineers at the time, asserting: "The field effect having been verified again, the concept had its moment of glory and returned to obscurity," Ross, "Invention of the Transistor," 21.

43. James Early, who had the good fortune never to work on the point-contact transistor, noted that it hurt the careers of many who worked on it. Early, interview. One further measure of the difference between the epitaxial transistor work and Atalla and Kahng's device is that in March 1963, Ian Ross of Bell Labs received the IEEE Morris Liebmann Award, in part for his work on the epitaxial transistor. This was just one month after Atalla and Kahng's device was first made known to the engineering public, through RCA's announcement of its work on such a device. By 1972 the IEEE still had not given Atalla and Kahng a formal award for their work, although that year the Awards Committee recommended they be recognized. IEEE Electron Device Group, Minutes of the Ad Com, 4 December 1972, IEEE Electron Device Society Records, IEEE Headquarters, Piscataway, New Jersey. A. E. Anderson stated that "Atalla was extremely ambitious and would not have chosen to follow in others' footsteps." A. E. Anderson, note to author, 16 May 1995; Frank Orpe, "Man Behind the News," *Electronic News,* 27 March 1963, 63; Joe McLean, "Use as Digital Circuit Seen for New Transistor," *Electronic News,* 11 February 1963, 1.

The very different, but very accomplished careers of Ross and Atalla suggest the following proposition: An engineer in a managerial position at Bell Labs in the late 1950s and early 1960s who is conceiving or working on something like an MOS transistor is not following a career path that will lead to the presidency of Bell Labs; one conceiving and pushing work on an epitaxial transistor might be. Ross spent his whole career within the Bell System, serving as president of Bell Labs from 1979 to 1991. Atalla left Bell Labs in 1962 to join hp associates, the semiconductor affiliate of Hewlett-Packard; joined Fairchild Semiconductor in 1969 to head its microwave semiconductor division; and then left in 1972 to form a start-up working on data security for electronic banking (at the time, the typical thing to do would have been to found a semiconductor company). In the 1990s Atalla founded Tri-Strata, an Internet data security company. For a biographical sketch of Ross, see "Invention of the Transistor," 28. Atalla's career can be followed through Don C. Hoefler, "Atalla Will Head New Fairchild Microwave Div.," *Electronic News*, 28 April 1969, 2; "The Antenna," *Electronic News*, 24 January 1972, 10; "Atalla Resigns from Fairchild," *Electronic News*, 31 January 1972, 2; "The Antenna," *Electronic News*, 28 February 1972, 18; Randall E. Stross, "When Money Goes Down the Drain," *Fortune*, 17 April 2000, 535–48; Atalla, interview.

44. All work done at Bell Labs was done under a case number, which was an administrative and accounting tool. A case authorized work in some general area. A later memo in this file, from June 1966, noted that Bell Labs' Allentown facility was just starting a program to fabricate MOS transistors. G. T. Cheney, "Proposal to Offer an Out-of-Hours Course in Field-Effect Transistors," 23 June 1966, Filecase 38589–35. Atalla and Kahng next did pioneering work on hot carrier devices, which used what is now called a Schottky barrier. Atalla said that he considered this work an extension of his MOS work. Atalla, interview; M. M. Atalla and D. Kahng, "A New 'Hot Electron' Triode Structure with Semiconductor-Metal Emitter," *IRE Transactions on Electron Devices* 9 (November 1962): 507–8.

45. Atalla, interview. Atalla and Kahng filed for and received patents on two alternative versions of their device. This move had to have been supported by Bell Labs management. Atalla and Kahng were unable to patent the MOS transistor as they described it because of previous patents by Lilienfeld and Bardeen. J. E. Lilienfeld, "Method and Apparatus for Controlling Electric Currents," U.S. Patent 1,745,175, 28 January 1930; J. E. Lilienfeld, "Device for Controlling Electric Current," U.S. Patent 1,900,018, 7 March 1933; J. Bardeen, "Three-Electrode Circuit Element Utilizing Semiconductive Materials," U.S. Patent 2,524,033, 3 October 1950. In a 1973 paper Shockley identifies the Bardeen patent as covering the MOS transistor. William Shockley, "The Invention of the Transistor: An Example of Creative-Failure Methodology," *National Bureau of Standards Special Publication 388: Proceedings of Conference on the Public Need and the Role of the Inventor (11–14 June 1973), Monterey, Calif.,* 50. Gordon Moore observed that the patents Atalla and Kahng filed were different from the paper they gave at the SSDRC. Moore, interview by author, 15 May 1996.

46. McWhorter to Pritchard, 9 October 1960; Early, "Semiconductor Devices," 1006–10; Petritz, "Contributions of Materials Technology," 1025–38. The programs for the conferences are given in "Program of Meeting [1960 Electron Device]," *IRE Transactions on Electron Devices* 8 (March 1961): 174–83; "Program of Meeting [1961 Electron Device]," *IRE Transactions on Electron Devices* 9 (January 1962): 109–19; "Solid-State Device Research Conference, Stanford University, Stanford, Calif.," *IRE Transactions on Electron Devices* 8 (September 1961): 420–30; "Solid-State Device Research Conference, July 9–11, 1962, Durham, New Hampshire," *IRE Transactions on Electron Devices* 9 (November 1962): 503–12.

47. Francis Bello, "Color TV: Who'll Buy a Triumph," *Fortune*, November 1955, 136. For further details on Sarnoff and RCA history, see Margaret B. W. Graham, *The Business of Research: RCA and the VideoDisc* (Cambridge: Cambridge University Press, 1986), 30–75;

Kenneth Bilby, *The General: David Sarnoff and the Rise of the Communications Industry* (New York: Harper & Row, 1986), 38–228; Robert Sobel, *RCA* (New York: Stein and Day, 1986), 36–168.

48. Quoted in Graham, *Business of Research*, 49. In 1928 Sarnoff gave a speech at Harvard Business School stating that business needed a new type of manager, knowledgeable about technology and its future possibilities. Bilby, *General*, 122.

49. Graham, *Business of Research*, 48–75; William B. Harris, "R.C.A. Organizes for Profit," *Fortune*, August 1957, 110–15. At the 1951 celebration, Sarnoff asked the laboratory to present him with three presents in 1956, when he would celebrate his fiftieth anniversary in radio: an electronic air conditioner, a light amplifier, and a videotape recorder. Research dutifully worked on these projects and was able to report progress on each by his golden anniversary. RCA Laboratories, *Research Report, 1956*, 1, David Sarnoff Library, Princeton, New Jersey (hereafter DSL).

50. Graham, *Business of Research*, 48–81; Harris, "R.C.A. Organizes for Profit," 232.

51. Quoted in RCA Laboratories, *Research Report, 1948*, 2, DSL.

52. Ibid., 70; RCA Laboratories, *Research Report, 1951*, i, DSL.

53. Quoted in Francis Bello, "The Year of the Transistor," *Fortune*, March 1953, 162. Further details of RCA's transistor symposium are provided in RCA Laboratories, *Research Report, 1952*, 27–29.

54. RCA Laboratories, *Transistors I* (Princeton, N.J.: RCA Laboratories, 1956).

55. Graham, *Business of Research*, 80–81; Harris, "R.C.A. Organizes for Profit," 232; Philip Querido, "Trade Ponders RCA's Patent Licensing Plans," *Electronic News*, 4 November 1957, 14; Philip Querido, "Patent Refund Battle Looms before RCA," *Electronic News*, 25 November 1957, 1, 19. The former *Electronic News* piece estimated that RCA had made $20 million annually through patent licensing.

56. RCA Laboratories, *Research Report, 1956*, 9. The 1950 report, noting that the Korean War made it "inevitable" that there should be "more effort into military research," asserted: "It has always been our experience that projects which originated as a result of our research have been carried to the most fruitful conclusions." Other statements of RCA's ambivalence towards military research contracts include RCA Laboratories, *Research Report, 1949*, 3, DSL; RCA Laboratories, *Research Report, 1950*, iii, 23, DSL; RCA Laboratories, *Research Report, 1951*, ii.

57. RCA saw Bell Labs as its main rival, and one can see in RCA's work a certain emulation of Bell Labs, with its transistor symposium and its publication of its collected papers on the transistor. In 1952, 1953, 1954, and 1956, RCA was in second place behind Bell Labs in the number of semiconductor patents received. In 1955, it was in third place behind Bell Labs and General Electric, while in 1957 it was in first place. RCA's reports on the Solid-State Device Research Conferences show that researchers from RCA considered Bell Labs their main competition. For example, of the 1955 conference, RCA researchers wrote: "The 'big three' [Bell Labs, RCA, and General Electric] has dwindled to the 'big two,' BTL [Bell Telephone Laboratories] (14 papers) and RCA (9 papers). They presented not only the most, but by far the best." J. Kurshan, ed., "Report on the Conference on Semiconductor Device Research, Philadelphia, Pennsylvania, June 20–22, 1955," Princeton Engineering Memo-

randum 565, 15 July 1955, 1, RCA-Hagley. Information on patents is from John E. Tilton, *International Diffusion of Technology: The Case of Semiconductors* (Washington, D.C.: Brookings Institution, 1971), 57.

58. RCA Laboratories, *Research Report, 1957,* 29, DSL.

59. Ibid., 5, 29.

60. RCA Laboratories, *Research Report, 1956,* 5; RCA Laboratories, *Research Report, 1957,* 5. In 1961 an executive at Bell Labs said: "RCA has five times as many people working on it [gallium arsenide] as we do. This is a somewhat special situation. They were given a large government contract and decided to place most of their eggs in this one basket." J. A. Hornbeck, "Devices for Digital Circuits," 2 May 1961, Location 64 06 03 05, AT&T Archives. Further historical background on gallium arsenide is provided by Heinrich J. Welker, "Discovery and Development of III–V Compounds," *IEEE Transactions on Electron Devices* 23 (July 1976): 664–74. The movement of RCA Laboratories out of silicon and germanium and into gallium arsenide was also confirmed by William Webster, interview by author, 27 May 1994.

61. RCA Laboratories, *Research Report, 1958,* 6, DSL. Although the gallium arsenide RCA had been able to produce was sufficient to build diodes, the report admitted, its "perfection is still not good enough for the gallium arsenide transistor." Ultimately, as researchers at RCA were gradually realizing, the material properties of gallium arsenide are so dramatically different from those of silicon or germanium that junction transistor types do not work well.

62. RCA Laboratories, *Research Report, 1957,* 7. Further information about Project Lightning is included in Samuel S. Snyder, "Computer Advances Pioneered by Cryptological Organizations," *Annals of the History of Computing* 2 (January 1980): 60–70.

63. J. T. Wallmark, "Semiconductor Logical Tree," 12 August 1957, RCA Laboratory Notebook 1745, 60, DSL. For RCA's earlier work, see Lester Hogan, "Reflections on the Past and Thoughts about the Future of Semiconductor Technology," *Interface Age,* March 1977, 24–26; Harwick Johnson, "Semiconductor Phase Shift Oscillator and Device," U.S. Patent 2,816,228, 10 December 1957. G. W. A. Dummer, of Britain's Royal Radar Establishment, had spoken in 1952 of building electronic equipment from a "solid block with no connecting wires." Quoted in Michael F. Wolff, "The Genesis of the Integrated Circuit," *IEEE Spectrum,* August 1976, 45. For further biographical details on Wallmark, see *IEEE Spectrum,* March 1964, 238. The following paragraphs have also benefitted from J. Torkel Wallmark, interview by author, 2 March 1994.

64. The basic structure of this device seems to be that of J. T. Wallmark and Harwick Johnson, "Shift Register Utilizing Unitary Multi-Electrode Semiconductor Device," U.S. Patent 3,038,085, 5 June 1962. It is described in more detail in J. T. Wallmark and S. M. Marcus, "Semiconductor Devices for Microminiaturization," *Electronics,* 26 June 1959, 35–37.

65. RCA Laboratories, *Research Report, 1957,* 6.

66. Quoted in Jerome P. Frank, "Shift Register Transistor Now under Development," *Electronic News,* 31 March 1958, 1, 15 (quote on 15). "RCA Experimental Device May Combine Entire Electronic Circuit in a Single Unit," Radio Corporation of America, Department of Information, 28 March 1958, DSL. I thank Alex Magoun of the David Sarnoff Library for providing me with this and other RCA press releases.

67. J. T. Wallmark, "Integrated Electronic Devices: New Approach to Microminiaturization," in *Aviation Age Research and Development Technical Handbook 1958–1959* (New York: Conover-Mast, 1958), F-6. Harvard Business School's Baker Library has a copy of this volume with a date stamp of 9 September 1958. I thank Janet Engstrand of the Baker Library for verifying this. I thank J. Torkel Wallmark for sending me a copy of this article.

68. Ibid., F-8. Wallmark also wrote: "In a sense, the integrated device concept shifts part of the responsibility for and cost of producing a reliable system from the system assembler to the component manufacturer."

69. RCA researchers claimed that the unipolar transistor was simpler and "its geometry and operation" was "more amenable to mass interconnection or integration techniques, particularly in planar arrays." RCA Laboratories, *Research Report, 1958*, 41–42. Later work was reported in RCA Laboratories, *Research Report, 1959*, 65–67, DSL; RCA Laboratories, *Research Report, 1960*, 63–65, DSL.

70. "RCA Scientists Developing New Type of Miniature Device So Small That 100,000,000 May Be Crammed into One Cubic Foot," RCA News, for release 22 March 1959, DSL. This announcement was reported in "Logic Element in Works," *Electronic News*, 23 March 1959, 93. Texas Instruments' announcement of its work is reported in "Gallium Arsenide Diode, S'ductor Solid Circuits Bow," *Electronic News*, 25 March 1959, 21.

71. *Electronic News* reported that Texas Instruments expected to have circuits available for "certain special applications" in 1959 and that although samples were not available, "depending on circuit complexity" a sample might be produced in a week. "Gallium Arsenide Diode, S'ductor Solid Circuits Bow," 21. RCA's 1959 *Research Report* claimed that RCA's work offered a "better approach to integrated device design than competitive approaches known at the present time." RCA Laboratories, *Research Report, 1959*, 65. RCA's micromodule put tiny individual components onto a ceramic wafer, which was then stacked onto other similar wafers. It is described in S. F. Danko, W. L. Doxey, and J. P. McNaul, "The Micro-Module: A Logical Approach to Microminiaturization," *Proceedings of the IRE* 47 (May 1959): 894–903. Jack Kilby's work at TI is described in Jack S. Kilby, "Invention of the Integrated Circuit," *IEEE Transactions on Electron Devices* 23 (July 1976): 648–54; Riordan and Hoddeson, *Crystal Fire*, 256–61; T. R. Reid, *The Chip: How Two Americans Invented the Microchip and Launched a Revolution* (New York: Simon & Schuster, 1985), 54–67, 79–80, 96–97. Reid presents evidence that TI had some awareness of RCA's work. Later published accounts of RCA's work include J. T. Wallmark and S. M. Marcus, "Semiconductor Devices for Microminiaturization," 35–37; J. T. Wallmark and S. M. Marcus, "Integrated Devices Using Direct-Coupled Unipolar Transistor Logic," *IRE Transactions on Electronic Computers* 8 (June 1959): 98–107; J. T. Wallmark, "Design Considerations for Integrated Electronic Devices," *Proceedings of the IRE* 48 (March 1960): 293–300. Wallmark later achieved a certain amount of notoriety in the industry for a paper given at the 1961 Electron Device Meeting that analyzed the maximum density and minimum feature size in semiconductor devices, claiming: "The end of the road to smaller size has already been reached." J. T. Wallmark and S. M. Marcus, "Maximum Packing Density and Minimum Size of Semiconductor Devices," *IRE Transactions on Electron Devices* 9 (January 1962): 112.

72. Webster had been one of six authors not affiliated with Bell Labs among the forty-five contributors to the second volume of the Bell Labs–sponsored *Transistor Technology*.

Webster received a Ph.D. in electrical engineering from Princeton University in 1954 while working for RCA. "Eight RCA Men Elected IRE Fellows," *RCA Engineer,* June–July 1960, 13. William Webster, interview by author, 31 May 1994. Webster's position is shown in the organization chart of RCA Laboratories in RCA Laboratories, *Research Report, 1959,* 7. Webster recounted that when William Pietenpol, one of the senior managers in Bell's transistor program, left Bell Labs, he was one of the few people who were not Bell Labs employees invited to the farewell dinner.

73. Atalla remembers Webster showing more interest in his work than Bell Labs management did. Webster recalled that Atalla did not have the disciplinary background one would expect from someone working on semiconductors: Atalla "wasn't an electrical engineer as I recall, he was a civil engineer or architect or unlikely thing." (Atalla was a mechanical engineer.) Atalla, interview; Webster, interview, 31 May 1994.

74. Karl Zaininger, RCA Laboratory Notebook 9294, 29–51, DSL; Karl Zaininger, RCA Laboratory Notebook 11917, 47–58, DSL; C. W. Mueller, memo to H. Weisberg, "Early MOS Transistor History," 20 February 1973 (in author's possession); Karl Zaininger, interview by author, 21 August 1994; Charles Mueller, interview by author, 4 May 1994. The exact roles of Zaininger, Mueller, and Moonan are not clear. The work was recorded in Zaininger's notebook, but there are pages pasted in in which Mueller and Moonan detailed procedures for making the device. Mueller signed the notebook as a coinventor rather than a witness, but one might suspect that if the work was seen as having important implications at the outset, it would have been recorded in Mueller's notebook. The device Zaininger, Mueller, and Moonan fabricated was similar to, but had substantial differences from, the device that Atalla and Kahng had presented at the 1960 Solid-State Device Research Conference. Later work at RCA was on devices much more similar to Atalla and Kahng's. Zaininger, Mueller, and Moonan's work was reported in RCA Laboratories, *Research Report, 1960,* 60.

75. Webster, interview, 31 May 1994; Steve Hofstein, note to author, 6 July 2000; S. R. Hofstein and F. P. Heiman, "The Silicon Insulated-Gate Field-Effect Transistor," *Proceedings of the IEEE* 51 (September 1963): 1190–1202.

76. RCA Laboratories, *Research Report, 1961,* 60, DSL. Other evidence of close collaboration with Somerville can be found in S. R. Hofstein, RCA Laboratory Notebook 12877, 17, 22, DSL; F. Heiman, "Insulated Gate Field-Effect Transistors," Princeton Technical Report 1207, 26 April 1962, RCA-Hagley.

77. Thomas Stanley, interview by author, 31 March 1994; RCA Laboratories, *Research Report, 1960,* 1; RCA Laboratories, *Research Report, 1963,* 50, 71–72, DSL. For details on the BIZMAC and RCA's entry into computing, see Paul E. Ceruzzi, *A History of Modern Computing* (Cambridge, Mass.: MIT Press, 1998), 55–57; Sobel, *RCA,* 170–80. Researchers at RCA used a number of different names at different times for what is now known as the MOS transistor. Since "MOS transistor" was one of the most common names used, and the term that endured, I will use it as well.

78. RCA Laboratories, *Research Report, 1961,* 3.

79. Thomas O. Stanley, "The Validity of Scaling Principles for Field-Effect Transistors," Princeton Technical Report 1282, 13 August 1962, RCA-Hagley. While there were ways to improve the performance of junction transistors, as the epitaxial junction transistor showed, they were not as straightforward as scaling all dimensions. Although Robert Den-

nard of IBM is typically credited with laying out the principles of MOS transistor scaling, based on his published work, a number of people in the industry had made informal studies well before Dennard's work. Robert H. Dennard, Fritz H. Gaensslen, Hwa-Nien Yu, V. Leo Rideout, Ernest Bassous, and Andre R. LeBlanc, "Design of Ion-Implanted MOSFET's with Very Small Physical Dimensions," *IEEE Journal of Solid-State Circuits* 9 (October 1974). 256–68; Moore, interview by author, 15 May 1996; Dale Critchlow, note to author, 15 May 1995.

80. The papers given at the 1962 Electron Device Conference are given in "Program of Meeting," *IEEE Transactions on Electron Devices* 10 (March 1963): 98–110. *Electronic News's* summary of the conference highlights is John Rhea and Nat Snyderman, "Newer Interest Seen in GaAs Applications," *Electronic News,* 29 October 1962, 1, 4. The quotations are from the RCA press release "RCA Develops Revolutionary Solid-State Element Combining Best Properties of Transistors and Vacuum Tubes," RCA News, for release 11 February 1963, DSL. I thank Alex Magoun of the David Sarnoff Library for providing me with this document.

81. Ibid., Joe McLean, "Use as Digital Circuit Seen for New Transistor," *Electronic News,* 11 February 1963, 1, 39; "Easier-to-Make Transistor Evolved by RCA; It May Affect Design, Cost of Computers," *Business Week,* 16 February 1963, 70; "RCA Makes New Type of Small Microcircuit," *Aviation Week and Space Technology,* 11 February 1963, 104.

82. The formation of Shockley Semiconductor Laboratory is detailed in Riordan and Hoddeson, *Crystal Fire,* 225–48; Christophe Lécuyer, "Making Silicon Valley: Engineering Culture, Innovation, and Industrial Growth, 1930–1970" (Ph.D. diss., Stanford University, 1999), 146–56; Moore, "Role of Fairchild," 53–54; Michael S. Malone, *The Big Score: The Billion Dollar Story of Silicon Valley* (Garden City, N.Y.: Doubleday, 1985), 68–70. For further details on Shockley, see Gordon Moore, interview by Rob Walker, 18 September 1995, Silicon Genesis: Oral History Interviews with Silicon Valley Scientists, Collection M0741, Department of Special Collections, Stanford University Libraries, Stanford, California (hereafter any reference to an interview in this collection will be identified as Silicon Genesis interviews); Harry Sello, interview by Rob Walker, 8 April 1995, Silicon Genesis interviews; Gordon Moore, interview by Christophe Lécuyer and Ross Bassett, 18 February 1997.

83. The most detailed account of the founding of Fairchild is Lécuyer, "Making Silicon Valley," 146–219. For more on the eight's departure from Shockley and the founding of Fairchild, see Riordan and Hoddeson, *Crystal Fire,* 248–62; Moore, "Role of Fairchild," 54–57; Malone, *Big Score,* 68–98; "Eight Leave Shockley to Form Coast Semiconductor Firm," *Electronic News,* 21 October 1957, 13; "West Coast Makers Set Transistor Push," *Electronic News,* 2 February 1959, 12; Moore, interview by Walker, 18 September 1995, Silicon Genesis interviews; Moore, interview by Lécuyer and Bassett, 18 February 1997. The eight founders of Fairchild were Julius Blank, Victor Grinich, Jean Hoerni, Eugene Kleiner, Jay Last, Gordon Moore, Robert Noyce, and Sheldon Roberts.

84. The details of Fairchild's early transistor work, including the development of the planar process and the integrated circuit, are discussed in Moore, "Role of Fairchild," 54–60; Lécuyer, "Making Silicon Valley," 146–219; Riordan and Hoddeson, *Crystal Fire,* 262–65. Other sources on this work include "West Coast Makers Set Transistor Push," *Electronic News,* 2 February 1959, 12; Jean A. Hoerni, "Planar Silicon Transistors and Diodes," Fairchild Semiconductor, Technical Articles and Papers TP-14, Bruce Deal Papers, Collection M1051, Department of Special Collections, Stanford University Libraries, Stanford, California;

Jean A. Hoerni, "Method of Manufacturing Semiconductor Devices," U.S. Patent 3,025,589, March 20, 1962; R. N. Noyce, "Semiconductor Device and Lead Structure," U.S. Patent 2,981,877, April 25, 1961; Ernest Braun and Stuart Macdonald, *Revolution in Miniature: The History and Impact of Semiconductor Electronics,* 2d ed. (Cambridge, England: Cambridge University Press, 1982), 73–104; Michael F. Wolff, "The Genesis of the Integrated Circuit," *IEEE Spectrum,* August 1976, 45–53.

85. "1960 Solid-State Device Research Conference, Attendance List," IEEE Electron Device Society Records, IEEE Headquarters, Piscataway, New Jersey; "Progress Report, Physics Section," 1 July 1960, Box 5, Folder 3, 1–2, Fairchild Camera and Instrument Technical Reports and Progress Reports, Collection M1055, Department of Special Collections, Stanford University Libraries, Stanford, California (hereafter FCI). In my interview with him in 1996, Gordon Moore confirmed that he heard Atalla and Kahng's paper. He recalled that "it kind of worked the way that people had tried to initially build a transistor." He also stated that he did not remember feeling that it was revolutionary.

86. Gordon Moore, interview by Allen Chen, 6 January 1993, 12, Intel Museum, Santa Clara, California.

87. "Noyce Appointed V-P at Fairchild Semiconductor," *Electronic News,* 11 May 1959, 12; "Palo Alto Firm Names Moore R&D Director," *Electronic News,* 1 June 1959, 24; Moore, interview by Lécuyer and Bassett, 18 February 1997; Gordon Moore, interview by Walker, 3 March 1995, Silicon Genesis interviews; Gordon Moore, interview by Allen Chen, 9 July 1992, 6 January 1993, Intel Museum, Santa Clara, California. In early 1963, Moore prepared a chart showing the progress that Fairchild R&D had made against milestones established at the beginning of 1962. For its work in gallium arsenide technology, Moore listed "many milestones," but "no accomplishments." One can easily imagine that in situations such as this, many a manager redefined "accomplishment" so as to have been able to report some progress. G. E. Moore to R. N. Noyce, "Research and Development Progress Report for 1962," 24 January 1963, Box 7, Folder 6, FCI.

88. G. E. Moore and V. H. Grinich to R. N. Noyce, "Research and Development Progress Report 1/1/61," 9 January 1961, 1, Box 5, Folder 9, FCI; G. E. Moore and V. H. Grinich to R. N. Noyce, "R&D Progress Report for Month Ending 1/31/61," 8 February 1961, 2, Box 5, Folder 11, FCI; G. E. Moore and V. H. Grinich to R. N. Noyce, "R&D Progress Report," 11 August 1961, 5, Box 6, Folder 7, FCI. I have identified Moore as the author by the typeface and the writing style—Moore was more expressive. In late 1960 C. T. Sah, a refugee from Shockley with a Ph.D. in electrical engineering from Stanford, conceived what he called a "surface-potential controlled tetrode," a four-terminal device that was a combination of a bipolar transistor and an MOS device. Work proceeded on it for some time, but it suffered from the same problems of inversion layers, with Moore reporting to Noyce in February of 1962: "We have been fooling with this problem for a long time and have no hope of an early solution. In fact, I'm not sure I have a strong hope of an eventual solution." Gordon E. Moore to R. N. Noyce, "Research & Development Progress Report," 15 February 1962, Box 7, Folder 2, FCI. For more on Sah's device, see C. T. Sah, "A New Semiconductor Tetrode—The Surface Potential Controlled Transistor," *Proceedings of the IRE* 49 (November 1961): 1623–34; C. T. Sah, "Evolution of the MOS Transistor—From Conception to VLSI," *Proceedings of the IEEE* 76 (October 1988): 1293–94.

89. V. H. Grinich and G. E. Moore to R. N. Noyce, "Progress Report, May 1962, R&D Laboratory," 12 June 1962, 7, Box 7, Folder 4, FCI. Fairchild was interested in an MOS capacitor in its efforts to implement the conventional circuit elements on its integrated circuits. On the four firms started by defectors from Fairchild (Rheem Semiconductor, Amelco, Signetics, and Molectro), see "Five Leave Firm to Form Rheem Unit," *Electronic News*, 9 March 1959, 15; "Drs. Hoerni, Last Resign Posts at Fairchild to Join Teledyne," *Electronic News*, 13 February 1961, 15; Richard Russack, "4 at Fairchild Resign; Plan to Form Firm," *Electronic News*, 4 September 1961, 4; Don C. Hoefler, "Silicon Valley—USA-Part II," *Electronic News*, 18 January 1971, 1. See also the poster put out by Semiconductor Equipment and Materials International, "Silicon Valley Genealogy" (Mountain View, Calif.: S.E.M.I., 1995).

90. "Fairchild Semiconductor Opens R&D Center," *Electronics*, 19 October 1962, 134; "Progress Report, Basic Physics Section," 1 February 1962, 1, Box 7, Folder 1, FCI.

91. G. E. Moore to R. N. Noyce, "Research and Development Progress Report for 1962," 24 January 1963.

92. Frank Wanlass, interview by author, 18 October 1994. Wanlass's idea for work on the MOS transistor came from an analogy to Paul Weimer's thin-film transistor work done at RCA. Weimer's thin-film transistor was made by evaporating thin films of cadmium sulfide onto a glass plate. The transistors were extremely unstable, and Wanlass believed that the planar process was more likely to yield stable devices. For other accounts of Wanlass's MOS work at Fairchild, see Michael J. Riezenman, "Wanlass's CMOS Circuit," *IEEE Spectrum*, May 1991, 44, and Clifford Barney, "He Started MOS from Scratch," *Electronics*, 8 October 1984, 64. On Weimer's thin-film transistor, see Ross Knox Bassett, "New Technology, New People, New Organizations: The Rise of the MOS Transistor, 1945–1975" (Ph.D. diss., Princeton University, 1998), 94–97.

93. While Wanlass claimed that Fairchild had a policy that new Ph.D.'s were allowed to choose their own research projects, Gordon Moore denied this was so, saying to me: "I hope not . . . but if we could make them feel that way, it was fantastic." Management and senior researchers had a number of ways to encourage work on some projects and not others. When Wanlass's initial efforts failed, he received counsel from a number of his coworkers that what he was trying to do was not a good way to make a transistor. Moore, interview by author, 15 May 1996; Wanlass, interview. Stanford's collection of Fairchild progress reports has nothing for Wanlass's department between August 1962 and January 1963; I have been forced to rely on an interview with him for details about this period, and it is unclear exactly what he did to achieve a reasonable level of progress in an area that had been so troublesome. What is clear is that in the July 1962 progress report (which does not list Wanlass as a contributor), there is no work on what would now be called an MOS transistor; however, in February 1963, such work was going on and Wanlass was the one doing it. "Progress Reports, Solid-State Physics Section," 1 August 1962, Box 7, Folder 5, FCI; "Progress Reports, Solid-State Physics Section," 1 February 1963, 5–7, Box 7, Folder 6, FCI.

94. "Progress Reports, Solid-State Physics Section," 1 February 1963, 5–7; "Progress Reports, Solid-State Physics Section," 1 June 1963, 10–11, Box 8, Binder, FCI; "Progress Reports, Solid-State Physics Section," 1 September 1963, 8, Box 8, Folder 2, FCI; Wanlass, interview.

95. F. M. Wanlass, "Novel Field-Effect Device Provides Broadband Gain," *Electronics*, 1 November 1963, 30–33; F. M. Wanlass, "Metal-Oxide-Semiconductor Field-Effect Transistors and Microcircuitry," *Wescon 1963 Technical Papers*, 13.2. Gordon Moore said that Wanlass was "really an inventor, more ideas than you could keep track of." Moore, interview by author, 15 May 1996.

96. "Progress Reports, Solid-State Physics Section," 1 February 1963, 6; D. Kahng and S. M. Sze, "A Floating Gate and Its Application to Memory Devices," *Bell System Technical Journal* 46 (July–August 1967): 1288–95; D. Frohman-Bentchkowsky, "Memory Behavior in a Floating-Gate Avalanche-Injection MOS (FAMOS) Structure," *Applied Physics Letters* 18 (1971): 332–34. C. T. Sah provides an account of this work in "Evolution of the MOS Transistor," 1295–96.

97. "Progress Reports, Solid-State Physics Section," 1 February 1963, 5. The complexities of isolating bipolar transistors and fabricating bipolar integrated circuits in general are described in G. E. Moore, "Semiconductor Integrated Circuits," in *Microelectronics: Theory, Design, and Fabrication*, ed. Edward Keonjian (New York: McGraw Hill, 1963), 262–359. See also Wolff, "Genesis of the Integrated Circuit," 45–53.

98. F. M. Wanlass and C. T. Sah, "Nanowatt Logic Using Field-Effect Metal-Oxide Semiconductor Triodes," *1963 International Solid-State Circuits Conference: Digest of Technical Papers*, 32–33; "Device Development Section, Progress Report," 1 April 1963, 1, Box 7, Folder 8, FCI.

99. G. E. Moore, C. T. Sah, and F. M. Wanlass, "Metal-Oxide-Semiconductor Field-Effect Devices for Micropower Logic Circuitry," in *Micropower Electronics*, ed. Edward Keonjian (New York: Macmillan, 1964), 41–56; Moore, "Semiconductor Integrated Circuits," 321. As a demonstration of the technology's immaturity, when Moore went to Europe, he took a number of MOS transistors with him to distribute as samples. However, they all failed to work, victims of electrostatic discharge, a peril MOS transistors are much more susceptible to than bipolar devices. Gordon Moore, note to author, 21 April 2001.

100. Moore, interview by Chen, 6 January 1993, 20. Gordon Moore's 1963 article "Semiconductor Integrated Circuits" shows his broad understanding of how integrated circuits had changed the electronics industry. He noted that the costs of components were determined primarily by the area they occupied on a silicon wafer (325). This article also contains an anticipation of Moore's 1965 *Electronics* article: "As the complexity is increased, microcircuits are favored more and more strongly, until one reaches the point where the yield because of the complexity falls below a production-worthy value. The point at which this occurs will continue to push rapidly in the direction of increasing complexity. As the technology advances, the size of circuit function that is practical to integrate will increase rapidly for a variety of reasons" (353). Compare this to Gordon E. Moore, " Cramming More Components onto Integrated Circuits," *Electronics*, 19 April 1965, 114–17.

101. In May 1961, John Hornbeck, a senior manager in Bell Labs' transistor effort, gave a speech to the Bell System Presidents' Conference reporting on devices for digital electronics. He gave an evaluation of the planar transistor's place that suggested he was still groping to understand it. He stated: "The planar is flat, rather than mesa. It is diffused like the others. It does not appear to be a major step forward. We are still evaluating it." He then went on to say that while future transistors might be planar, they would "certainly be epitaxial"—Bell's

innovation that led to a significant improvement in speed. J. A. Hornbeck, "Devices for Digital Circuits," 2 May 1961, Location 64 06 03 05, AT&T Archives. In September 1961 a joint Bell Labs–Western Electric evaluation cited four supposed advantages of the planar transistor over the Bell-developed mesa transistor: reliability, low current gain, saturation current, and noise. It then went on to highly qualify those so-called advantages, concluding. "It is not entirely obvious that the best planar transistor is markedly superior to the best mesa transistor." It somewhat petulantly asserted that a "well conceived publicity program" had made many believe that planar transistors were "the only acceptable type of silicon transistor." H. J. Patterson and H. E. Talley, "Joint Western Electric–BTL Program for Planar Silicon Transistor Development," Filecase 38589–51, 19 September 1961, AT&T Archives.

102. A complete analysis of Bell's transistor program in the 1960s is beyond the scope of this book. The most significant innovation Bell Labs made in the 1960s in semiconductors for applications for digital electronics was the silicon gate process, an innovation also initially rejected at Bell Labs, which became the technological foundation of Intel. (See chapter 6.) In 1969 Jack Morton gave the keynote address at the IEEE International Solid-State Circuits Conference. In his speech, made roughly six months after Robert Noyce and Gordon Moore had started Intel to exploit LSI (large-scale integration), the economies resulting from integrating many components onto a single chip, Morton denigrated this approach, saying it resulted in low yields and high costs. Jack A. Morton, "Strategy and Tactics for Integrated Electronics," *IEEE Spectrum*, June 1969, 26–33. Bell's position as an equipment manufacturer put them in a similar position to IBM, which I discuss in chapter 7. Raymond M. Warner, a former Bell Labs engineer who moved on to Motorola and Texas Instruments, presents an evaluation of Bell's semiconductor work that is consistent with mine in Warner and Grung, *Transistors*, 37, 52–53.

TWO • BACK FROM THE FRONTIER

1. A detailed history of IBM's entry into computing and electronics is provided by Charles J. Bashe, Lyle R. Johnson, John H. Palmer, and Emerson W. Pugh, *IBM's Early Computers* (Cambridge, Mass.: MIT Press, 1986), and Emerson W. Pugh, *Building IBM: Shaping an Industry and Its Technology* (Cambridge, Mass.: MIT Press, 1995), 131–261. The assertion that IBM's computing revenue surpassed its electric accounting revenue in only 1962 is from Emerson W. Pugh, Lyle R. Johnson, and John H. Palmer, *IBM's 360 and Early 370 Systems* (Cambridge, Mass.: MIT Press, 1991), 1. For an argument that IBM's transition from accounting machines to computers was in many ways not as radical as it might appear, see Steven W. Usselman, "IBM and Its Imitators: Organizational Capabilities and the Emergence of the International Computer Industry," *Business and Economic History* 22 (winter 1993): 1–35.

2. Bashe et al., *IBM's Early Computers*, 6–21, 523–27; Pugh, *Building IBM*, 37–87.

3. International Business Machines Corporation, "Research and Development Facilities," 18 December 1950, 10, 12, 17, 23, Box TBR 108, IBM Technical History Project, IBM Archives, Somers, New York (hereafter IBM THP). One might get a sense of how far IBM lagged Bell Labs in scientific capability by comparing the eight college graduates in IBM's

Physics Laboratory with the physicists listed on one column of one page of the Bell Labs phonebook in 1949. On this column one sees the names of Shockley, Bardeen, and Brattain, the inventors of the transistor and winners of the 1956 Nobel Prize in Physics; Charles Kittel, whose classic text taught generations solid-state physics; Lester Germer, collaborator with Nobel Prize–winning physicist Clinton Davisson; Karl Darrow, former student of Robert Millikan and longtime secretary of the American Physical Society; and Philip Anderson, a recent Harvard Ph.D. who would win the 1977 Nobel Prize in Physics. Bell Telephone Laboratories, *Telephone Directory, 1949*, 43, AT&T Archives, Warren, New Jersey.

4. Bashe et al., *IBM's Early Computers*, 373–78. Another feature of this work was that Thomas J. Watson's chauffeur was pressed into service to assist as a technician in his spare time. M. Loren Wood, interview by Charles J. Bashe, John H. Palmer, and Lyle R. Johnson, 12 November 1980, 7, IBM Technical History Project Oral Histories (*IBM's Early Computers*), Hagley Museum and Library, Wilmington, Delaware (hereafter these interviews will be cited as IBM THP interviews, with the initial citation for each interview identifying the volume for which the interview was conducted). When I discussed this work in a 1995 interview with Rolf Landauer, one of the first top-flight Ph.D. physicists hired at IBM (see below), it was still embarrassing to him and he insisted that I turn the tape recorder off while he described this work. Rolf Landauer, interview by author, 23 February 1995. Landauer did mention this work briefly in Rolf Landauer, interview by C. J. Bashe, J. H. Palmer, and L. R. Johnson, 9 October 1980, 3, IBM THP interviews (*IBM's Early Computers*).

5. Bashe et al., *IBM's Early Computers*, 378; A. L. Samuel to W. W. McDowell, 3 July 1951, Box TBR 107, IBM THP; A. L. Samuel to J. C. McPherson, 18 September 1951, Box TBR 107, IBM THP.

6. Bashe et al., *IBM's Early Computers*, 378–90; L. P. Hunter and H. Fleisher, "Graphical Analysis of Some Transistor Switching Circuits," *Proceedings of the IRE* 40 (November 1952): 1559–61. IBM's early transistor papers at the SSDRC (with the conference's variant names during this period) are summarized in J. Kurshan, ed., "Report on Semiconductor Device Research Conference, University of Illinois, June 19–20, 1952," Princeton Engineering Memorandum 231, 9 July 1952, RCA Collection, Hagley Museum and Library, Wilmington, Delaware (hereafter RCA-Hagley); J. Kurshan, ed., "Report on the Conference on Transistor Research Pennsylvania State College, July 6–8, 1953," Princeton Engineering Memorandum 325, 27 July 1953, RCA-Hagley; J. Kurshan, ed., "Report on the Conference on Semiconductor Device Research, Minneapolis, Minn., June 28–30, 1954," Princeton Engineering Memorandum 452, 17 August 1954, RCA-Hagley. Landauer, one of the physicists Hunter hired, provides his perspective on Hunter in Landauer, interview by author, and Landauer, IBM THP interviews, 4–5. Hunter provides information on his background and work at IBM in Lloyd P. Hunter, interview by E. W. Pugh, 17 December 1982, 1–17, IBM THP interviews (*IBM's Early Computers*). In his interview with me, Landauer asserted that in 1952 Hunter said that the future of IBM was in semiconductors, but IBM did not know it yet.

7. Bashe et al., *IBM's Early Computers*, 183–84, 544–49; Pugh, *Building IBM*, 237–44. A broader perspective on the development of the decentralized, multidivision corporate structure is provided by Alfred D. Chandler Jr., *Strategy and Structure: Chapters in the History of the American Industrial Enterprise* (Cambridge, Mass.: MIT Press, 1962).

8. Thomas J. Watson Jr. and Peter Petre, *Father, Son & Co.: My Life at IBM and Beyond* (New York: Bantam Books, 1990), 246–48.

9. Bashe et al., *IBM's Early Computers,* 555–63. Although IBM's facility opened first, Saarinen designed it and Bell Labs' Holmdel facility concurrently. Aline B. Saarinen, ed., *Eero Saarinen on His Work* (New Haven, Conn.: Yale University Press, 1962), 70. G. Robert Gunther-Mohr, a Research manager who sat on an advisory committee to assist in establishing the new building, asserts that Saarinen initially had a different design in mind for IBM, but difficulties with this plan made him turn to a plan that looked much like the Bell Labs design. G. Robert Gunther-Mohr, interview by Emerson Pugh, 22 December 1983, 4–5, IBM THP interviews (*IBM's Early Computers*). For a discussion of Saarinen's research laboratories within the context of 1950s ideas about industrial research, see Scott G. Knowles and Stuart W. Leslie, "'Industrial Versailles,' Eero Saarinen's Corporate Campuses for GM, IBM and AT&T," *Isis* 92 (March 2001): 1–33.

10. David A. Hounshell, "The Evolution of Industrial Research in the United States," in *Engines of Innovation: U.S. Industrial Research at the End of an Era,* ed. Richard S. Rosenbloom and William J. Spencer (Boston: Harvard Business School Press, 1996), 38–51; "Dr. E. R. Piore," transcript of speech at Williamsburg Conference, 3–4, Box TMB 119, IBM THP. George Wise has called it an assembly line model of the relationship between science and technology. "Science and Technology," *Osiris* 2d ser., 1 (1985): 229–46.

11. "Dr. E.R. Piore," 5–6.

12. "IBM Research Organizational Charts," 1 January 1958, Box TBR 109, IBM THP; G. Robert Gunther-Mohr, interview by author, 21 March 1995; Gunther-Mohr, IBM THP interviews, 4 October 1983 and 22 December 1983.

13. G. R. Gunther-Mohr, "A Program of Research for the Field of Semiconductor Science," 3, in "Committee Reports Program Research Conference, 1957," Box TBR 109, IBM THP; Vannevar Bush, *Science, the Endless Frontier: A Report to the President on a Program for Postwar Scientific Research* (Washington, D.C.: GPO, 1945), 19.

14. Gunther-Mohr, "Program of Research," 3. In a description of the operations of Bell Labs given in 1950, its executive vice president, Mervin Kelly, had explained its sharp distinction between research and development by asserting that scientists would lose their productivity if they were to become too concerned with development. Mervin J. Kelly, "The Bell Telephone Laboratories—An Example of an Institute of Creative Technology," *Proceedings of the Royal Society* 203A (10 October 1950): 292–93.

15. G. R. Gunther-Mohr, "The IBM Semiconductor Research Program: A Report for the 1958 Research Conference," 2, Box TAR 156, IBM THP. Another example of this ideal of the autonomy of the industrial researcher at that time comes from Henry Ehrenreich: "When I arrived at the General Electric Research Laboratory at the beginning of 1955, fresh from a Ph.D. at Cornell, I was greeted by my supervisor, Leroy Apker, who looked after the semiconductor section of the general physics department. I asked him to suggest some research topics that might be germane to the interests of the section. He said that what I did was entirely up to me. After recovering from my surprise, I asked, 'Well, how are you going to judge my performance at the end of the year?' He replied, 'Oh, I'll just call up the people at Bell and ask them how they think you are doing.'" Henry Ehrenreich, "Strategic Curiosity: Semiconductor Physics in the 1950s," *Physics Today,* January 1995, 28.

16. Gunther-Mohr, interview by author; R. W. Landauer to Emerson W. Pugh, 17 July 1979, Box TAR 242, IBM THP. Esaki's arrival at IBM is discussed in Bashe et al., *IBM's Early Computers*, 560–61. Leo Esaki describes his discovery of the tunnel diode in "Discovery of the Tunnel Diode," *IEEE Transactions on Electron Devices* 23 (July 1976): 644–47. J. B. Gunn describes his discovery of the Gunn effect in "The Discovery of Microwave Oscillations in Gallium Arsenide," *IEEE Transactions on Electron Devices* 23 (July 1976): 705–13.

17. "How to Win at Research," *Fortune*, October 1950, 118; "The Age of Research," *Time*, 9 July 1956, 74.

18. "How to Win at Research," 118.

19. David C. Mowery and Nathan Rosenberg, *Technology and the Pursuit of Economic Growth* (Cambridge: Cambridge University Press, 1989), 123–68; Harvey M. Sapolsky, *Science and the Navy: The History of the Office of Naval Research* (Princeton, N.J.: Princeton University Press, 1990).

20. "G.M. Technical Center," *Fortune*, December 1951, 82. In his discussion of the rationale for the design of IBM's research facility, Eero Saarinen stated: "It has always seemed to me that many scientists in the research field are like university professors—tweedy, pipe-smoking men." Saarinen, *Eero Saarinen on His Work*, 70.

21. George Wise, "Science at General Electric," *Physics Today*, December 1984, 52–61; C. E. Kenneth Mees and John A. Leermakers, *The Organization of Industrial Scientific Research*, 2d ed. (New York: McGraw-Hill, 1950), 136; H. H. Race, "The Research Laboratory Builds a House," *General Electric Review*, November 1948, 11–22; John A. Miller and John Horn, "New Laboratory Facilities to Improve Engineering Effort," *General Electric Review*, December 1949, 21–27. At the end of their article, Miller and Horn state: "The Research Laboratory will be well equipped to discharge effectively its prime responsibility—the task of pushing forward the frontiers of scientific knowledge."

22. Harold Mott-Smith, interview by George Wise, 1 March 1977, quoted in Leonard Reich, *The Making of American Industrial Research: Science and Business at GE and Bell, 1876–1926* (Cambridge: Cambridge University Press, 1985), 105.

23. Although Bell Labs hired William Shockley with the idea that he would work on a solid-state amplifying device, before World War II he spent time on loan to the Vacuum Tube Development Department. William Shockley, "The Path to Conception of the Junction Transistor," *IEEE Transactions on Electron Devices* 23 (July 1976): 601–2.

24. David A. Hounshell and John Kenly Smith Jr., *Science and Corporate Strategy: Du Pont R&D, 1902–1980* (Cambridge: Cambridge University Press, 1988), 350–64.

25. "The Atomic Airplane: 'This Program Has Had a Very Irregular History,'" *Science* 132 (9 September 1960): 658–59; "The Atomic Airplane: Its Death Has Been Mourned by Few," *Science* 133 (21 April 1961): 1223–25; W. Henry Lambright, *Shooting Down the Nuclear Plane* (Indianapolis, Ind.: Bobbs-Merrill, 1967); Steven L. Del Sesto, "Wasn't the Future of Nuclear Engineering Wonderful," in *Imagining Tomorrow: History, Technology, and the American Future*, ed. Joseph J. Corn (Cambridge, Mass.: MIT Press, 1986), 58–76.

26. Bashe et al., *IBM's Early Computers*, 567–68; B. L. Havens, "Report on Microwave Program," 1 June 1958, Box TAR 156, IBM THP; Landauer, IBM THP interviews, 2–3.

27. Bashe et al., *IBM's Early Computers*, 568–70; Robert Henkel, "Cryogenic Computer Seen by IBM in '64," *Electronic News*, 27 February 1961, 29.

28. Melville H. Hodge Jr., "Rate Your Company's Research Productivity," *Harvard Business Review*, November–December 1963, 109–22.

29. Joan Lisa Bromberg, *The Laser in America, 1950–1970* (Cambridge, Mass.: MIT Press, 1991), 73–96, 143–54; Bashe et al., *IBM's Early Computers*, 563–64. The IBM semiconductor laser was initially reported in Marshall I. Nathan, William P. Dumke, Gerald Burns, Frederick H. Dill Jr., and Gordon Lasher, "Stimulated Emission of Radiation from GaAs p-n Junctions," *Applied Physics Letters* 1 (1 November 1962): 62–64.

30. Bashe et al., *IBM's Early Computers*, 387, 402; "IBM-Texas Instruments, Inc. Agreement," 20 January 1958, Box TBR 107, IBM THP.

31. E. W. Pugh, "Report of Second Meeting," 4 November 1959, 7, Box TBR 109, IBM THP; Bashe et al., *IBM's Early Computers*, 399–415.

32. Gunther-Mohr, IBM THP interviews, 22 December 1983, 2–3; G. R. Gunther-Mohr, "Vapor Growth," *IBM Journal of Research and Development* 4 (July 1960): 247.

33. Pugh et al., *IBM's 360 and Early 370 Systems*, 59–76.

34. Ibid., 59–76. In October 1960 Harding wrote a memo describing the direction he believed IBM's semiconductor efforts should go. Although he acknowledged that if a semiconductor technology had to be implemented immediately, it would have to be done in germanium, he clearly stated: "The future of silicon device technology holds much greater promise." The advantages of silicon over germanium included greater temperature range, greater reliability, and the possibility of use in integrated circuits. Harding then outlined a proposal for implementing the silicon technology. The next April, Harding was on the study committee that proposed that IBM develop the planar silicon technology that later came to be known as SLT. W. E. Harding, "Semiconductor Devices Required for COMPACT," 13 October 1960, Box TBR 110, IBM THP. Harding's assessments of Fairchild and Texas Instruments are given in W. E. Harding, "Visit to Fairchild Corporation, Palo Alto, California," 22 April 1959, Box TBR 110, IBM THP; W. E. Harding, "Visit to Texas Instruments to Discuss Silicon High Current, High Speed Switching Transistors," Box TBR 110, IBM THP.

35. The term "$5 billion gamble" is from T. A. Wise, "IBM's $5,000,000,000 Gamble," *Fortune*, September 1966, 118. Pugh et al., *IBM's 360 and Early 370 Systems*, 68–105.

36. The establishment of the East Fishkill facility is described in "IBM Div. to Start Work on Plant in Fishkill, N.Y.," *Electronic News*, 10 September 1962, 40. It was common practice at IBM to refer to the Components Division as "CD," and I will follow this practice.

37. L. W. Atwood et al., "Report of an Integrated Circuit Study," 25 January 1963, Box TAR 242, IBM THP; William E. Harding, interview by Emerson W. Pugh, 24 January 1990, 8–9, IBM THP interviews (*IBM's 360 and Early 370 Systems*); Pugh et al., *IBM's 360 and Early 370 Systems*, 60; Rolf Landauer, "Integrated Circuitry," 11 September 1964, 3–4, Box TAR 242, IBM THP. Landauer quoted CD's 1963 long-range plan: "It is expected that SLT and its natural derivatives will be the mainstay of IBM logic for the next seven to ten years. . . . No revolutionary change at the manufacturing level is expected in this period. . . . It is not anticipated that IBM will follow what is now known as Fairchild Circuit Diffused Technology." For a discussion of leasing's effect on the computer industry, see Katherine Davis Fishman, *The Computer Establishment* (New York: Harper & Row, 1981), 15–18.

38. L. O. Hill, interview by E. W. Pugh, 1 October 1986, 1–3, IBM THP interviews (*IBM's 360 and Early 370 Systems*).

39. A. H. Eschenfelder, "Research-Components Division Interface," 27 June 1962, Box TAR 242, IBM THP; G. R. Gunther-Mohr, "Meeting to Discuss the Coordination of Research and Components Programs on Solid State," 23 April 1962, Box TAR 242, IBM THP.

40. As one example of the expectations for large jumps in technology, see "Ultra High Speed Computers Key AIEE Meeting Topic," *Electronic News*, 5 February 1962, 1, 5. Research's programs to make such a discrete jump in semiconductor technology were centered around vapor growth (or epitaxy) and gallium arsenide. Although Fairchild was pressing ahead with its integrated circuits based primarily on the diffusion process, a 1962 Research proposal by G. R. Gunther-Mohr expressed skepticism that the diffusion process was sufficient to realize a "new art of computer assembly." The superiority of vapor growth was premised on the possibilities of much greater physical control of structures, making it possible "in principle to program at will the composition of a crystal" on a molecular level. A 1962 description of Research's work by Rolf Landauer suggested: "If a really high degree of control were available, we could make larger device arrays at one time, and eliminate some of the hierarchy of chips, modules, small boards, large boards." Gallium arsenide was seen as having a wide range of applications, particularly for the highest-speed computer logic, and a number of novel device structures were proposed in this area. G. R. Gunther-Mohr, "Program Motivation," document fragment marked "I–F," Box TAR 242, IBM THP; Rolf Landauer to W. J. Pietenpol, 11 October 1962, Box TAR 242, IBM THP. Landauer offers his own commentary on these documents in "History of Large Scale Integration," 8 October 1971, 2, Box TAR 242, IBM THP.

41. Landauer, interview by author; Landauer, "History of Large Scale Integration," 3–4.

42. R. W. Landauer, "Solid State Science, Five-Year Plan, 1963–67," 1, Box TAR 242, IBM THP.

43. R. W. Landauer, "Solid State Science 2 Year Plan, 1964–65," 1, Box TAR 242, IBM THP. Landauer identified the relative dates of this document and the preceding reference in his memo "History of Large Scale Integration," 1.

44. R. W. Landauer, "Research Effort in Large Scale Integration," n.d., Box TAR 242, IBM THP.

45. Rolf Landauer, "Integrated Circuitry," 3.

46. Landauer, interview by author; Rolf Landauer, "FET NDRO Memory Matrices," 10 December 1963, Box TAR 242, IBM THP. Landauer's biography is traced in his obituary written by colleagues at IBM Research, who concluded: "He was the heart and soul of IBM research." Alan Fowler, Charles H. Bennett, Seymour P. Keller, and Yosef Imry, "Rolf William Landauer," *Physics Today*, October 1999, 104–5.

47. Rolf Landauer, "History of Large Scale Integration," 3–4; Landauer, "Solid State Science 2 Year Plan, 1964–65," 6–10, unpaginated attachment. The term *large-scale integration*, which later became common in the semiconductor industry, appears to have originated with IBM Research and spread throughout the industry through a request for proposal from the Air Force. (See below.) Originally, the term meant roughly two hundred to a thousand transistors per chip. As Landauer planned the program he noted that some Research personnel were bound by an Air Force contract to pursue a specific type of semiconductor research (low-temperature vapor growth) and were thus unavailable for the program. Rolf Landauer, "Research Effort in Integrated Circuitry," n.d., Box TAR 242, IBM THP.

48. Hill, THP interviews, 3–6; Herb Lehman, interview by author, 6 November 1995.

49. Sol Triebwasser, Robert Dennard, and Lewis Terman, interview by author, 1 September 1994.

50. Landauer, "History of Large Scale Integration," 3–4; Landauer, "Solid State Science, Five-Year Plan, 1963–67," 3.

51. "Minutes of the Fifty-Third Meeting of the R&D Board," 12 September 1963, E-42, Box TAR 130, IBM THP.

52. Ibid., E-36.

53. Landauer, "Integrated Circuitry," 6–8; Landauer, interview by author. Although IBM generally called the MOS transistor a "field-effect transistor" (FET) or an "insulated-gate field-effect transistor" (IGFET), for clarity I will use "MOS transistor" throughout.

54. Pugh et al., *IBM's 360 and Early 370 Systems*, 105–12. John Haanstra, whose fierce opposition to System/360 cost him his position as president of a division, used the denigration of SLT and the boosting of monolithic circuits as a springboard to another division presidency. One of Thomas Watson Jr.'s outside advisors was Robert Galvin of Motorola, who was attempting to sell IBM its version of integrated circuitry. Gunther-Mohr, interview by author.

55. T. J. Watson to A. K. Watson, 19 November 1964, Box TBR 114, IBM THP. Around this time, due to a combination of Watson's dissatisfaction with Research and the pressing needs for System/360, Research's budget was held flat for several years. M. J. Schiller, "Interview of Dr. Gardiner L. Tucker," 31 October 1977, Box TBR 109, IBM THP. Landauer recalled that in 1964, while he was attending a class for managers, Thomas Watson Jr. made his traditional appearance before the class and asked Landauer why IBM Research was so unsuccessful. Landauer, interview by author.

56. Gardiner Tucker to A. K. Watson, 21 August 1964, Box TAR 242, IBM THP.

57. S. P. Keller to E. R. Piore and J. A. Haddad, "Research Activities in NGT and Y," 15 September 1964, Box TAR 242, IBM THP.

58. H. G. Cohen, S. P. Keller, and D. N. Streeter, "Technology Transfer," 5 August 1975, Box TAR 251, IBM THP.

59. In September 1964 Landauer wrote: "In view of the length of time it took the corporation to initiate an integrated circuitry program aimed at IBM's over-all needs . . . it would have been better for Research to embark, in 1963, on a program using p-n junctions instead of field effect transistors." Rolf Landauer, "Integrated Circuitry," 7.

60. Rolf Landauer to D. L. Critchlow, E. W. Pugh, and S. Triebwasser, November 1979, 2, Box TAR 242, IBM THP.

61. Rolf Landauer to Emerson Pugh, 17 July 1979, 2, Box TAR 242, IBM THP; IBM Research, "A Proposal for a Large Scale Integrated Circuit Array," 9 August 1965, Box TAR 242, IBM THP; Richard L. Petritz, "Current Status of Large Scale Integration Technology," *IEEE Journal of Solid-State Circuits* 2 (December 1967): 130–47.

62. The Corporate Technical Board was a staff function that reviewed technical strategies and oversaw IBM's technical direction. Its members in October 1965 were B. O. Evans, president of the Federal Systems Division; J. W. Haanstra, president of the Systems Development Division; J. A. Haddad, director of Technology and Engineering and chair of the board; E. R. Piore, IBM vice president and chief scientist; and G. L. Tucker, director of Research.

63. D. T. Doody, "Semiconductor Device, Circuit, and Packaging Strategy," October 1965, 5, 7, Box TBR 110, IBM THP.

64. Ibid, 10–11, 35–37.

65. "CTB Minutes, October 29, 1965," 4 November 1965, 3, Box TAR 242, IBM THP.

66. C. J. Bashe, "Interview with Dr. Gardiner L. Tucker, October 31, 1977," Box TBR 109, IBM THP; A. G. Anderson to G. L. Tucker, "FET's," 4 February 1966, Box TAR 242, IBM THP. The interview conducted with Tucker includes the following summary: "The committee [CTB] decided that the IGFET work that by then had started in Research should be discontinued. Tucker kept it going. Haddad [the chair of the CTB] said, 'You're in trouble.' Tucker said, 'I work for Tom Watson. The IGFET work stops when he says it should.' He didn't."

THREE · DEVELOPMENT AT RESEARCH

1. For details of the electroluminescent photoconductor program, which attempted to deposit electroluminescent photoconductive material on a glass substrate to make digital logic elements, see Sol Triebwasser and S. P. Keller, interview by E. W. Pugh, 17 October 1983, IBM Technical History Project Oral Histories (*IBM's Early Computers*), Hagley Museum and Library, Wilmington, Delaware (hereafter these interviews will be cited as IBM THP interviews, with the initial citation for each interview identifying the volume for which the interview was conducted). On the tunnel diode, Merlin Smith, note to author, 12 and 13 September 1995; Dale Critchlow, note to author, 13 September 1995. Through their work on MOS technology, many of the early members of IBM's MOS program have received major recognition both inside IBM and in the profession at large. For example, both Dale Critchlow and Robert Dennard are IBM fellows, fellows of the IEEE, and members of the National Academy of Engineering. Frank Fang is an IEEE fellow and a member of the National Academy of Engineering and has received the Buckley Prize from the American Physical Society and the Sarnoff Award from the IEEE. Alan Fowler is an IBM fellow and an IEEE fellow as well as a member of the National Academy of Engineering, the National Academy of Science, and the American Academy of Arts and Sciences; he has received the Buckley Prize of the American Physical Society. *IEEE Membership Directory*, 1997.

2. Emerson W. Pugh, Lyle R. Johnson, and John H. Palmer, *IBM's 360 and Early 370 Systems* (Cambridge, Mass.: MIT Press, 1991), 48–112.

3. Rolf Landauer, "Solid State Science 2 Year Plan, 1964–65," 9, Box TAR 242, IBM Technical History Project, IBM Archives, Somers, New York (hereafter IBM THP). The final report on the LSI program credited Cheroff and Hochberg with "bringing the science and art of the Silicon Planar Technology into Research." S. Triebwasser, "Large Scale Integration, Final Program Report, November 22, 1966," RC 1980, 1, IBM Research, Yorktown Heights, New York. George Cheroff, interview by author, 21 August 1995; Sol Triebwasser, Robert Dennard, and Lewis Terman, interview by author, 1 September 1994.

4. T. C. Ainslie memo to R. T. Miller, 5 December 1969, Box TAR 242, IBM THP; H. S. Lehman, "Chemical and Ambient Effects on Surface Conduction in Passivated Silicon Semiconductors," *IBM Journal of Research and Development* 8 (September 1964): 422–26; Herb Lehman, interview by author, 6 November 1995.

5. K. H. Zaininger, "What the Competition Knows about the MOS-Structure," Princeton Engineering Memorandum 2483, 12 August 1964, 2, RCA Collection, Hagley Museum and Library, Wilmington, Delaware. Abstracts of the conference papers are given in "Solid-State Device Research Conference," *IEEE Transactions on Electron Devices* 11 (November 1964): 530–37.

6. Zaininger, "What the Competition Knows," 2.

7. William E. Harding, "Semiconductor Manufacturing in IBM, 1957 to the Present: A Perspective," *IBM Journal of Research and Development* 25 (September 1981): 651; D. R. Kerr, J. S. Logan, P. J. Burkhardt, and W. A. Pliskin, "Stabilization of SiO_2 Passivation Layers with P_2O_5," *IBM Journal of Research and Development* 8 (September 1964): 376–84.

8. Donald Seraphim, interview by author, 12 November 1995; Cheroff, interview; Dale Critchlow, note to author, 5 May 1995. A discussion of IBM's early concentration on *n*-channel MOS transistors can be found in E. W. Pugh, D. L. Critchlow, R. A. Henle, and L. A. Russell, "Solid State Memory Development in IBM," *IBM Journal of Research and Development* 25 (September 1981): 594. Further details are provided in a letter from George Cheroff to Dale Critchlow dated 30 April 1980, which was a review of a draft of the above article (in author's possession). I thank George Cheroff for giving me this document.

9. G. Cheroff, "Notes on Threshold and Substrate Bias Parameters in FETS," 16 December 1964, Box TAR 243, IBM THP; Cheroff, interview. *N*-channel devices could not be turned off because positively charged sodium ions were the main contaminant in the oxide. They induced an *n*-channel in the transistor even without the application of an external voltage. The substrate bias effect was discovered earlier by researchers at RCA. F. P. Heiman, "Integrated Insulated-Gate Field-Effect Transistor Circuit on a Single Substrate Employing Substrate-Electrode Bias," U.S. Patent 3,233,123, 1 February 1966.

10. Cheroff, interview; Seraphim, interview; Alan Fowler, interview by author, 30 October 1995; Alan Fowler, note to author, 17 March 1999.

11. Biographical details on Hochberg are given in *IBM Journal of Research and Development* 8 (September 1964): 479. Fowler, interview; Seraphim, interview; Cheroff, interview; Jerome Eldridge, interview by author, 4 January 1996.

12. Dale Critchlow, note to author, 15 May 1995; Joe Shepard, interview by author, 20 June 1995; Eldridge, interview. Even in its final report on LSI, Research cited wafer 3717 as an example of its best devices. S. Triebwasser, "Large Scale Integration, Final Program Report," 58.

13. H. S. Lehman, "Chemical and Ambient Effects on Surface Conduction in Passivated Silicon Semiconductors," 422–26; P. Balk, "Effects of Hydrogen Annealing on Silicon Surfaces," presented at the 1965 Electrochemical Society Spring Meeting, San Francisco, California, Extended Abstracts of the Electronics Division, vol. 14, abstract 109, 237–40; P. Balk, P. J. Burkhardt, and L. V. Gregor, "Orientation Dependence of Built-in Surface Charge on Thermally Oxidized Silicon," *Proceedings of the IEEE* 53 (December 1965): 2133–34.

14. D. E. Rosenheim to G. L. Tucker, "Large Scale Integration Program," 19 June 1964, Box TAR 243, IBM THP. Only point three of Rosenheim's four fundamentals remained in the final transformation of Research's MOS/LSI program. Rosenheim went on to write: "Although at this point we are certainly not counting out the FET as the best element for large

scale integration, we do recognize the possibility that the control of the Si-SiO$_2$ interface may not be satisfactorily obtained within our schedule for showing the feasability of large scale integration."

15. D. E. Rosenheim to G. L. Tucker, "The Role of IGFET in IBM," 19 November 1965, Box TAR 243, IBM THP. IBM Research and Bell Labs called MOS transistors "IGFETs" (insulated-gate field-effect transistors). The "MOS transistor" name was used by most of the rest of the industry.

16. "CTB Minutes, October 29, 1965," 4 November 1965, 3, Box TAR 242, IBM THP.

17. I. M. Mackintosh and D. Green, "Programmed Interconnections—A Release from Tyranny," *Proceedings of the IEEE* 52 (December 1964): 1648–51.

18. Don Rosenheim, interview by author, 4 January 1996; M. S. Axelrod, A. S. Farber, and D. E. Rosenheim, "Some New High-Speed Tunnel-Diode Logic Circuits," *IBM Journal of Research and Development* 6 (April 1962): 158–69. Biographical details on Smith are found in *IEEE Spectrum,* May 1967, 40. Dale Critchlow, note to author, 12 May 1995; Dale Critchlow, note to author, 27 November 1995. The other key manager of the program, Sol Triebwasser, while a physicist, also had engineering sensibilities.

19. Merlin Smith, note to author, 12 September 1995. In January 1966, Smith wrote: "We have an important responsibility to assure the Corporation and the potential user that the technology is applicable to real machine environments." M. G. Smith, "LSI Engineering Resources," 28 January 1966, Box TAR 243, IBM THP.

20. Merlin Smith and Dale Critchlow, interview by author, 17 May 1995.

21. D. L. Critchlow, "Some Comparisons between Fixed Interconnection Patterns and Programmed Interconnection Patterns for Planar Technology with IGFET's," 17 March 1965, Box TAR 243, IBM THP; Smith and Critchlow, interview.

22. Charles J. Bashe, Lyle R. Johnson, John H. Palmer, and Emerson W. Pugh, *IBM's Early Computers* (Cambridge, Mass.: MIT Press, 1986), 547–48; Harlow Frietag, interview by Richard T. Miller, 25 July 1968, Box TAR 242, IBM THP.

23. Triebwassser, "Large Scale Integration, Final Program Report," 95–104; R. H. Dennard, "A Cost Study for Integrated Circuits with Many Logical Decisions per Chip," 3 February 1966, RC 1552, IBM Research, Yorktown Heights, New York; Smith and Critchlow, interview.

24. Merlin Smith, "Cost Studies," 17 October 1966, Box TAR 243, IBM THP; Smith and Critchlow, interview. In my interview, Smith said one of the things he was proudest of was his focus on costs. He asserted: "It was almost unheard of in the climate of early Research, it was even a bad thing to do, to ask a Research guy what it might cost."

25. Ibid.; Triebwasser, "Large Scale Integration, Final Program Report," 4; James E. Hull et al., "Large Scale Integrated Circuit Arrays," Technical Report AFAL-TR-70–55, April 1970. The discretionary wiring approach did not appeal to everyone. Fairchild never pursued this approach, and Gordon Moore stated that he used to give conference talks on the Fairchild approach to LSI (see chapter 4), while Texas Instruments gave its talks on discretionary wiring. It was "the easiest argument I ever had to make, why one was a much more reasonable approach then the other." He asserted that the discretionary wiring approach had all of the disadvantages of the Fairchild approach, but none of the advantages. Gordon Moore, interview by author, 15 May 1996.

26. Triebwasser, "Large Scale Integration, Final Program Report," 150. This section of the report was written by Harlow Freitag.

27. In 1967 Eugene Fubini and Merlin Smith of IBM estimated that with LSI the industry might require as many as 100,000 unique part numbers. E. G. Fubini and M. G. Smith, "Limitations in Solid-State Technology," *IEEE Spectrum*, May 1967, 55–59; R. T. Miller, "History and Impact of LSI in IBM," 14 August 1969, 1–11, Box TAR 242, IBM THP (a report prepared by IBM Research in 1968 and 1969, but never formally issued, describing its LSI work).

28. Arnold Weinberger, "Large Scale Integration of MOS Complex Logic: A Layout Method," *IEEE Journal of Solid-State Circuits* 2 (December 1967): 182–90; D. L. Critchlow, "Layout and Mask Generation for LSI," *1967 IEEE International Convention Record*, pt. 6, 62–70; Triebwasser, "Large Scale Integration, Final Program Report," 35–49.

29. W. A. Notz, "Interests within IBM in an IGFET [MOS] Technology," 27 June 1966, Box TAR 243, IBM THP.

30. S. Triebwasser, "Large Scale Integration, Final Program Report," 6.

31. One of the harshest but most perceptive critics of IBM's semiconductor operations was Americo DiPietro, a member of the corporate staff during this period (see below). In 1971 he wrote a lengthy memo analyzing IBM's semiconductor operations compared to the semiconductor industry. "Unless an extremely large volume production is predicted, and perhaps not even then," he noted, "it is likely that a different component design will not be introduced into CD manufacturing, regardless of its desirability. Its cost would be prohibitively high since it would have to carry the full expense of a different manufacturing technology and tooling. . . . We have not been as versatile as the typical vendor. We do not develop as many different components nor do we produce our components as early or as easily as the typical vendor." DiPietro also noted that while the semiconductor industry was oriented towards the continuous development of new products and improvement of existing products, IBM's new semiconductor technology, which was tightly linked to new IBM computer systems, was much more nearly a process of stepwise development, with discrete jumps over long intervals. A. R. DiPietro, "IBM's Semiconductor Components," 19 May 1971, Box TAR 242, IBM THP.

32. S. Triebwasser, "Phaseover of LSI Program to CD," 26 July 1966, Box TAR 243, IBM THP. According to Triebwasser, some of the potential customers for a low-cost technology expected the low-cost objective to "very rapidly be swallowed up in the drive for performance."

33. Gardiner L. Tucker, "LSI Transfer to the Components Division," 24 August 1966, Box TAR 243, IBM THP.

34. The audit committees' members are listed in S. Triebwasser, "Large Scale Integration, Final Program Report," 172.

35. A. R. DiPietro, interview by E. W. Pugh, 12 November 1982, 1–2, Box TAR 243, IBM THP; A. R. DiPietro, interview by author, 1 March 1996.

36. DiPietro, interview by Pugh, 2–3.

37. Ibid., 2–7; DiPietro, interview by author.

38. DiPietro, interview by Pugh, 7–10; A. R. DiPietro, "Corporate SLT Task Force," 25 January 1966, Box TAR 243, IBM THP.

39. Pugh et al., *IBM's 360 and Early 370 Systems*, 105–12; "Why System/360 Is the Computer with a Future," IBM Advertisement, *Time*, 27 November 1964, 80–81.

40. Rolf Landauer, "Integrated Circuitry," 11 September 1964, 4, Box TAR 242, IBM THP.

41. Watson's later belief that SLT was a proper choice and the restoration of John Gibson to his former position are discussed in Pugh et al., *IBM's 360 and Early 370 Systems*, 109–12. RCA's strategy and performance are discussed in Franklin M. Fisher, James W. McKie, and Richard B. Mancke, *IBM and the U.S. Data Processing Industry: An Economic History* (New York: Praeger, 1983), 202–13. For a perceptive analysis of the factors that contributed to IBM's success in the computer industry as compared to companies such as RCA that had greater expertise in component technology, see Steven W. Usselman, "IBM and Its Imitators: Organizational Capabilities and the Emergence of the International Computer Industry," *Business and Economic History* 22 (winter 1993): 1–35.

42. One of the major points of Alfred Chandler's work is that in the large corporation, administrative coordination replaced market coordination. *The Visible Hand: The Managerial Revolution in American Business* (Cambridge, Mass.: Harvard University Press, 1977), 7–8. The fact that IBM's semiconductors were incompatible with merchant producers' gave IBM's semiconductor operations an added degree of freedom. It became very difficult to make any meaningful comparison of costs or prices between IBM and outside producers. Because IBM's semiconductors were different, it was always possible to say they had some extremely desirable characteristics not found in commercially available devices, and therefore any cost comparisons were not valid.

43. DiPietro, interview by Pugh; A. R. DiPietro, "Integrated Circuit Technology," 11 July 1966, Box TAR 243, IBM THP; J. W. Gibson, "Integrated Circuit Technology," 21 July 1966, Box TAR 243, IBM THP. Haddad's background is discussed in Pugh et al., *IBM's 360 and Early 370 Systems*, 43.

44. DiPietro, interview by Pugh; A. R. DiPietro, "Report on Trip to General Microelectronics, Inc.," 27 December 1963, Box TAR 243, IBM THP; J. G. Beavan, "Trip Report," 4 September 1964, Box TAR 243, IBM THP. Whatever data Wanlass showed DiPietro, General Microelectronics had not completely solved the MOS stability problems. (See chapter 5.) DiPietro was the first advocate of MOS and thus potential ally of the Research LSI program at the corporate level. When the Corporate Technical Board voted to instruct Research to halt its MOS work in October 1965, all of its members, with the exception of the director of Research, were in agreement. This included Haddad. When DiPietro took his position under Haddad, he was able to convince him of the merit of MOS technology.

45. Triebwasser, "Large Scale Integration, Final Program Report," 174.

46. Ibid., 173.

47. Ibid., 172; DiPietro, interview by Pugh.

48. Cheroff, interview; Fowler, interview; Shepard, interview.

49. Eldridge, interview; E. H. Snow and B. E. Deal, "Polarization Phenomena and Other Properties of Phosphosilicate Glass Films on Silicon," *Journal of the Electrochemical Society* 113 (March 1966): 269.

50. J. M. Eldridge and P. Balk, "Formation of Phosphosilicate Glass Films on Silicon Dioxide," *Transactions of the Metallurgical Society of the AIME* 242 (March 1968): 539–45; J. M. Eldridge, "A Thermochemical Evaluation of the Doping of SiO_2/Si with P_2O_5 and B_2O_3,"

Research Report, 8 December 1966, IBM Research, Yorktown Heights, New York; Eldridge, interview.

51. Shepard, interview; Eldridge, interview; F. Montillo and D. Dong, "Progress Report for December 15, 1966," 21 December 1966, 1–5, Box TAR 243, IBM THP. Dong was an extremely talented and observant technician who had previously played an important but largely unnoticed part in suggesting a solution to a problem of cracked metal stripes in IBM's circuit modules that had delayed deliveries of IBM's most powerful computer, the Series 360/Model 91.

52. Cheroff, interview; Dale Critchlow, note to author, 18 July 1996; P. Balk, "Progress Report, 15 July 1966," Box TAR 243, IBM THP.

53. Pieter Balk, interview by author, 23 August 1995; Triebwasser, "Large Scale Integration, Final Program Report," 64.

54. Montillo and Dong, "Progress Report for December 15, 1966," 1–5; Triebwasser, "Large Scale Integration, Final Program Report," 62–64.

55. S. Triebwasser, "FET's vs Bipolars: The SMT Decision," 28 December 1966, Box TAR 243, IBM THP. The audience for this memo is unclear. In November 1966, after the audit, Triebwasser had written to Pugh, the audit chair: "It is always difficult for people who have committed their energies to an area for some time to be able to maintain the kind of objectivity which your committee brought to the problem." S. Triebwasser, "LSI Audit," 14 November 1966, Box TAR 243, IBM THP.

56. R. W. Landauer, untitled, 17 July 1979, Box TAR 242, IBM THP. Landauer made this assertion in response to a query as to why Research did not concentrate on MOS memories earlier.

57. Triebwasser, Dennard, and Terman, interview; R. E. Matick, P. Pleshko, C. Sie, and L. M. Terman, "A High-Speed Read Only Store Using Thick Magnetic Films," *IBM Journal of Research and Development* 10 (July 1966): 333–41; L. M. Terman, note to author, 5 March 1996.

58. Ibid.; P. Pleshko and L. M. Terman, "An Investigation of the Potential of MOS Transistor Memories," *IEEE Transactions on Computers* 15 (August 1966): 423–27; John D. Schmidt, "Integrated MOS Transistor Random Access Memory," *Solid State Design*, January 1965, 21–25.

59. R. L. Elfant, "Integrated Circuits for Memories," 3 November 1965, Box TAR 243, IBM THP.

60. Pugh et al., *IBM's 360 and Early 370 Systems*, 210–19; Smith and Critchlow, interview.

61. Smith and Critchlow, interview; F. G. Carey, "Poughkeepsie Support," 16 August 1966, Box TAR 243, IBM THP; S. Triebwasser, "Contract for G-Machine Memory," 14 September 1966, Box TAR 243, IBM THP.

62. M. Smith, "Reorientation of the IGFET/LSI Program," 8 July 1966, Box TAR 243, IBM THP; Smith and Critchlow, interview.

63. Pugh et al., *IBM's 360 and Early 370 Systems*, 458–66; G. Constantine et al., "Potential for Monolithic Megabit Memories," 27 January 1965, Box TBR 112, IBM THP. Henle discusses his career and his work on semiconductor memories in Robert A. Henle, interview by E. W. Pugh, 16 April 1987, IBM THP interviews (*IBM's 360 and Early 370 Systems*).

64. J. W. Gibson, "CD Response to T&E Recommendations in Concurrence Letter," 30 November 1966, Box TAR 242, IBM THP. The Data Processing Group consisted of five divisions that formed the heart of IBM: CD, the Systems Development Division, the Systems Manufacturing Division, the Field Engineering Division, and the Data Processing Division (marketing). Pugh et al., *IBM's 360 and Early 370 Systems,* 664–65.

65. E. R. Piore, "Semiconductor Device, Circuit, and Packaging Strategy," 9 January 1967, Box TAR 242, IBM THP; J. W. Gibson, "Semiconductor Device, Circuit, and Packaging Strategy," 8 February 1967, Box TAR 242, IBM THP. A copy of the last memo reveals that the CTC and the corporate staff were working in concert. Gibson's memo was sent to Piore, the head of the CTC. Although no one on the corporate staff was copied on the letter, the secretary of the CTC made a copy of the memo and sent it on to A. R. DiPietro.

66. Robert Meade, interview by author, 25 February 1995; Smith and Critchlow, interview; "History and Impact of LSI in IBM," 16–17.

67. On the capacity of IBM's East Fishkill manufacturing line, see William E. Harding, interview by E. W. Pugh, 24 January 1990, 1, IBM THP interviews (*IBM's 360 and Early 370 Systems*). Harding claimed that in 1966 IBM was "producing more semiconductor logic circuits than all other companies combined."

68. S. Triebwasser, "Notes of Meeting in Fishkill on FET's 2/3/67," 8 February 1967, Box TAR 243, IBM THP.

69. Ibid.

70. R. E. Markle and D. E. Rosenheim, "FET Program," 9 May 1967, Box TAR 242, IBM THP; R. M. Meade, "Field-Effect Transistor Program," 4 April 1967, Box TAR 242, IBM THP.

FOUR • MOS IN A BIPOLAR COMPANY

1. On the state of MOS transistor stability at this time, see Bruce E. Deal and James M. Early, "The Evolution of Silicon Semiconductor Technology: 1952–1977," *Journal of the Electrochemical Society* 126 (January 1979): 24C. In a 1996 interview, Gordon Moore told me that when he put together this group, he was more concerned with improving bipolar transistors than with MOS transistors as such. Gordon Moore, interview by author, 15 May 1996.

2. Bruce Deal, interview by Henry Lowood, 9 June 1988, 12 July 1988, Stanford Oral History Project, Department of Special Collections, Stanford University Libraries, Stanford, California.

3. Deal, interview by Lowood, 12 July 1988. For details on Grove's early life, see Joshua Cooper Ramo, "A Survivor's Tale," *Time,* 29 December 1997/5 January 1998, 54–72; Linda Geppert, "Profile: Andy Grove," *IEEE Spectrum,* June 2000, 34–38; George Gilder, *Microcosm: The Quantum Revolution in Economics and Technology* (New York: Simon & Schuster, 1989), 83–85.

4. Ed Snow, interview by author, 9 January 1996. One part of Snow's dissertation research was published as E. H. Snow and P. Gibbs, "Dielectric Loss Due to Impurity Cation Migration in α Quartz," *Journal of Applied Physics* 35 (August 1964): 2368–74.

5. Deal, interview by Lowood, 12 July 1988; Snow, interview; "Progress Reports—Solid State Physics Section," 1 July 1963, 1–2, Box 8, Binder, Fairchild Camera and Instrument

Technical Reports and Progress Reports, Collection M1055, Department of Special Collections, Stanford University Libraries, Stanford, California (hereafter FCI).

6. Frank Wanlass, interview by author, 18 October 1994; "Progress Reports—Solid State Physics Department," 1 February 1963, 5–7, Box 7, Folder 6, FCI; "Progress Reports—Solid State Physics Section," 1 June 1963, 10–11, Box 8, Binder, FCI; "Progress Reports—Solid State Physics Section," 1 October 1963, 9, Box 9, Binder, FCI; "Progress Reports—Solid State Physics Section," 1 December 1963, 9–11, Box 9, Binder, FCI; "Wanlass, Gattuso Join GME," *Electronic News*, 13 January 1964, 8.

7. John L. Moll, "Variable Capacitance with Large Capacity Change," *Wescon Conference Record*, pt. 3 (1959): 32–36; Lewis Madison Terman, "An Investigation of Surface States at a Silicon/Silicon-Oxide Interface Employing M-O-S Diodes" (Ph.D. diss., Stanford University, 1961); R. Lindner, "Semiconductor Surface Varactor," *Bell System Technical Journal* 41 (May 1962): 803–31; K. Lehovec, A. Slobodskoy, and J. L. Sprague, "Field-Effect Capacitance Analysis of Surface States on Silicon," *Physica Status Solidi* 3 (1963): 447–64.

8. A. S. Grove, B. E. Deal, E. H. Snow, and C. T. Sah, "Investigation of Thermally Oxidized Silicon Surfaces Using Metal-Oxide-Semiconductor Structures," *Solid State Electronics* 8 (1965): 145–63.

9. "Progress Reports—Solid State Physics Section," 1 August 1963, 13–14, Box 8, Folder 2, FCI; "Progress Reports—Solid State Physics Section," 1 October 1963, 17–18; "Progress Reports—Solid State Physics Section," 1 December 1963, 23–24.

10. "Progress Reports—Solid State Physics Section," 1 June 1963, 23; "Progress Reports—Solid State Physics Department," 1 October 1963, 18.

11. "Progress Reports—Solid State Physics Section," 1 November 1963, 10, 21–23, Box 9, Binder, FCI (quotation on 23); "Progress Report—Device Development Section," 1 November 1963, 24, Box 9, Binder, FCI. Although the significance of electron beam–evaporated aluminum clearly appears to have come from Snow's work, it was reported in November by Wanlass, Deal, the Device Development Department, and the engineer in charge of running the electron beam apparatus. For a discussion of methods for evaporating metals, see Leon I. Maissel and Reinhard Glang, eds., *Handbook of Thin Film Technology* (New York: McGraw-Hill, 1970), 1–36–55. One of Gordon Moore's great strengths has been his ability to see the ultimate implications of trends. The methods of applying aluminum also used a vacuum system, and as more and more groups requested funding for their own vacuum system, Moore joked that it would be more cost-effective to issue everyone spacesuits and evacuate the building. Snow, interview.

12. "Progress Reports—Solid State Physics Section," 1 January 1964, 21, Box 9, Binder, FCI; "Progress Reports—Solid State Physics Department," 1 February 1964, 15–16, Box 9, Folder 1, FCI; Snow, interview. The idea that sodium was a source of instability in MOS transistors was in the air at the time. Gordon Moore recalls that when he interviewed Snow, he thought his work on sodium migration in quartz might prove applicable because there was suspicion about sodium's role in MOS instabilities. The fact that Snow did not jump directly to sodium suggests that the general suspicions were not strong enough to lead him in that direction. Gordon Moore reported to Robert Noyce in February 1961: "Our field effect measurements on oxide covered surfaces seem to be explainable on the basis of ion migration. This is interesting in itself and important with respect to long term reliability of

the planar devices. I think we have our teeth in something here now." G. E. Moore and V. H. Grinich to R. N. Noyce, "R&D Progress Report for Month Ending 1/31/61," 8 February 1961, 3, Box 5, Folder 11, FCI. While this appears tantalizingly close to what Snow found over two and a half years later, it did not result in a definitive implication of sodium's role in transistor instabilities. Moore, interview by author.

13. J. Earl Thomas, interview by author, 5 October 1995; "Progress Reports—Materials and Processes Department," 3 August 1964, 2, Box 10, Folder 4, FCI; "Progress Reports—Materials and Processes Department," 6 October 1964, 13, Box 10, Folder 7, FCI; "Progress Reports—Materials and Processes Department," 5 November 1964, 2, 14, Box 11, Folder 1, FCI; "Progress Reports—Materials and Processes Department," 30 April 1965, 5, Box 11, Folder 6, FCI. In my interview, Thomas cited John Strong, *Procedures in Experimental Physics* (Englewood Cliffs, N.J.: Prentice-Hall, 1938), 168–87, as providing the standard techniques for the evaporation of metal.

14. B. E. Deal and A. S. Grove, "General Relationship for the Thermal Oxidation of Silicon," *Journal of Applied Physics* 36 (December 1965): 3770–78. As of 1988, this paper had been cited an average of 30 times a year since its publication. Deal, interview by Lowood, 12 July 1988.

15. The early history of the work on surface states is reviewed in Lillian Hoddeson et al., *Out of the Crystal Maze: Chapters from the History of Solid-State Physics* (New York: Oxford University Press, 1992), 467–70. William Shockley, "On the Surface States Associated with a Periodic Potential," *Physical Review* 56 (15 August 1939): 317–23; John Bardeen, "Surface States and Rectification at a Metal Semi-Conductor Contact," *Physical Review* 71 (15 May 1947): 717–27; A. S. Grove, B. E. Deal, E. H. Snow, and C. T. Sah, "Investigation of Thermally Oxidized Silicon Surfaces Using Metal-Oxide-Semiconductor Structures," 145–63; B. E. Deal, M. Sklar, A. S. Grove, and E. H. Snow, "Characteristics of the Surface-State Charge (Q_{ss}) of Thermally Oxidized Silicon," *Journal of the Electrochemical Society* 114 (March 1967): 266–74.

16. A review of the group's work is provided in B. E. Deal, A. S. Grove, E. H. Snow, and C. T. Sah, "Recent Advances in the Understanding of the Metal-Oxide-Silicon System," *Transactions of the Metallurgical Society of the AIME* 233 (March 1965): 524–29. For the papers by Fairchild R&D during this time, see "Index to Published Technical Papers," Bruce Deal Papers, Collection M1051, Department of Special Collections, Stanford University Libraries, Stanford, California (hereafter Bruce Deal Papers).

17. "Digital Integrated Electronics, Monthly Technical Summary," September 1967, 18, Box 14, Folder 3, FCI; Snow, interview; A. S. Grove, *Physics and Technology of Semiconductor Devices* (New York: John Wiley and Sons, 1967); James Downey, interview by Rob Walker, 20 May 1995, Silicon Genesis: Oral History Interviews with Silicon Valley Scientists, Collection M0741, Department of Special Collections, Stanford University Libraries, Stanford, California (hereafter any reference to an interview in this collection will be cited as Silicon Genesis interviews). Prior to the book's publication, Grove taught a graduate course at Berkeley based on the manuscript.

18. "Progress Reports, Physics Department," 1 June 1965, 2, Box 11, Folder 8, FCI.

19. "Progress Reports, Physics Department," 1 September 1965, 1, Box 12, Folder 3, FCI.

20. Bruce E. Deal, "The XIX Drifts of Silicon Oxide Technology" (in author's possession).

This paper was given at the 1974 Electrochemical Society Meeting, but Deal asserts it was presented internally prior to this.

21. "Progress Report, Device Development Section," 1 August 1963, 2, Box 8, Binder, FCI. Diodes are two terminal devices that were made from a subset of the process used to make bipolar transistors.

22. Fairchild Semiconductor, "Product Manual FT-0028," April 1965 (in author's possession). I thank Jim Kelley for this document.

23. "Progress Reports—Solid State Physics Section," 1 June 1963, 22; "Progress Report, Device Development Section," 1 October 1963, 40–41, Box 9, Binder, FCI. In November 1966 the Digital Integrated Electronics Department reported: "We continue to make unstable MOS devices. We continue to hope for stable devices when we have a separate set of tubes for MOS." "Digital Integrated Electronics, Monthly Technical Summary," November 1966, 1, Box 13, Folder 9, FCI. The tubes held the wafers as they went into the diffusion or oxidation furnaces.

24. "Progress Report, Device Development Department," 1 September 1964, 31, Box 9, Binder, FCI; "Progress Report, Device Development Department," 1 November 1964, 29, Box 10, Folder 7, FCI; "Progress Report, Device Development Department," 1 January 1965, 30, Box 11, Folder 2, FCI.

25. "Progress Report, Device Development Department," 1 October 1964, 6–7, Box 10, Folder 5, FCI. The overvoltage problem, the same problem that had affected the transistors Moore took with him to Europe, was most likely what is now known as electrostatic discharge, where high voltages produced by static electricity destroy the MOS transistor's thin gate oxide.

26. Fairchild Advertisement, *Electronic News*, 5 October 1964, 20–21; "Fairchild Markets First Device Using Planar II Process," *Electronic News*, 12 October 1964, 35; "Fairchild Semiconductor Announces New Development in Semiconductor Manufacturing Process," Fairchild Semiconductor Press Release, 21 August 1964 (in author's possession); "Progress Report, Device Development Department," 1 November 1964, 29; Larry Blaser and Earl Cummins, "Designing FET's and MOST's into A-M Radios," Fairchild Technical Paper 29, October 1964, Bruce Deal Papers; John Barrett, Larry Blaser, and Harry Suzuki, "An FM Tuner Using MOS-FET's and Integrated Circuits," Fairchild Technical Paper 31, November 1965, Bruce Deal Papers. The General Microelectronics MOS transistor is described in the advertisement "GME 1004 P-Channel MOS Field Effect Transistor," *Electronic News*, 4 May 1964, 43. I thank Bruce Deal for providing me with a copy of the press release announcing the Planar II process.

27. "Progress Report, Device Development Department," 1 November 1964, 29; "Progress Report, Device Development Department," 1 January 1965, 30. On Fairchild's previous use of R&D for manufacturing, see Moore and Grinich, "R&D Progress Report for Month Ending 1/31/61," 1.

28. An AND gate is a basic digital logic function that provides a 1 output only if all the inputs are 1. See Gordon Moore's description of customers' reaction to the integrated circuit in chapter 1.

29. Quoted in Walter Mathews, "Fairchild Semicon to Expand Activity," *Electronic News*, 16 December 1963, 1, 33. On the establishment of the High Speed Memory Engineering

Department at Fairchild, see G. E. Moore and V. H. Grinich, "R&D Progress Report Month of March 1962," 13 April 1962, 3, Box 7, Folder 3, FCI; "High Speed Memory Engineering Progress Report," 1 June 1962, 1, Box 7, Folder 4, FCI.

30. "Fairchild Semiconductor Opens R&D Center," *Electronics*, 19 October 1962, 134; "Fairchild Semiconductor, Research and Development Laboratory, Organization Chart," 21 August 1968 (in author's possession); Gordon Moore to R. N. Noyce, "Monthly Progress Report June 1962, R&D Laboratories," 6 July 1962, 7, Box 7, Folder 5, FCI; "Progress Report, High Speed Memory Engineering," 1 June 1962, 2, Box 7, Folder 5, FCI.

31. What Moore was proposing, and what Fairchild would later work on, was a memory plane, a plane of flip-flops (memory elements), analogous to a plane of magnetic cores. Other chips would send signals to and receive signals from the memory plane. Moore's statements were made in a question-and-answer period during a lecture series on electronics held in Europe under the sponsorship of the North Atlantic Treaty Organization's Advisory Group for Aeronautical Research and Development. "Panel Discussion," *Micropower Electronics*, ed. Edward Keonjian (New York: MacMillan, 1964), 200; Moore, interview by author. The term "MOST," metal-oxide-semiconductor transistor, was sometimes used at Fairchild instead of "MOS transistor."

32. "Progress Report, High Speed Memory Engineering," 28 February 1964, 2, 7, Box 9, Folder 2, FCI; John D. Schmidt, "Integrated MOS Transistor Random Access Memory," *Solid State Design*, January 1965, 21–25. The comparison to production bipolar circuits is based on "Fairchild DTL Integrateds Due by July 1," *Electronic News*, 22 June 1964, 39. Although the focus of comment on Gordon Moore's 1965 *Electronics* paper has been on semiconductor makers' ability to put more and more components onto integrated circuits, the pace of development he suggested in that article serves as a heuristic delimiting what is attainable. Schmidt's IMM chip was well beyond the curve that Moore's article defined in 1965, suggesting that this heuristic was not recognized in early 1964. Gordon E. Moore, "Cramming More Components onto Integrated Circuits," *Electronics*, 19 April 1965, 114–17.

33. G. B. Herzog, "Report on Computer Memory Workshop—UCLA Conference Center, Lake Arrowhead, September 10–11–12, 1964," RCA Princeton Engineering Memorandum 2506, 9 February 1965, RCA Collection, Hagley Museum and Library, Wilmington, Delaware.

34. "Progress Report, High Speed Memory Engineering," 6 May 1964, 2, 9–10, Box 10, Folder 1, FCI; "Progress Report, High Speed Memory Engineering," 1 November 1964, 2, Box 10, Folder 7, FCI; "Progress Report, High Speed Memory Engineering," 1 January 1965, 9, Box 11, Folder 2, FCI. The last report I have found mentioning work on the IMM is "Progress Report, Device Development Department," 1 April 1965, 30, Box 11, Folder 5, FCI.

35. Les Vadasz, interview by author, 11 July 1996; C. D. Root and L. Vadasz, "Design Calculations for MOS Field Effect Transistors," *IEEE Transactions on Electron Devices* 11 (June 1964): 294–99.

36. "Progress Report, Device Development Department," 1 October 1964, 22.

37. "Progress Report, Device Development Department," 1 January 1965, 23–24; "Progress Report, Device Development Department," 1 February 1965, 36, Box 11, Folder 8, FCI; "Progress Report, Device Development Department," 1 March 1965, 23, Box 11, Folder 4,

FCI; "Progress Report, Device Development Department," 1 July 1965, 33, Box 12, Folder 1, FCI. NAND and NOR gates are basic digital logic blocks. A NAND gate returns a logical zero only if all inputs are logical ones. A NOR gate returns a logical one only if all inputs are logical zeros.

38. "Digital Integrated Electronics, Progress Report," 1 November 1965, 13, Box 12, Folder 5, FCI; "Digital Integrated Electronics, Monthly Technical Summary," October 1966, 5–6, Box 13, Folder 9, FCI; "Digital Integrated Electronics, Monthly Technical Summary," August 1966, 15, Box 13, Folder 7, FCI; "Digital Integrated Electronics, Monthly Technical Summary," November 1966, 8–9; "Digital Integrated Electronics, Monthly Technical Summary," December 1966, 1, Box 13, Folder 10, FCI. On the problems of the two-level metal MOS parts, see also Rob Walker, *Silicon Destiny: The Story of Application Specific Integrated Circuits and LSI Logic Corporation* (Milpitas, Calif.: CMC Publications, 1992), 18–20. In August 1967 the bipolar group working on two-level metal reported success: "THEY SAID IT COULDN'T BE DONE. But last month, FAIRCHILD delivered the first two layer metal arrays to a customer (ten octal decoder parts to Lockheed). At last count, we had found well over 300 good decoder parts at wafer-sort and were wondering what to do with all these devices. Scribing and packaging have, however, done much towards solving this problem and we are no longer considering stocking distributors." "Digital Integrated Electronics, Monthly Technical Summary," August 1967, 1, Box 14, Folder 3, FCI. Fairchild's two-level bipolar parts are described in "LSI, Fairchild Style," *Electronics*, 5 February 1968, 45–46.

39. "Digital Integrated Electronics, Monthly Technical Summary," November 1966, 5; Walker, *Silicon Destiny*, 18–22.

40. Lee Boysel, interview by author, 23 January 1996. Unless otherwise noted, all material in this section and the following section is from this source.

41. Bill Holmes, "10 Channel MOS Multiplexer," 18 June 1965 (in author's possession). I thank Lee Boysel for providing me with this document, as well as others detailing his MOS work. Boysel also showed me a notebook full of early MOS articles he had collected.

42. For a general overview of IBM's role in the Apollo program, see *IBM News, Federal Systems Division, Lunar Landing Special*, 24 July 1969.

43. The joint IBM-GI work is described in James Lydon, "LSI Cost Cuts Spur Microcircuit Research," *Electronic News*, 7 November 1966, 32; and L. L. Boysel and D. R. Hill, "Large-Scale Integrated Metal Oxide Semiconductor Applied to Data Acquisition," *NEREM Record*, 1966, 74–75.

44. L. L. Boysel, "Trip Report," 17 February 1966 (in author's possession); L. L. Boysel, "Trip Report," 22 March 1966 (in author's possession).

45. Digital differential analyzers are discussed in chapters 5 and 8. IBM's 360 architecture based on the use of control stores is discussed in Emerson W. Pugh, Lyle R. Johnson, and John H. Palmer, *IBM's 360 and Early 370 Systems* (Cambridge, Mass.: MIT Press, 1991), 128–35, 210–19.

46. The MOS circuits in Fairchild's catalog were a four- and a five-channel switch, very simple integrated circuits designed by R&D. *Fairchild Semiconductor Integrated Circuits* (Mountain View, Calif.: Fairchild Semiconductor, 1966), 55.

47. Gordon Moore remembered Boysel as being extremely secretive and recalled being

annoyed at one meeting where Boysel suddenly presented the complete design of an integrated circuit no one else knew he had been working on. Moore, interview by author; Moore, interview by Christophe Lécuyer and Ross Bassett, 18 February 1997.

48. Lee Boysel, untitled document, 13 September 1967 (in author's possession).

49. Ibid.

50. Two news pieces in *Electronic News* describe the status quo as it existed in the semiconductor industry. In 1965 Fairchild announced an order for 500,000 integrated circuits worth $2 million from Sperry Rand. In 1966 Fairchild received an order from Burroughs for an estimated 10 million integrated circuits. Both orders were entirely for bipolar circuits containing a few logic gates per chip. "Fairchild Semicon Gets Sperry Order," *Electronic News*, 5 April 1965, 6; "Fairchild Semicon Gets 20 Million Ics Order," *Electronic News*, 7 November 1966, 2.

51. Gilder, *Microcosm*, 247–61; "The 30th Anniversary of the Integrated Circuit: Thirty Who Made a Difference," *Electronic Engineering Times*, September 1988, 60–62; Michael S. Malone, *The Big Score: The Billion Dollar Story of Silicon Valley* (Garden City, N.Y.: Doubleday, 1985), 98–99; John Hulme, interview by author, 16 November 1996. A biographical sketch of Widlar is given in *IEEE Journal of Solid State Circuits* 6 (February 1971): 58.

52. Hulme, interview. Hulme told me that he and others suspected that Boysel's work was not intended for the benefit of Fairchild.

53. Wanlass, interview; J. Donald Trotter, note to author, 5 December 1996; L. Cohen, R. Rubenstein, and F. Wanlass, "MTOS Four Phase Clock Systems," *NEREM Record*, 1967, 170–71. A suggestion of Booher's role in the development of four-phase logic is provided in R. K. Booher, "MOS GP Computer," *Proceedings of the Fall Joint Computer Conference*, 1968, 877.

54. One clear point of origin of dynamic memory is the one-transistor memory cell conceived by Robert Dennard of IBM Research (see chapter 7). Dennard and others at IBM Research suspected it spread to GI through the National Security Agency, but I have been unable to confirm this. Sol Triebwasser, Robert Dennard, and Lewis Terman, interview by author, 1 September 1994.

55. Lee L. Boysel to R. B. C. Newcomb, 1 May 1966 (in author's possession). Boysel's articles include Lee Boysel, "Memory on a Chip: A Step toward Large-Scale Integration," *Electronics*, 6 February 1967, 93–97; Lee L. Boysel, "Adder on a Chip: LSI Helps Reduce Cost of Small Machine," *Electronics*, 18 March 1968, 119–21; Lee Boysel and Joe Murphy, "Multiphase Clocking Achieves 100-Nsec MOS Memory," *EDN*, 10 June 1968, 50–53; Lee Boysel, "Cutting Systems Costs with MOS," *Electronics*, 20 January 1969, 105–7.

56. "Fairchild Has Introduced 52 New Products in the Last 52 Weeks," advertising copy in author's possession; Marshall Wilder, interview by author, 26 November 1996. For the advertisement of the nineteenth week of the campaign, showing a 25-bit MOS shift register, see *Electronic News*, 12 February 1968, 3.

57. "Progress Report, Device Development Department," 1 March 1965, 21; "Digital Integrated Electronics, Monthly Technical Summary," August 1966, 15; "Digital Integrated Electronics, Monthly Technical Summary," December 1966, 5; "Digital Integrated Electronics, Monthly Technical Summary," August 1967, 7.

58. In 1970 *Datamation* gave the lease price of an IBM System/360 Model 40 at $6,500

per month. Angeline Pantages, "RCA's New Line: Just Enough to Check Migration to 370?" *Datamation*, 1 October 1970, 30–31.

59. "Digital Integrated Electronics, Monthly Technical Summary," February 1967, 23, Box 13, Folder 11, FCI; "Digital Integrated Electronics, Monthly Technical Summary," December 1967, 13, Box 14, Folder 5, FCI; "Digital Integrated Electronics, Monthly Technical Summary," April 1968, 11, Box 14, Folder 6, FCI; "Digital Integrated Electronics, Monthly Technical Summary," March 1968, 8, Box 14, Folder 6, FCI.

60. Fairchild Semiconductor, *MOS/LSI* (Mountain View, Calif.: Fairchild Semiconductor, 1968), 1.

61. Jim Kelley, interview by author, 8 January 1996; Wilder, interview; Bob Cole, interview by author, 13 December 1995. Gordon Moore has made the following observations about transferring the MOS technology out of R&D at Fairchild: "It would seem that the more technically competent a receiving organization becomes, the more difficult it is to transfer technology to it. When it clearly had all the technical capability in the company, the Fairchild lab had no trouble transferring technology to production. As the production organization became more successful and began to recruit more technical people, however, technology transfer became more difficult. Production, it seemed, had to kill a technology and reinvent it in order to get it into manufacturing." Gordon Moore, "Some Personal Perspectives on Research in the Semiconductor Industry," in *Engines of Innovation: U.S. Industrial Research at the End of an Era,* ed. Richard S. Rosenbloom and William J. Spencer (Boston: Harvard Business School Press, 1996), 167. When I mentioned Mountain View's use of a vapox process in MOS manufacturing, while Gordon Moore did not remember the details, one could see a flash of the tension between Mountain View and Palo Alto in his comment that that was "an excellent example of them trying to use the wrong technology to make something work." Moore, interview by author.

62. "Digital Integrated Electronics, Monthly Technical Summary," May 1967, 1, Box 14, Folder 1, FCI.

63. "Digital Integrated Electronics, Monthly Technical Summary," July 1967, 9, Box 14, Folder 2, FCI.

64. Boysel, interview, 23 January 1996; Cole, interview.

65. "Digital Integrated Electronics, Monthly Technical Summary," July 1967, 1. The February 1967 report stated: "The main newsworthy event was the production of stable MOS devices. A clean MOS furnace and strict cleaning procedures paid off. A process manual is in the mill." "Digital Integrated Electronics, Monthly Technical Summary," February 1967, 7.

66. "Digital Integrated Electronics, Monthly Technical Summary," October 1967, 11, Box 14, Folder 4, FCI; "Digital Integrated Electronics, Monthly Technical Summary," November 1967, 9, Box 14, Folder 4, FCI.

67. Moore, interview by author; "Fairchild Semiconductor MOS Micro Matrix Design Handbook," 15 October 1966, 1, Box 6, Rob Walker Papers, Collection M0742, Department of Special Collections, Stanford University Libraries, Stanford, California. Later the document refers to "the rather enigmatic interface between the system builder and the semiconductor device manufacturer." It further states: "As the technology evolves and becomes

more sophisticated, it appears that a means must be found to merge the fabricator and the systems builder into a coherent design team. This is a very difficult problem and a great deal of cut and try experimentation will be required during the coming year to develop a common language, information formats, document controls, and all the other implements needed to bridge the user-manufacturer interface" (7).

68. "Fairchild Semiconductor MOS Micro Matrix Design Handbook," 34–36; Moore, interview by author; Gordon E. Moore, "Customized LSI," *IEEE International Convention Digest*, 1968, 79. Initially, the computer-aided-design approach seemed to work best in vertically integrated companies, where in spite of all the division of labor, the work was done within one corporation. E. J. Rymaszewski, J. L. Walsh, and G. W. Leehan, "Semiconductor Logic Technology in IBM," *IBM Journal of Research and Development* 25 (September 1981): 603–16.

69. Fairchild Semiconductor, *MOS/LSI*, 1–38.

70. Although tension existed between Mountain View and R&D in bipolar work, a much greater level of cooperation and communication existed between the two groups. Moore, interview by author; Hulme, interview.

71. In April 1967 Fairchild essentially restarted the integrated MOST memory project, now called the SAM (semiconductor advanced memory), which was transferred to the Physics Department in July 1967. The project was still attempting to build a sixty-four-bit memory chip, the same density Fairchild R&D had been working on in 1964. "Monthly Technical Summary—Physics," July 1967, 27, Box 14, Folder 2, FCI.

72. Moore, interview by Lécuyer and Bassett. In a 1996 interview, Gordon Moore acknowledged that Boysel had a better idea of how MOS technology could be used than Moore's R&D group did. He also said of Boysel: "The way he did his work at Fairchild, you would be hard pressed to find a better way to do it to get it rejected." In retrospect, Moore said that Boysel should have been part of the R&D group. Moore, interview by author.

FIVE • IT TAKES AN INDUSTRY

1. Bruce Deal, interview by Henry Lowood, 9 June 1988, Stanford Oral History Project, Stanford University Libraries, Stanford, California; Gordon Moore, interview by Rob Walker, 18 September 1995, Silicon Genesis: Oral History Interviews with Silicon Valley Scientists, Collection M0741, Department of Special Collections, Stanford University Libraries, Stanford, California (hereafter any reference to an interview in this collection will be cited as Silicon Genesis interviews). The issue of researchers publishing their work has been around since the beginning of the industrial research lab. Leonard S. Reich, *The Making of American Industrial Research: Science and Business at GE and Bell, 1876–1926* (Cambridge: Cambridge University Press, 1985), 110, 118–20, 189, 195.

2. The RCA Collection at the Hagley Museum and Library in Wilmington, Delaware (hereafter RCA-Hagley), contains a number of reports on the SSDRC and its precursors. Many are riddled with staple holes, suggesting that they were bucked around many offices. For example, J. Kurshan, ed., "Report on Semiconductor Device Research Conference, University of Illinois, June 19–20, 1952," RCA Princeton Engineering Memorandum 231, 9 July

1952; J. Kurshan, ed., "Report on the Conference on Transistor Research, Pennsylvania State College, July 6–8, 1953," RCA Princeton Engineering Memorandum 326, 27 July 1953; J. Kurshan et al., "Report on the Conference on Semiconductor Device Research, Minneapolis, Minn., June 28–30, 1954," RCA Princeton Engineering Memorandum 452, 17 August 1954.

3. K. H. Zaininger, "What the Competition Knows about the MOS-Structure," RCA Princeton Engineering Memorandum 2483, 12 August 1964, 1, RCA-Hagley. Abstracts of the conference papers appear in "Solid-State Device Research Conference, July 1–3, 1964, Boulder, Colorado," *IEEE Transactions on Electron Devices* 11 (November 1964): 530–37.

4. Zaininger, "What the Competition Knows," 1.

5. Donald Young, interview by author, 13 September 1994. In August 1960, about a month after the SSDRC that dealt heavily with the subject of epitaxy, Ian Ross wrote a memo on the subject that included the phrase "a number of workers in the field now suspect." I. M. Ross, "Some Transistor Structures Using a Combination of Wide Gap and Narrow Gap Semiconductors," 25 August 1960, Case 19881–4, AT&T Archives, Warren, New Jersey. These conferences had their own moral economy, with the expectation that there would be an equal exchange of information among firms. In the 1950s, conferees at the SSDRC groused about Texas Instruments, which sent people but did not present papers. It is probably not coincidental that Texas Instruments, which had previously been primarily a geophysical services company, was a newcomer to industrial research. E. O. Johnson, ed., "Report on the Conference on Semiconductor Device Research, University of Colorado, Boulder, Colorado, July 15–17, 1957," RCA Princeton Engineering Memorandum 925, 12 August 1957, RCA-Hagley; E. Johnson, ed., "Report on the Conference on Semiconductor Device Research, Purdue University, Lafayette, Indiana, June 25–27, 1956," RCA Princeton Engineering Memorandum 748, 17 August 1956, RCA-Hagley. On the use of E. P. Thompson's term "moral economy" in a scientific setting, see Robert E. Kohler, *Lords of the Fly: Drosophila Genetics and the Experimental Life* (Chicago: University of Chicago Press, 1994), 11–13, 133–35.

6. George Warfield, "Editorial," *IEEE Transactions on Electron Devices,* 13 (February 1966): 221. Abstracts of the conference papers appear in "Solid-State Device Research Conference, June 21–23, 1965, Princeton, New Jersey," *IEEE Transactions on Electron Devices* 12 (September 1965): 502–11.

7. "Silicon Interface Specialists Conference," n.d. (in author's possession); "Attendance List, 1965 Silicon Interface Specialist Conference," n.d. (in author's possession); Karl Zaininger, interview by author, 21 August 1994. The letter of invitation to the second Silicon Interface Specialists Conference stated: "The purpose of this conference will be to bring together a small group of specialists, actively working in the field in an informal, workshop-like atmosphere where the free interchange of ideas and discussion will lead to an increase in our knowledge and understanding of silicon interface phenomena. . . . To maintain ease of communication and interchange of ideas, the size of the meeting is being planned with an attendance level of roughly sixty to seventy people." S. R. Hofstein to Bruce E. Deal, 22 November 1966 (in author's possession).

Among the seventy-one people registered for the first Silicon Interface Specialists Conference, only five had academic positions. Sixty-one were affiliated with corporations, while five were affiliated with government or other laboratories, such as the Research Triangle Institute or Lincoln Laboratories. I thank Bruce Deal for providing me with these documents.

8. E. H. Snow, A. S. Grove, B. E. Deal, and C. T. Sah, "Ion Transport in Insulating Films," *Journal of Applied Physics* 36 (May 1965): 1664–73; A. S. Grove, E. H. Snow, and B. E. Deal, "Stable MOS Transistors," *Electro-Technology,* December 1965, 40–43; B. E. Deal, A. S. Grove, E. H. Snow, and C. T. Sah, "Recent Advances in the Understanding of the Metal-Oxide-Silicon System," *Transactions of the Metallurgical Society of the AIME* 233 (March 1965): 524–29.

9. N. Goldsmith and F. Zaber, "Clean Oxide Growth," 27 April 1966 (in author's possession); Norman Goldsmith, interview by author, 4 August 1994. For evidence of others who suspected the role of sodium prior to the Fairchild work, see Zaininger, interview. I thank Norman Goldsmith for providing me with this document.

10. Gordon Moore, interview by author, 15 May 1996; Zaininger, interview; Deal, interview by Lowood, 12 July 1988.

11. George Warfield, "Editorial," *IEEE Transactions on Electron Devices* 12 (March 1965): 87; George Warfield, "Editorial," *IEEE Transactions on Electron Devices* 13 (February 1966): 221; George Warfield, "Introduction," *IEEE Transactions on Electron Devices* 14 (November 1967): 727. In the 1967 introduction Warfield also said: "The history of the device field is replete with announcements of revolutionary new devices, later followed by obituaries containing the usual phrase of condolence 'still a powerful new tool for research.' The MIS system is one of the few which has essentially lived up to its announcement and whose obituary seems a long way off."

12. Earl S. Schlegel, "A Bibliography of Metal-Insulator-Semiconductor Studies," *IEEE Transactions on Electron Devices* 14 (November 1967): 728–49.

13. Jean A. Hoerni, "Planar Silicon Transistors and Diodes," Fairchild Technical Articles and Papers, TP-14, Bruce E. Deal Papers, Collection M1051, Department of Special Collections, Stanford University Libraries, Stanford, California; Jean A. Hoerni, "Method of Manufacturing Semiconductor Devices," U.S. Patent 3,025,589, 20 March 1962; R. Dale Painter, "FC&I to Drop Mesa Method for Planar," *Electronic News,* 31 October 1960, 1, 5; Fairchild C&I Shift to Planar Stirs Controversy in Field," *Electronic News,* 7 November 1960, 50; "Why Fairchild Converted to Planar" (Fairchild Advertisement), *Electronic News,* 5 December 1960, 3; Jerome P. Frank, "Sperry Div. to Shift to Planar Transistors," *Electronic News,* 30 January 1961, 43; "Fairchild Planar Process License Given to Philco," *Electronic News,* 20 August 1962, 98; Jack Robertson, "Shakeout of Makers Forecast in Planar Transistors Field," *Electronic News,* 21 October 1963, 36; "Industry Eyes Planar Pacts by Fairchild," *Electronic News,* 27 April 1964, 1, 29; Roger Borovoy, interview by Christophe Lécuyer and Ross Bassett, 14 June 1996.

14. Christophe Lécuyer, "Making Silicon Valley: Engineering Culture, Innovation, and Industrial Growth, 1930–1970" (Ph.D. diss., Stanford University, 1999), 237–38; "Offers $120 Flip-Flop Element at Fairchild Semiconductor," *Electronic News,* 20 March 1961, 128; Walter Mathews, "Fairchild Semicon Ics Drop to $2.55," *Electronic News,* 4 May 1964, 1, 30. Lécuyer argues convincingly that Fairchild responded to a dropoff in its military business by entering the much more price-sensitive commercial market.

15. II–VI compounds are made from elements from columns II and VI from the periodic table.

16. Paul K. Weimer, "The TFT—A New Thin-Film Transistor," *Proceedings of the IRE* 50 (June 1962): 1462–69; "Microelectronics Research, Progress Report," 1 February 1962, 1, Box 7, Folder 1, Fairchild Camera and Instrument Technical Reports and Progress Reports, Collection M1055, Department of Special Collections, Stanford University Libraries, Stanford, California (hereafter FCI); Alan Fowler, interview by author, 30 October 1995; Rolf Landauer to Emerson Pugh, 17 July 1979, Box TAR 242, IBM Technical History Project, IBM Archives, Somers, New York (hereafter IBM THP). Weimer's original work on the thin-film transistor can be followed in Paul K. Weimer, Laboratory Notebook 12266, David Sarnoff Library, Princeton, New Jersey (hereafter DSL) and Paul K. Weimer, Laboratory Notebook 13268, DSL.

17. T. Peter Brody, "The Thin Film Transistor—A Late Flowering Bloom," *IEEE Transactions on Electron Devices* 31 (November 1984): 1615.

18. Ibid.

19. H. G. Cohen, S. P. Keller, and D. N. Streeter, "Technology Transfer," 5 August 1975, Box TAR 251, IBM THP.

20. In their assessment of IBM Research's germanium program and the problems it faced, the authors noted: "Absence of work outside of IBM was an inhibiting factor on the [product] divisions." Ibid.

21. Frank Wanlass, interview by author, 18 October 1994, 8 February 1996; Walter Matthews, "Hear Lowell, 3 FC&I Men to Form Firm," *Electronic News*, 15 July 1963, 1, 36; "Wanlass, Gattuso Join GME," *Electronic News*, 13 January 1964, 8.

22. Wanlass, interview, 18 October 1994; Lee Boysel, interview by author, 23 January 1996; Bob Cole, interview by author, 13 December 1995; Keith Eaton, interview by author, 18 January 1996.

23. "GME 1004 P-Channel MOS Field Effect Transistor" (Advertisement), *Electronic News*, 4 May 1964, 43; Walter Mathews, "MOS Technology Being Used in Developing Family of Ics," *Electronic News*, 24 August 1964, 84; Wanlass, interview, 18 October 1994, 8 February 1996; Boysel, interview, 23 January 1996; A. R. DiPietro, interview by E. W. Pugh, 12 November 1982, Box TAR 243, IBM THP.

24. Wanlass, interview, 18 October 1994, 8 February 1996; Joel Karp, interview by author, 29 January 1999; Mary Bubb, "Growing Trend to IC Mosfet Use," *Electronic News*, 4 July 1966, 28.

25. Walter Mathews, "Hear GM Backing New Bay Area Firm," *Electronic News*, 29 July 1963, 1, 28; "Bulletin," *Electronic News*, 26 August 1963, 1; Walter Mathews and James Lydon, "GME Plans Set for Wide Diversity," *Electronic News*, 26 August 1963, 1, 10; Wanlass, interview, 18 October 1994; Eaton, interview; "GME Negotiating Deal with Victor," *Electronic News*, 21 September 1964, 8; "Feel Victor Deal with GME Nearer," *Electronic News*, 12 October 1964, 36.

26. "Victor MOS Calculator Delivery Due," *Electronic News*, 25 August 1966, 26; "Electronics Newsletter," *Electronics*, 18 October 1965, 26; "Victor Calculator out of Woods Soon," *Electronics*, 20 February 1967, 20; Lewis H. Young, "Uncalculated Risks Keep Calculator on the Shelf," *Electronics*, 6 March 1967, 231–34.

27. "Philco Giving Up on MOS Calculator," *Electronics*, 24 June 1968, 25; "Down the Drain," *Electronics*, 8 July 1968, 66.

28. Walter Mathews, "Gattuso, Wanlass Leaving Coast to Join GI Semicon," *Electronic News*, 7 December 1964, 6; James Lydon, "GI Realigns Top Staff of Semicon Group," *Electronic News*, 15 February 1965, 1, 26; "Takes New Post at Gen'l Instr.," *Electronic News*, 6 July 1964, 29; "Dr. C. H. Sutcliffe Leaving Philco," *Electronic News*, 6 July 1964, 28; Wanlass, interview, 18 October 1994; untitled (undated prospectus in author's possession from J. Leland Seely); "General Instrument Bides Its Time," *Business Week*, 29 November 1969, 118–20. As an example of how little capital was required to start a semiconductor firm in the early 1960s (or how optimistic Seely and Wanlass were), their prospectus estimated that $63,000 would be required to get their proposed new company through the first six months. This included, among other things, buying all capital equipment (estimated at $45,200), incorporating the company (estimated at $600), and paying the monthly rent and utilities (estimated at $500 a month)!

29. James Lydon, "Monolithic Circuitry Use in Subsystems Available," *Electronic News*, 24 March 1965, 1, 55; "General Instrument Adds 90-Bit MOS Shift Register," *Electronic News*, 3 January 1966, 26; Wanlass, interview, 18 October 1994.

30. "Gen'l Instrument to Show Monolithic Digital Analyzer," *Electronic News*, 22 August 1966, 92; "Integrator on a Chip," *Electronics*, 22 August 1966, 38; Wanlass, interview, 18 October 1994.

31. Wanlass, interview, 18 October 1994, 8 February 1996; Roy Johns, "Gen'l Instrument Sights '67 Gains," *Electronic News*, 6 February 1967, 42; "General Instrument Dedicates R&D Center," *Electronic News*, 7 August 1967, 46; Walter Mathews, "R&D Unit Thrives at GI's 'Ivory Tower' in Utah," *Electronic News*, 3 February 1969, 36.

32. Don Hoefler, "MOS—Wide-Awake Sleeper," *Electronic News*, 23 June 1969, 2:12, 44, 47. In its 1972 review of the MOS business, *Electronic News* estimated that General Instrument had been fourth in MOS sales in 1971. In 1976 the consulting firm Integrated Circuit Engineering estimated that GI had been eighth in MOS sales in 1974. John Rhea and James Lydon, "MOS: A Red-Only Business," *Electronic News*, 12 June 1972, 1, 4–5, 52; Integrated Circuit Engineering, *Status 1976: A Report on the Integrated Circuit Industry* (Scottsdale, Ariz.: Integrated Circuit Engineering, 1976), 24. After Wanlass left, he started a new company, which was acquired within months. McDonnell Douglas acquired the successor company, and Wanlass left by September 1972, at which point *Electronic News* quoted him as saying: "There are big company people and there are little company people. I guess I'm one of the latter." "The Antenna," *Electronic News*, 16 February 1970, 34; Neil Kelley, "Varadyne Agrees to Buy Wanlass Firm, Others," *Electronic News*, 22 June 1970, 1, 40; "Wanlass Leaves Post at Nitron," *Electronic News*, 2 October 1972, 1, 48.

33. Raymond Warner, conversation with author; Wanlass, interview, 18 October 1994, 8 February 1996; Boysel, interview, 23 January 1996; Cole, interview; "Device Development Progress Report," June 1966, 21, Box 13, Folder 5, FCI.

34. Moore, interview by author; L. L. Vadasz, "G.I.'s MOS Seminar," 10 December 1968, Intel Museum, Santa Clara, California. In my interview with him, Gordon Moore said he felt that Wanlass could have been more useful to the industry if he had not left Fairchild. He stated: "I was very sorry when he was attracted away to go off and try to set up an MOS

company, that was the wrong use of him. A guy with so many ideas like that is awfully hard to keep on one track. You've got to narrow to one path to make a successful company. He could have been a really great contributor to the industry, but he faded in a useful capacity." One can imagine that had Wanlass been hired at Intel, he might have played a role similar to Ted Hoff, coming up with new ways to use MOS technology.

35. On tacit knowledge, see H. M. Collins, "The TEA Set: Tacit knowledge and scientific networks," *Science Studies* 4 (1974): 165–86; and "Building a TEA Laser: The Caprices of Communication," *Social Studies of Science* 5 (November 1975): 441–50. Wanlass has received little recognition within the field for his work. It was only in 1991 that he received the IEEE Solid State Circuits award, which was for his patent on complementary MOS (CMOS). Because Wanlass was primarily affiliated with start-up companies after he left Fairchild, and most of these were not successful, he had no institution pushing for his recognition.

36. Donald Seraphim, interview by author, 12 November 1995; Dale Critchlow, note to author, 27 November 1995, 12 May 1995; D. P. Seraphim, "Meeting with General Instrument Co. Staff, March 16, 1965," 17 March 1965, Box TAR 242, IBM THP. Both Seraphim and Critchlow cited the importance of Wanlass's counsel without prompting. Seraphim, who managed the LSI program, credited Wanlass with encouraging him to drop the programmable interconnect. Critchlow credited Wanlass for encouraging the move to smaller gate lengths.

Wanlass, who believed that people were too secretive with technical information, had previously served in the Army Special Services working with atomic weapons, where much of what he worked on was top-secret. Wanlass, interview, 8 February 1996.

37. J. E. Thomas, "Telephone Conversation with William Webster of RCA 2/20/63," Box TAR 242, IBM THP. Although Thomas's position in CD may appear to contradict the previous paragraph, it does not. Thomas was a researcher at heart and was hired to head an Advanced Development group in CD in 1962 and played no role in the actual implementation of SLT, IBM's production semiconductor technology. In 1964 he left to join GI, in apparent recognition of the fact that with SLT, IBM had set off on its own course and that Thomas's broad industry experience was more than offset by his lack of involvement in SLT. J. Earl Thomas, interview by author, 5 October 1995.

38. Dale Critchlow and Merlin Smith, interview by author, 17 May 1995; G. Cheroff, "Monthly Progress Report for September, MOS Devices and Surface Studies," 29 September 1966, 5, Box TAR 243, IBM THP.

39. Cheroff, "Monthly Progress Report for September," 3. IBM Research's final report on its LSI program carried detailed results of stability tests comparing devices made by Research with those made by other firms. S. Triebwasser, "Large Scale Integration, Final Program Report, November 22, 1966," RC 1980, 58–61, IBM Research, Yorktown Heights, New York.

40. Cheroff, "Monthly Progress Report for September," 7; D. L. Critchlow, "General Instruments IGFET Layout Rules," 27 February 1968, Box TAR 243, IBM THP.

41. J. F. Shepard, Progress Report, 21 November 1968, 2, Box TAR 243, IBM THP; Joe Shepard, interview by author, 20 June 1995; Goldsmith, interview; Goldsmith and Zaber, "Clean Oxide Growth," 27 April 1966; Werner Kern and David A. Puotinen, "Cleaning Solutions Based on Hydrogen Peroxide for Use in Silicon Semiconductor Technology," *RCA Review* 31 (June 1970): 187–206. As late as 1994 the "RCA clean" was still widely used in the

industry. Pieter Burggraaf, "Keeping the 'RCA' in Wet Chemistry Cleaning," *Semiconductor International,* June 1994, 86–90.

42. W. A. Notz, "Interests within IBM in an IGFET [MOS] Technology," 27 June 1966, Box TAR 243, IBM THP.

43. Alfred D. Chandler Jr., *The Visible Hand: The Managerial Revolution in American Business* (Cambridge, Mass.: Harvard University Press, 1977), 6–7; Dale Critchlow, note to author, 30 June 1996.

44. S. Triebwasser, "FET's vs. Bipolars: The SMT Decision," 28 December 1966, Box TAR 243, IBM THP.

45. Erich Bloch to G. L. Tucker, "MOS Devices for Memory Application," 5 November 1965, Box TAR 243, IBM THP.

46. The most notable example of considering military R&D funding a positive force for economic development is Kenneth Flamm, *Creating the Computer: Government, Industry and High Technology* (Washington, D.C.: Brookings Institution, 1988), and *Targeting the Computer* (Washington, D.C.: Brookings Institution, 1987). Richard C. Levin presents a fairly positive view of military R&D support in "The Semiconductor Industry," in *Government and Technical Progress: A Cross Industry Analysis,* ed. Richard R. Nelson (New York: Pergamon Press, 1982), 9–100. To say that military support did not translate into commercial success for RCA is not to say that the military's support was not a success by its standards. It was able to keep two technologies alive that were potentially useful to the military at a time when they might have died without military support. RCA's ranking in MOS sales comes from Rhea and Lydon, "MOS: A Red-Only Business," 5.

47. On gallium arsenide, see chapter 1. On molecular electronics, see Ernest Braun and Stuart Macdonald, *Revolution in Miniature: The History and Impact of Semiconductor Electronics,* 2d ed. (Cambridge, England: Cambridge University Press, 1982), 95–96. Earl Arthur, "Ge, Si Research Emphasis Hit at AF Conclave," *Electronic News,* 18 September 1961, 39.

48. William Webster, note to author, 29 August 1994. For the sake of comparison, prior to its LSI program, IBM Research was charged with the technologies to be used in products five years out and further. G. R. Gunther-Mohr, "Meeting to Discuss the Coordination of Research and Components Programs on Solid State," 23 April 1962, Box TAR 242, IBM THP.

49. G. C. Sziklai, "Symmetrical Properties of Transistors and Their Applications," *Proceedings of the IRE* 41 (June 1953): 717–24; Paul K. Weimer, "An Earlier CMOS Patent," *IEEE Spectrum,* February 1992, 11; F. M. Wanlass and C. T. Sah, "Nanowatt Logic Using Field-Effect Metal-Oxide Semiconductor Triodes," *1963 International Solid-State Circuits Conference: Digest of Technical Papers,* 32–33; F. M. Wanlass, "Low Stand-By Power Complementary Field Effect Circuitry," U.S. Patent 3,356,858, 5 December 1967; Israel Khalish, interview by author, 19 January 1994; Gerald Herzog, interview by author, 24 May 1994; Moore, interview by author.

50. G. B. Herzog, "COS/MOS from Concept to Manufactured Product," *RCA Engineer,* Dec.–Jan. 1975–76, 38. RCA's CMOS work is described in G. B. Herzog et al., "Large-Scale Integrated Circuit Arrays," Technical Report AFAL-TR-69-352, January 1970. I thank Paul Ceruzzi for providing me with a copy of this report.

51. Khalish, interview; Don C. Hoefler, "Feathers Fly over MOS Semicon Merits," *Electronic News,* 24 August 1966, 30; F. Ambler, "A Timing Fuze Employing p-Channel MOS

Arrays," *RCA Engineer,* June–July 1968, 22–24. RCA engineers seem to have had a persistent attraction to complex technologies. As RCA Laboratories considered various possibilities for video recording, researchers considered magnetic tape "spent technology" and favored more exotic approaches. Margaret B. W. Graham, *The Business of Research: RCA and the VideoDisc* (Cambridge: Cambridge University Press, 1986), 92.

52. Herzog, "COS/MOS from Concept to Manufactured Product," 36–38; "RCA Metal Oxide Semicons to Be Exhibited at Wescon," *Electronic News,* 19 August 1968, 79; Hoefler, "MOS—Wide-Awake Sleeper," 47; Rhea and Lydon, "MOS: A Red-Only Business," 5; "RCA Invests $2M in AMS; Set RAM Deal," *Electronic News,* 20 May 1974, 1, 20. In what might be seen as emblematic of RCA's semiconductor efforts, RCA entered the one-kilobit DRAM market four years after Intel. Just months later the semiconductor industry went into a severe downturn.

53. William Webster, interview by author, 31 May 1994; C. W. Mueller and P. H. Robinson, "Grown-Film Silicon Transistors on Sapphire," *Proceedings of the IEEE* 52 (December 1964): 1487–90; RCA Laboratories, *Research Report, 1964,* 5, DSL; RCA Laboratories, *Research Report, 1965,* 4, DSL.

54. Jack Wessling, "SOS for Computer Designs," *Electronic News,* 23 May 1966, 1; "People," *Electronics,* 11 October 1971, 14. Arthur Lowell, the promoter behind GME, was one of the key figures advocating SOS at Autonetics.

55. "People," 14. Sevin is quoted in Martin Gold and Andy McCue, "SOS: Fact or Fancy?" *Electronic News,* 30 December 1974, 24. In 1976 a three-inch sapphire substrate cost fifty dollars while a silicon substrate cost five dollars. Sapphire substrate suppliers estimated that if they could get 5 percent of the silicon market, they could bring the substrate cost down to twelve dollars. For a four-kilobit RAM the fifty-dollar sapphire substrate would represent a doubling of the die manufacturing cost, even assuming manufacturers could achieve yields with the SOS process that matched those of the silicon MOS process. Nat Snyderman, "SOS—The Substrate Hangup," *Electronic News,* 20 September 1976, 14; Integrated Circuit Engineering, *Status 1977: A Report on the Integrated Circuit Industry* (Scottsdale, Ariz.: Integrated Circuit Engineering, 1977), 4-1–4-5.

56. "Inselek Enters Chap. XI: Liability $2M; Assets $1.1M," *Electronic News,* 20 January 1975, 2. An assessment of SOS in 1977 noted that an assortment of difficulties had kept most semiconductor manufacturers away, with RCA being the only firm to offer standard products. It concluded that a "fundamental breakthrough" was necessary before SOS could compete with MOS. Integrated Circuit Engineering, *Status 1977,* 5–8.

57. In a matter of technological style, large East Coast R&D laboratories, with their vast resources and long-term outlook, seemed to prefer complicated technologies to simple ones. West Coast start-ups, which needed to get a product to market quickly, preferred simpler technologies. In discussing why Fairchild did not pursue the more complicated MOS technologies, Gordon Moore said: "I believe you go the way nature takes you." Moore, interview by author.

58. Webster, interview. Nathan Rosenberg has written that because of uncertainties in the research process, "It is more likely to prove successful, and far more likely to proceed efficiently, if decision making is sequential in nature." "Economic Experiments," in *Exploring the Black Box: Technology, Economics, and History* (Cambridge: Cambridge University

Press, 1994), 93–94. He continues this way: "The trouble with a War on Cancer, as with Supersonic transport, is that large amounts of money are committed to a particular project, or to a particular design, long before sufficient information is available to permit an intelligent evaluation of the research options. Where government funds are involved the problem is likely to be compounded by the failure to terminate a project, or a design, even after compelling evidence has accumulated that the thrust of R&D expenditures is in a direction that is unlikely to yield acceptable results."

While RCA did not have a large financial commitment to CMOS and SOS, it made a large intellectual commitment to these technologies very early on. This commitment was then reinforced by military funding.

SIX • THE END OF RESEARCH

1. An award-winning history of post–World War II America calls 1968 "the most turbulent year." James T. Patterson, *Grand Expectations: The United States, 1945–1974* (Oxford: Oxford University Press, 1996), 678–709. The comments of leaders in the electronics industry on the assassination of Kennedy are reported in "Industry Leaders Join in Kennedy Tributes," *Electronic News*, 10 June 1968, 1, 13.

2. In 1973, Gordon Moore said: "We are really the revolutionaries in the world today— not the kids with the long hair and beards who were wrecking the schools a few years ago." Quoted in Gene Bylinsky, "How Intel Won Its Bet on Memory Chips," *Fortune*, November 1973, 143–44. Moore has said that this comparison was in his mind for several years before this article. Gordon Moore, interview by Ross Bassett and Christophe Lécuyer, 18 February 1997.

3. Alfred Chandler defines first movers as "the pioneers or other entrepreneurs who made the three interrelated sets of investments in production, distribution, and management required to achieve the competitive advantages of scale, scope or both." Alfred D. Chandler Jr., *Scale and Scope: The Dynamics of Industrial Capitalism* (Cambridge, Mass.: Harvard University Press, 1990), 35.

4. Gordon E. Moore, "Some Personal Perspectives on Research in the Semiconductor Industry," in *Engines of Innovation: U.S. Industrial Research at the End of an Era*, ed. Richard S. Rosenbloom and William J. Spencer (Boston: Harvard Business School Press, 1996), 171; AnnaLee Saxenian, *Regional Advantage: Culture and Competition in Silicon Valley and Route 128* (Cambridge, Mass.: Harvard University Press, 1994).

5. Gordon E. Moore, "Cramming More Components onto Integrated Circuits," *Electronics*, 19 April 1965, 114–17.

6. Christophe Lécuyer, "Making Silicon Valley: Engineering Culture, Innovation, and Industrial Growth, 1930–1970" (Ph.D. diss., Stanford University, 1999), 166; Chih-Tang Sah, "Evolution of the MOS Transistor—From Conception to VLSI," *Proceedings of the IEEE* 76 (October 1988): 1293; "Offer $120 Flip-Flop Element at Fairchild Semiconductor," *Electronic News*, 20 March 1961, 128; "The 30th Anniversary of the Integrated Circuit: Thirty Who Made a Difference," *Electronic Engineering Times*, September 1988, 60–62.

7. "Offer $120 Flip-Flop Element at Fairchild Semiconductor," 128; Richard Russack, "4 at Fairchild Resign: Plan to Form Firm," *Electronic News*, 4 September 1961, 4; James

Lydon, "Integrated Circuits Spur Semicon Makers Race," *Electronic News*, 27 March 1962, 1, 8; Walter Mathews and James Lydon, "GME Plans Set for Wide Diversity," *Electronic News*, 26 August 1963, 1, 10. A 1962 Signetics price sheet appears in "Thirty Who Made a Difference," 32.

8. Semiconductor Equipment and Materials International, "Silicon Valley Genealogy," (Mountain View, Calif.: S.E.M.I., 1995). This is a poster based on Don Hoefler's work. Fairchild's movement into commercial markets is traced in Lécuyer, "Making Silicon Valley," 220–64. In 1964 Fairchild Semiconductor made substantial price cuts in its integrated circuits, reducing its prices as low as $2.55. "Market Crowding Presages Squeeze," *Electronic News*, 3 August 1964, 1, 25.

9. "Nat'l Semiconductor Moving, Realigning Top Management," *Electronic News*, 6 March 1967, 54; "2d Firm Agrees to Second-Source TI Series 54 Line," *Electronic News*, 16 October 1967, 38. For further details of Sporck's takeover at National, see Michael S. Malone, *The Big Score: The Billion Dollar Story of Silicon Valley* (Garden City, N.Y.: Doubleday, 1985), 129–35; Charles Sporck, interview by Rob Walker, 21 February 2000, Silicon Genesis: Oral History Interviews with Silicon Valley Scientists, Collection M0741, Department of Special Collections, Stanford University Libraries, Stanford, California (hereafter any reference to an interview in this collection will be cited as Silicon Genesis interviews). The former Fairchild managers who followed Sporck to National Semiconductor were Floyd Kvamme, marketing manager; Pierre Lamond, IC production manager; Roger Smullen, manufacturing manager for Ics; and Fred Bialek, overseas IC operations manager.

10. Sporck, Silicon Genesis interviews; Regis McKenna, interview by Rob Walker, 22 August 1995, Silicon Genesis interviews, Malone, *Big Score* 129–31.

11. Sporck, Silicon Genesis interviews; Malone, *Big Score*, 108–9; Tom Rowe, interview by author, 14 March 1997. Stock options were typically reserved for engineers and managers, although Sporck asserted that National gave stock options to long-serving operators.

12. A 1971 report by a consulting firm asserts that in mid-1968 Fairchild's "production yields were substantially below industry norms. In some instances, they were achieving only 20% of their best competitor." Integrated Circuit Engineering, *Status of Integrated Circuits, Midyear 1971* (Phoenix, Ariz.: Integrated Circuit Engineering, 1971), 37.

13. "Fairchild Has Introduced 52 New Products in the Last 52 Weeks," advertising copy in the author's possession. Fairchild ran weekly ads in *Electronic News* touting its product of the week. One such advertisement (week 19 of the program) is *Electronic News*, 12 February 1968, 3. The program started in October 1967.

14. "FC&I Head Resigns; Earnings Plummet," *Electronic News*, 23 October 1967, 2; Don Hoefler, "FC&I, Mountain View Breathes Easier," *Electronic News*, 30 October 1967, 16; "A High Flier Runs into Turbulent Air," *Business Week*, 23 December 1967, 76.

15. Fairchild Camera and Instrument, *Annual Report, 1967;* "The Antenna," *Electronic News*, 27 May 1968, 8; "Noyce Leaves V-P, Director Post at FC&I," *Electronic News*, 1 July 1968, 1–2; "Parent and Child," *Electronics*, 8 July 1968, 52–53; Gordon E. Moore, "Intel—Memories and the Microprocessor," *Daedalus* 125 (spring 1996): 58–59; Gordon Moore, interview by author, 15 May 1996; Gordon Moore, interview by Rob Walker, 3 March 1995, Silicon Genesis interviews.

16. Mimi Real, *A Revolution in Progress: A History of Intel to Date* (Santa Clara, Calif.: Intel

Corporation, 1984), 4–7; Tom Wolfe, "The Tinkerings of Robert Noyce," *Esquire*, December 1983, 364–67. Jean Hoerni, one of the Fairchild eight, invested in Intel even as he publicly complained that its name was too similar to the name of his company, Intersil. Hoerni said: "Just as he [Robert Noyce] will not be second-sourcing new products, we would hope that he would not second-source his company name." Gordon Moore later said in a note confirming the investment in Intel of the other six Fairchild founders that Hoerni lamented that he earned more from his investment in Intel than the companies he had founded himself. Hoerni is quoted in Don Hoefler, "Dr. Noyce Delivers Baby, Calls it Intel," *Electronic News*, 5 August 1968, 39. Gordon Moore, note to author, 23 April 2001.

17. Malone, *Big Score*, 124; "Fairchild Ends Manhunt, Elects C. L. Hogan President and Chief Executive Officer," *Electronic News*, 12 August 1968, 1, 6; "Musical Chairs," *Electronics*, 19 August 1968, 45–47; C. Lester Hogan, interview by Rob Walker, 24 January 1995, Silicon Genesis interviews. The details of Hogan's pay package at Fairchild are given in the misleadingly titled "The Fight That Fairchild Won," *Business Week*, 5 October 1968, 106.

18. Don C. Hoefler, "Fairchild Hoganized; Tom Bay Leaves Firm," *Electronic News*, 19 August 1968, 1, 14; Don C. Hoefler, "Hear Hogan Forces Grab More Bodies," *Electronic News*, 26 August 1968, 12; "Rudin, Linear IC Specialist, Resigns Position at Fairchild," *Electronic News*, 11 November 1968, 36; "Five More Resign Posts at Fairchild," *Electronic News*, 18 November 1968, 48; "Sanders Resigns New Post at FC&I," *Electronic News*, 9 December 1968, 2; "Siegel Leaves Fairchild Semicon for Intersil," *Electronic News*, 9 December 1968, 2; "4 Fairchild Grads Enter Memory Mart," *Electronic News*, 3 February 1969, 8; "The Antenna," *Electronic News*, 17 March 1969, 18. Both Tom Wolfe and Michael Malone have colorful accounts of the arrival of Hogan and his team. Wolfe, "Tinkerings of Robert Noyce," 364; Malone, *Big Score*, 110–25.

19. "Core Assault," *Electronics*, 24 June 1968, 49; Walter Barney, "MOS Bandwagon Starts Rolling," *Electronics*, 18 March 1968, 173–86.

20. Walter Johnson, "IBM Reports Major Advance in EDP Memory Technology," *Electronic News*, 25 October 1965, 6; Jack Robertson, "IBM LSI Arrays Go into System 360 Soon," *Electronic News*, 16 October 1967, 37. By the time Intel was founded, IBM had already made the decision to essentially stop magnetic core memory development and use semiconductor memories in its next line of mainframe computers. Emerson W. Pugh, Lyle R. Johnson, and John H. Palmer, *IBM's 360 and Early 370 Systems* (Cambridge, Mass.: MIT Press, 1991), 466–69. See also chapter 7.

21. Moore, interview by author; Moore, interview by Lécuyer and Bassett; Gordon E. Moore, "Semiconductor Memories," *Abstracts of the 1968 Intermag Conference*, 4.5. Moore discussed a hybrid scheme where MOS integrated circuits stored the information and bipolar integrated circuits served as the "drive electronics," which interfaced between the MOS memory and the rest of the system. The core memories used in IBM's System/360 introduced in 1965 had manufacturing costs of between one and three cents per bit. Pugh et al., *IBM's 360 and Early 370 Systems*, 475.

22. Moore, "Cramming More Components onto Integrated Circuits," 114–17. The problems the semiconductor industry had in moving towards products at higher and higher densities are discussed in "LSI Still Chases Its Press Clippings," *Electronics*, 10 June 1968, 157–63.

23. Moore, interview by author.

24. The computer-aided-design approach to integrated circuits did not become successful within the semiconductor industry until the early 1980s. Rob Walker, *Silicon Destiny: The Story of Application Specific Integrated Circuits and LSI Logic Corporation* (Milpitas, Calif.: CMC Publications, 1992).

25. Raymond M. Warner Jr., ed., *Integrated Circuits: Design Principles and Fabrication* (New York: McGraw-Hill, 1965), 296–97; Ted Jenkins, interview by author, 18 February 1997. Gordon Moore discusses Jean Hoerni's development of the gold-doped transistor process in "The Role of Fairchild in Silicon Technology in the Early Days of 'Silicon Valley,' " *Proceedings of the IEEE*, 86 (January 1998): 57. Some early Intel employees have claimed that they always believed that Intel was formed to be an MOS company, with the bipolar work pursued as a temporizing measure. But Moore has asserted that this was not so: had the bipolar work been able to win Intel a competitive advantage, Intel would have been more bipolar oriented. Moore, interview by author.

26. Moore, interview by author; Jenkins, interview. Similar work was also done in Japan at the University of Tokyo and the Electrotechnical Laboratory. Kunio Tada and Jose Luis Laraya, "Reduction of Storage Time of a Transistor Using a Schottky-Barrier Diode," *Proceedings of the IEEE* 55 (November 1967): 2064–65; Y. Tarui et al., "Transistor Schottky-Barrier Diode Integrated Logic Circuit," *1968 International Solid-State Circuits Conference: Digest of Technical Papers*, 164–65. Fairchild R&D used the term "Jenkins Diode" and in February 1968 was doing paper studies investigating its use. "Digital Integrated Electronics, Monthly Technical Summary," February 1968, 7, Box 14, Folder 6, Fairchild Camera and Instrument Technical Reports and Progress Reports, Collection M1055, Department of Special Collections, Stanford University Libraries, Stanford, California (hereafter FCI). Although Ted Jenkins came to work at Intel, his initial assignment, possibly to avoid the appearance that he had been hired to bring over the Schottky process, was on a small project in Pasadena. The sixty-four-bit bipolar memory was Intel's first product. R. N. Noyce, R. E. Bohn, and H. T. Chua, "Schottky Diodes Make IC Scene," *Electronics*, 21 July 1969, 74–80.

27. J. C. Sarace, R. E. Kerwin, D. L. Klein, and R. Edwards, "Metal-Nitride-Oxide-Silicon Field-Effect Transistors, with Self-Aligned Gates," *Solid-State Electronics* 11 (1968): 653–60; Ed Snow, interview by author, 9 January 1996; Les Vadasz, interview by author, 11 July 1996. I have discussed Bell Labs' response to the silicon gate process informally with several Bell engineers. One indication of Bell's response is that the IEEE did not recognize the silicon gate process with a formal award until 1994, when Robert Kerwin, Donald Klein, and John Sarace (all previously of Bell Labs) were given the Jack A. Morton award twenty-five years after Intel commercialized the silicon gate process. General Microelectronics also did early silicon gate work and received a patent on it. Boyd G. Watkins, "IGFET Comprising N-Type Silicon Substrate, Silicon Oxide Gate Insulator and P-Type Polycrystalline Silicon Gate Electrode," U.S. Patent 3,576,478, 27 April 1971.

28. "Digital Integrated Electronics, Monthly Technical Summary," October 1967, 4, Box 14, Folder 4, FCI; "Digital Integrated Electronics, Monthly Technical Summary," November 1967, 2, Box 14, Folder 4, FCI; "Digital Integrated Electronics, Monthly Technical Summary," February 1968, 1, 4, Box 14, Folder 6, FCI; "Digital Integrated Electronics, Monthly Technical Summary," May 1968, 4–6, Box 14, Folder 7, FCI; Federico Faggin, interview by author, 6 June 1997.

29. Vadasz, interview; Edward Gelbach, Andrew Grove, and Ted Jenkins, interview, 24 October 1983, 10, Intel Museum, Santa Clara, California.

30. Eugene Flath, interview by author, 2 March 1997. Flath was the fifth person on board at Intel. Flath estimated that he had responsibility for roughly 70 percent of Fairchild's digital integrated circuits. The names of early Intel employees, their order hired, and in some cases their month of hire were provided in "Intel Corporation, Original Employee List," undated document and a CD consisting of badge photos, both provided by the Intel Museum, Santa Clara, California.

31. Flath, interview; Tom Rowe, interview, 14 March 1997; George Staudacher, interview by author, 9 March 1997.

32. Flath, interview; Snow, interview. For example, Ed Snow, who had done the work on the effect of sodium on MOS transistors, was interested in leaving Fairchild to go to Intel but was told by Grove, who had been another member of the Fairchild R&D group working on MOS stability problems, that the two of them duplicated each other and Intel could not afford any duplication.

33. Flath, interview; Tom Rowe, interview, 14 March 1997; Tom Rowe, telephone conversation with author, 24 February 1996; Bob Holmstrom, interview by author, 21 February 1997. The semiconductor industry recognized the ability to apply thin films of material onto silicon or silicon dioxide as a distinct craft. Although different materials might be used, the techniques and equipment were common. See, for example, Leon I. Maissel and Reinhard Glang, eds., *Handbook of Thin Film Technology* (New York: McGraw-Hill, 1970). Holmstrom was what would be called a thin films man.

34. Joel Karp, interview by author, 29 January 1997; Bob Abbott, interview by author, 3 March 1997; John Reed, note to author, 26 November 1996; Brian Santo, "1K-Bit DRAM," *IEEE Spectrum*, November 1988, 108–12.

35. Holmstrom, interview; Abbott, interview.

36. Staudacher, interview. One might speculate that had Fairchild simply gone out of business in August 1968, the effect would have been much less advantageous for Intel. It would not have been able to absorb all the people it could potentially use. These people would have had to find permanent jobs, perhaps outside the area, and would have likely been unavailable to Intel at a later date.

37. "Fairchild Fights Back," *Electronics*, 4 January 1971, 75.

38. "Musical Chairs," *Electronics*, 19 August 1968, 45.

39. Gordon Moore, Gerhard Parker, and Les Vadasz, interview, 17 October 1983, 50–51, Intel Museum, Santa Clara, California.

40. Gelbach, Grove, and Jenkins, interview, 11–12.

41. Moore, "Cramming More Components onto Integrated Circuits," 116.

42. Real, *Revolution in Progress*, 6–7.

43. Jenkins, interview; Tom Rowe, interview, 14 March 1997; Snow, interview.

44. Tom Rowe, interview, 14 March 1997.

45. Ibid.; Flath, interview; Real, *Revolution in Progress*, 6.

46. Tom Rowe, interview, 14 March 1997; Flath, interview; Real, *Revolution in Progress*, 6. The stakes were a bottle of champagne.

47. Tom Rowe, interview, 14 March 1997.

48. Ibid.; Flath, interview.

49. Ibid. Fairchild R&D experimented with this approach in August 1968, and it is possible that people from Intel knew of this. "Digital Integrated Electronics, Monthly Technical Summary," August 1968, 3, Box 15, Folder 2, FCI.

50. Tom Rowe, interview, 14 March 1997. My interview with Rowe was supplemented by a cross-sectional drawing of the silicon gate process dated 12 March 1997 and a note dated 13 March 1997 in which Rowe described the key elements of the process.

51. Ibid.

52. Flath, interview; Tom Rowe, interview, 14 March 1997. Flath recalled a conversation with Robert Noyce where Noyce stated the advantage of buying rather than making production equipment was that it enabled Intel to spend money to buy time. Intel's jig is consistent with Eric von Hippel's finding that in the semiconductor equipment business many of the innovations have originated with semiconductor producers, for they stand to benefit most by such work. Eric von Hippel, *The Sources of Innovation* (New York: Oxford University Press, 1988).

53. Tom Rowe, interview, 14 March 1997; Moore, interview by author; Jenkins, interview; Karp, interview.

54. Flath, interview; Tom Rowe, interview, 14 March 1997; Jenkins, interview. Describing these meetings, Flath stated: "The ideas flew so fast and furiously I had no idea where the ideas came from."

55. Tom Rowe, interview, 14 March 1997.

56. Ibid.

57. Ibid.; Real, *Revolution in Progress*, 10.

58. Tom Rowe, interview, 14 March 1997.

59. Ibid.

60. L. L. Vadasz, A. S. Grove, T. A. Rowe, and G. E. Moore, "Silicon-Gate Technology," *IEEE Spectrum*, October 1969, 28–35. The key features of Intel's silicon gate process are from Tom Rowe, note to author, 13 March 1997. For the sake of comparison, workers at Fairchild R&D published an article on their silicon gate process that provided much more operational information. F. Faggin and T. Klein, "Silicon Gate Technology," *Solid-State Electronics* 13 (1970): 1125–44. The article was submitted just after the Intel article was published.

61. A 1997 reference guide to semiconductor processing contains the following statements in its chapter on metal deposition: "Several methods are used to ensure good step coverage. Primary is a planetary domed wafer holder." Peter van Zant, *Microchip Fabrication: A Practical Guide to Semiconductor Processing*, 3d ed. (New York: McGraw-Hill, 1997), 404.

62. "Digital Integrated Electronics, Monthly Technical Summary," September 1969, 12, Box 16, Folder 2, FCI; "Digital Integrated Electronics, Monthly Technical Summary," December 1969, 2, Box 16, Folder 5, FCI. The "rock stable" quote is from September.

63. A 1969 tutorial article on solid-state memories by Ted Hoff makes no mention of dynamic RAM. Marcian E. Hoff, "Solid-State Memory Offers New Organizations," *EDN*, 15 October 1969, 55–61.

64. Karp, interview. An example of the Fairchild approach is given by Dov Frohman-Bentchkowsky and Leslie Vadasz, "Computer-Aided Design and Characterization of Digital MOS Integrated Circuits," *IEEE Journal of Solid-State Circuits* 4 (April 1969): 57–64.

65. "First Silicon Gate MOS," Intel Advertisement, *Electronic News,* 1 September 1969, 25; "2d Intel Product: LSI Memory Chip," *Electronic News,* 8 September 1969, 2. In January 1970 Intel reduced the price of the 1101 to $23.50 in large volumes, but this was still roughly 10 cents per bit. "Intel Slices 2 Circuit Prices," *Electronic News,* 26 January 1970, 14. Real, *Revolution in Progress,* 10–11.

66. Moore, Parker, and Vadasz, interview, 21–23; Intel Advertisement, *Electronics,* 8 June 1970, 56; "Intel Reduces Prices 65% on 8 MOS Shift Registers," *Electronic News,* 20 July 1970, 33.

67. Semiconductor Equipment and Materials International, "Silicon Valley Genealogy."

68. "If Cogar Can Keep from Being Suited Up," *Datamation,* October 1969, 147–48; "Where the Action Is in Electronics," *Business Week,* 4 October 1969, 86–98. Cogar is discussed in greater detail in chapter 7.

69. "Advanced Memory Systems, Corporate Summary Profile," Raytheon Company, Electronics Industry Files, Collection M0661, Box 6, Folder 2, Department of Special Collections, Stanford University Libraries, Stanford, California (hereafter Raytheon Collection); Advanced Memory Systems, "Preliminary Product Specification, AMS91600111," June 1970, Raytheon Collection; Advanced Memory Systems, *Annual Report, 1970.*

70. *Mostek's 15th Anniversary* (Carrollton, Tex.: Mostek, 1984); "Pair of TI Aces Form MOS Company," *Electronics,* 14 April 1969, 34; "Who's Who in Electronics," *Electronics,* 28 April 1969, 14; "Sprague, Mostek Link Up for Slice of MOS Market," *Electronic News,* 30 June 1969, 40. I thank Anne Millbrooke for providing me with Mostek's fifteenth anniversary newspaper.

71. Lee Boysel, interview by author, 28 February 1997; Lee Boysel, Wallace Chan, and Jack Faith, "Random-Access MOS Memory Packs More Bits to the Chip," *Electronics,* 16 February 1970, 109–15.

72. Blyth & Co., "Confidential Report on Four-Phase Systems, Inc.," 30 April 1971 (in author's possession); Boysel, interview, 28 February 1997. (I thank Lee Boysel for providing me with this document.)

73. "IBM Making Semicon Memory," *Electronic News,* 24 February 1969, 41; Martin Gold, "IC Producers Jam Memory Cost Talks," *Electronic News,* 3 November 1969, 1, 49.

74. "New MOS Firm Bows, Eyes Custom Wafers," *Electronic News,* 19 May 1969, 43; Bob Cole, interview by author, 13 December 1995; Karp, interview. For more on Cartesian, see chapter 8.

75. Santo, "1K-Bit DRAM," 108–12; Karp, interview; George Sideris, "The Intel 1103: The MOS Memory That Defied Cores," *Electronics,* 26 April 1973, 108–13. Real, *Revolution in Progress,* 10–11, also discusses the history of the 1103.

76. Karp, interview; Santo, "1K-Bit DRAM," 108–10; Sideris, "Intel 1103," 108–11.

77. Santo, "1K-Bit DRAM," 110–11; Karp, interview. One of the customers Intel had in mind for the 1103, Data General, very publicly backed out, stating that the 1103 could not meet its speed requirements. In what was seen as a setback for MOS memory in general, Data General shifted to a bipolar memory produced by Fairchild. Don C. Hoefler, "Data Gen'l Shifts Supernova to Bipolar," *Electronic News,* 7 June 1971, 1, 18.

78. Santo, "1K-Bit DRAM," 111.

79. Sideris, "Intel 1103," 110.

80. "MOS Firms Trying 1103 Market Again," *Electronic News*, 21 June 1971, 1; Intel Advertisement, *Electronic News*, 21 June 1971, 14–15; Martin Gold and James Moran, "MOS Sales Surge," *Electronic News*, 28 June 1971, 1, 23; "As 1024 RAMs Firm," *Electronic News*, 28 June 1971, 1, 14. Ed Gelbach recalled that although many 1103s were being sold, for a long time no computer systems using 1103s were being shipped to customers. Gelbach, Grove, and Jenkins, interview, 18.

81. Don C. Hoefler, "Second Sources Confident on 1103 Despite Setback," *Electronic News*, 14 June 1971, 1, 16; Advanced Memory Systems, "Preliminary Product Specification, 91600111," June 1970, Raytheon Collection; "Advanced Memory to Ship 1st MOS 360 Extension," *Electronic News*, 7 June 1971, 55; Advanced Memory Systems, *Annual Report, 1970*; Intel Corporation, *Annual Report, 1972*. An engineer from Burroughs, one of the leading consumers of the 1103, said: "The 1103 is like Listerine. Everyone hates its taste, but everyone uses it." J. Reese Brown, "1103 Semiconductor Memory Device Pattern Sensitivities and Error Modes," *1973 Symposium on Semiconductor Memory Testing, Digest of Papers*, 74. Among the works by economists on the establishment of technical standards and why the optimum technical solution does not always become the standard, see Paul A. David, "Clio and the Economics of QWERTY," *American Economic Review* (May 1985): 332–37; and "The Hero and the Herd in Technological History: Reflections on Thomas Edison and the Battle of the Systems," in *Favorites of Fortune: Technology, Growth, and Economic Development since the Industrial Revolution*, ed. Patrice Higonnet, David S. Landes, and Henry Rosovsky (Cambridge, Mass.: Harvard University Press, 1991), 72–119.

82. John Rhea, "One Recipe, Two Cooks. Or the Wafer That Wouldn't," *Electronic News*, 11 March 1974, 1, 54. An engineer with Burroughs estimated that it bought over 60 percent of its 1103 DRAMs from Intel. The primary reason was that Intel was able to consistently deliver its product. John Durkin, interview by author, 17 April 1997. Microsystems International Limited, after a strong start, suffered major problems in making the transition from two- to three-inch wafers.

83. Martin Gold, "Intel Describes 4096-Bit Chip," *Electronic News*, 21 February 1972, 1, 19. At this time some controversy in the industry existed over whether four-kilobit RAMs or two-kilobit RAMs should be the industry's next step. Had two-kilobit RAMs been produced it might have slowed the Moore's Law doubling time, as firms would have looked to keep that particular product long enough to earn a return on their investment.

84. Gordon Bell and William D. Strecker, "Computer Structures: What Have We Learned from the PDP-11?" *Proceedings of the Third Annual Symposium on Computer Architecture* (1976), 1–14; Gordon Bell and W. D. Strecker, "Retrospective: What Have We Learned from the PDP-11-What We Have Learned from the VAX and Alpha," *25 Years of the International Symposia on Computer Architecture (Selected Papers)* (1998): 6–10. John L. Hennessy and David A. Patterson observe that at the time the PDP-11 architecture was developed, magnetic core memories were slowly increasing in size. *Computer Architecture: A Quantitative Approach* (San Mateo, Calif.: Morgan Kaufmann, 1990), 485. C. Gordon Bell, a key figure in the development of the PDP-11 and the later VAX, asserted that he had not read Moore's 1965 paper, but "discovered" something analogous to it in 1972. C. Gordon Bell, note to author, 16 May

2001. Articles on the PDP-8, PDP-11, and the VAX are included in Daniel Siewiorek, C. Gordon Bell, and Allen Newell, *Computer Structures: Principles and Examples* (New York: McGraw-Hill, 1982).

85. Wolfe, "Tinkerings of Robert Noyce," 368.

86. Chandler Jr., *Scale and Scope*, 17. The number of products introduced by Intel is given in Intel Corporation, *Annual Report, 1975*, 12.

87. Intel Advertisement, *Electronic News*, 31 January 1972, 40–41; Intel Advertisement, *Electronic News*, 2 October 1972, 37.

88. Gelbach, Grove, and Jenkins, interview, 37–44; Moore, "Intel—Memories and the Microprocessor," 68–69; Real, *Revolution in Progress*, 21–22.

89. Gelbach, Grove, and Jenkins, interview, 37–44.

90. Ibid., 44; Moore, "Intel—Memories and the Microprocessor," 74–75.

91. Sideris, "Intel 1103," 112; Moore, "Intel—Memories and the Microprocessor," 65–67. Moore called the 1103 "one of the most difficult to use integrated circuits ever produced."

92. Moore, Parker, and Vadasz, interview, 90–93; Moore, "Intel—Memories and the Microprocessor," 67; Real, *Revolution in Progress*, 32–33; "The Antenna," *Electronic News*, 9 August 1971, 12; "Intel Markets 1103s on Cards in Subsystems Push," *Electronic News*, 18 October 1971, 36; "Intel to Widen IBM/370 MOS Line," *Electronic News*, 2 April 1973, 56. In the Intel interview, Moore estimated that IBM's products accounted for roughly 60 to 70 percent of the total market for semiconductor memory.

93. Bill Regitz, interview by author, 21 February 1997; Real, *Revolution in Progress*, 32–33; Moore, "Intel—Memories and the Microprocessor," 67.

94. "COS/MOS Pact Awarded to RCA by Electro/Data," *Electronic News*, 21 December 1970, 25; Don C. Hoefler, "Intersil Gets Seiko Pact," *Electronic News*, 25 January 1971, 22; "Motorola Entering Watch Parts Field," *Electronic News*, 17 January 1972, 6; "Solitron/Riker Firm Eyes Watch Market," *Electronic News*, 16 October 1972, 46; "GE, SSSI Plan Firm to Produce Watch Modules," *Electronic News*, 23 October 1972, 31.

95. Real, *Revolution in Progress*, 38–39; "Intel Acquires Microma, Watch Movement Producer," *Electronic News*, 12 June 1972, 21; Robert W. Wilson, Peter K. Ashton, and Thomas P. Egan, *Innovation, Competition, and Government Policy in the Semiconductor Industry* (Lexington, Mass.: D.C. Heath, 1980), 100–101; Ernest Braun and Stuart Macdonald, *Revolution in Miniature: The History and Impact of Semiconductor Electronics*, 2d ed. (Cambridge: Cambridge University Press, 1982), 192. Intel's watch experience demonstrates the difficulties of determining what constitutes economies of scope. Intel got into the watch business because it believed its MOS processing skills gave it a competitive advantage. After several years the price of the digital electronics component of the watch had fallen to the point where it represented a tiny fraction of the watch's cost. At this point Intel recognized that the key skills needed to succeed in the watch business were mass consumer marketing, where it had no special capabilities. Gelbach, Grove, and Jenkins, interview, 75–77.

96. Roger Borovoy, interview by Christophe Lécuyer and Ross Bassett, 14 June 1996; Tom Rowe, interview, 14 March 1997. A 1977 consulting report on the industry stated: "It is interesting to speculate whether Intel will ever resort to patent protection of its high technology." Integrated Circuit Engineering, *Status '77: A Report on the Integrated Circuit Industry* (Scottsdale, Ariz.: Integrated Circuit Engineering, 1977), 3–6.

97. Jenkins, interview; Tom Rowe, interview, 14 March 1997.

98. *Microelectronics News,* 17 January 1976, 1–2.

99. *Microelectronics News,* 17 January 1976, 3; *Microelectronics News,* 24 January 1976, 1–2; *Microelectronics News,* 31 January 1976, 2. The quote is from *Microelectronics News,* 17 January 1976, 3. Don Hoefler, the editor of *Microelectronics News* and the leading journalist of Silicon Valley—the person who coined the term—took up the cause of AMS because he believed Intel was violating basic Silicon Valley norms.

100. Fairchild R&D provided a number of Ph.D. semiconductor technologists who joined Intel in the years after it was founded and went on to become senior managers. This group included Albert Yu, Sunlin Chou, and Ron Whittier.

101. *International Electron Devices Meeting Technical Digest,* 1968–75.

102. Moore, "Some Personal Perspectives on Research in the Semiconductor Industry," 168.

103. In 1994, in commemoration of its fortieth anniversary, the IEDM published an anniversary volume listing the most important papers presented between 1955 and 1984. The only Intel papers were two industry overviews given by Gordon Moore. *Fortieth Anniversary of the International Electron Devices Meeting, Technical Survey.* I thank William Van Der Vort of the IEEE Electron Device Society for making this volume available to me. Moore has made the following comments about Intel's relationship to other R&D labs: "In its early years, the company [Intel] looked to Bell Labs for basic materials and science related to semiconductor devices; it looked to RCA's Princeton labs for consumer-oriented product ideas; and it sought insights into basic materials problems and metallurgy from the laboratories of General Electric. Over time, Intel found that most of the basic R&D relevant to its needs was being done by companies such as Fairchild and Texas Instruments, which had evolved into the product leaders in semiconductors." Moore, "Some Personal Perspectives on Research in the Semiconductor Industry," 170.

104. D. J. Fitzgerald, G. H. Parker, and P. Spiegel, "Reliability Studies of MOS Si-Gate Arrays," *Reliability Physics, 1971, 9th Annual Proceedings,* 57–61; R. T. Jenkins and D. J. Fitzgerald, "Reliability Studies of LSI Arrays Employing Al-Si Schottky Barrier Devices," *Reliability Physics 1970, 8th Annual Proceedings,* 34; G. H. Parker, J. C. Cornet, and W. S. Pinter, "Reliability Consideration in the Design and Fabrication of Polysilicon Fusable Link PROM's," *Reliability Physics 1974, 12th Annual Proceedings,* 88–89. Again, Intel had representation in the leadership of these meetings, with Gerhard Parker on the board of directors during much of this period.

105. Intel, *Annual Report, 1974,* 2; Intel, *Annual Report, 1975,* 1; Flath, interview. Flath estimated that in 1975 bipolar wafer starts were one-fifth of the MOS wafer starts. Integrated Circuit Engineering, a consulting firm, estimated that in 1974, Intel's MOS sales were five times its bipolar sales. Integrated Circuit Engineering, *Status 1976: A Report on the Integrated Circuit Industry* (Scottsdale, Ariz.: Integrated Circuit Engineering, 1976), 24.

106. Rob Walker, note to author, 22 April 2001; "The Antenna," *Electronic News,* 19 October 1970, 8. Some of the veterans from Bell Labs who joined Fairchild under Hogan include Jim Early, John Atalla, and Gil Amelio. James Early, interview by author, 13 January 1995. John Atalla, interview by author, 19 October 1994. Walker asserts that Hogan had his own private bathroom installed at Fairchild headquarters.

107. Walker, *Silicon Destiny*, 22–23. In 1975 *Business Week* asserted that Fairchild's MOS operations had had five managers in the previous seven years. "'Creative Tension' in a Growth Company," *Business Week*, 6 October 1975, 58–59. The case of Roy Pollack typifies Fairchild's MOS program during this period. In April 1971 Pollack came to Fairchild from RCA as the head of Fairchild's MOS operations. In September 1971 Pollack announced that Fairchild would not put a priority effort on large-scale standard products—precisely what Intel was focusing on. The next year Pollack announced: "After extensive evaluation of business alternatives, we have decided to redirect the efforts of our organization to much greater emphasis on the development and production of standard products." The next year Pollack was gone, back at RCA. Martin Gold, "Pollack Leaves RCA for FC&I," *Electronic News*, 26 April 1971, 1, 2; Roy H. Pollack and Bernie Marren, "Program Priority Relative to Large Scale Standards," 2 September 1971, Box 6, Rob Walker Papers, Collection M0742, Department of Special Collections, Stanford University Libraries, Stanford, California (hereafter Rob Walker Papers); Roy H. Pollack, "Organization Announcement," 7 September 1972, Box 6, Rob Walker Papers; "RCA TV Position Filled by Pollack," *Electronic News*, 14 May 1973, 83.

108. "FC&I Alters Forecast for Semicon Memories," *Electronic News*, 19 May 1971, 11; Integrated Circuit Engineering, *Status 1976*, 18–19; Integrated Circuit Engineering, *Status 1978: A Report on the Integrated Circuit Industry* (Scottsdale, Ariz.: Integrated Circuit Engineering, 1978), 2:8–9.

109. "Fairchild Bets on a Memory Chip," *Business Week*, 20 January 1975, 26; "FC&I Demonstrates CCDs," *Electronic News*, 26 March 1973, 72. One impetus to Fairchild's CCD work was Gil Amelio, who came from Bell Labs, where the CCD had been invented. Intel also worked on CCDs, but without the same enthusiasm as Fairchild. Real, *Revolution in Progress*, 37–38.

110. "Fairchild's Radical Process for Building Bipolar VLSI," *Electronics*, 4 September 1986, 55–59. On Fairchild's problems under Schlumberger, see David Sylvester, "Fairchild Flounders under Schlumberger," *San Jose Mercury News*, 10 March 1985, F1–2. For several perspectives on Fairchild's troubles in this period, see Charles Sporck, Silicon Genesis interviews; Gil Amelio, interview by Rob Walker, 24 March 2000, Silicon Genesis interviews; Richard B. Schmitt, "Schlumberger Reaches Accord to Sell Fairchild—Agreement With National Semiconductor Corp. Valued at $122 Million," *Wall Street Journal*, 1 September 1987, 1. The consulting firm Integrated Circuit Engineering estimated Fairchild's 1985 MOS sales at $25 million, accounting for roughly 5 percent of its total. This placed it in twenty-seventh place in MOS sales among the 50 companies tracked. Integrated Circuit Engineering, *Status 1987: A Report on the Integrated Circuit Industry* (Scottsdale, Ariz.: Integrated Circuit Engineering: 1987), 2-11–2-12.

SEVEN • IBM

1. Alfred D. Chandler Jr., *The Visible Hand: The Managerial Revolution in American Business* (Cambridge, Mass.: Harvard University Press, 1977); Alfred D. Chandler Jr., *Scale and Scope: The Dynamics of Industrial Capitalism* (Cambridge, Mass.: Harvard University Press, 1990); Alfred D. Chandler Jr., *Strategy and Structure: Chapters in the History of the American Industrial Enterprise* (Cambridge, Mass.: MIT Press, 1962).

2. Karl Ganzhorn, interview by author, 22 July 1997.

3. Wolfgang Liebmann, interview by author, 7 June 1996; L. L. Rosier, C. Turrel, and W. K. Liebmann, "Semiconductor Crosspoints," *IBM Journal of Research and Development* 13 (July 1969): 439–46.

4. Liebmann, interview; Robert Meade, interview by author, 25 February 1995. Liebmann's biographical details are given in *IBM Journal of Research and Development* 13 (July 1969): 483. Liebmann noted that with his background in compound semiconductor materials at RCA and IBM, the MOS transistor was a distinct step down in difficulty. This is much like the situation that occurred at RCA (chapter 1).

5. R. M. Meade, "Products Procedure," 11 May 1967, Box TAR 243, IBM Technical History Project, IBM Archives, Somers, New York (hereafter IBM THP).

6. Meade, interview; Sol Triebwasser, Robert Dennard, and Lewis Terman, interview by author, 1 September 1994.

7. R. T. Miller, "History and Impact of LSI in IBM," 14 August 1969, Box TAR 242, IBM THP; Dale Critchlow, note to author, 15 May 1995; George Cheroff, interview by author, 21 August 1995.

8. Dick Gladu, interview by author, 30 November 1995; Dick Linton, interview by author, 1 September 1994; W. B. Ittner III to R. P. Pecoraro, "SDD Memory Program," 5 June 1967, Box TAR 243, IBM THP. One young engineer who joined Gladu's group and whose career took off with the MOS work was Nick Donofrio, who has held a number of senior management positions within IBM and as of 2000 was a senior vice president with responsibility for technology and manufacturing. IBM Corporation, *Annual Report 2000*, 96.

9. Robert H. Dennard, "Evolution of the MOSFET Dynamic RAM—A Personal View," *IEEE Transactions on Electron Devices* 31 (November 1984): 1549–55; Triebwasser, Dennard, and Terman, interview.

10. Dennard, "Evolution of the MOSFET Dynamic RAM," 1550–51; Triebwasser, Dennard, and Terman, interview; Dale Critchlow, letter to author, 23 November 1997. Had IBM been doing *p*-channel MOS, it would have been easier to implement Dennard's idea.

11. Triebwasser, Dennard, and Terman, interview; Merlin Smith, letter to author, 16 January 1998; R. H. Dennard, "Field-Effect Transistor Memory," U.S. Patent 3,387,286, 4 June 1968; R. H. Dennard, M. G. Smith, and L. M. Terman, "Low Power FET Memories," ITL Component Development and Packaging Meeting, 26 April 1967. Merlin Smith, who was assigned to evaluate Dennard's disclosure and recognized its importance, asserted that one IBM patent attorney worried that the simplicity of Dennard's invention (using only one transistor and one capacitor) made it obvious and thus unpatentable. Smith, Critchlow, and Dennard successfully impressed upon the IBM patent attorneys the significance of the work so that it was quickly pushed through the system. The patent was issued less than eighteen months after Dennard conceived the idea, an extraordinarily short period of time. For a number of reasons (among them, the lack of a publication, the delay between the time of invention and the time of its exploitation, and IBM's position outside the semiconductor industry), Dennard did not get the recognition in the field that he might have otherwise. But the one-device cell has been the standard in dynamic RAMs for over twenty-five years. One engineer called it "the most abundant man-made object on this planet earth." C. T. Sah, "Evolution of the MOS Transistor—From Conception to VLSI," *Proceedings of the IEEE* 76

(October 1988): 1301. Dennard sent me figures from the Nikkei Market Access survey projecting the demand for DRAM in 2001 would reach 370×10^{15} bits. Robert Dennard, note to author, 6 June 2001.

12. Miller, "History and Impact of LSI in IBM"; Linton, interview.

13. Miller, "History and Impact of LSI in IBM"; R. W. Bartolus et al., "FET Cost-Performance Main Storage Design," 6 September 1968, Box TAR 243, IBM THP; W. F. Bankowski et al., "FET Technology Feasibility Model," 20 December 1968, Box TAR 243, IBM THP.

14. Triebwasser, Dennard, and Terman, interview. In December 1968 Dale Critchlow, a manager in the Research MOS group, wrote an impassioned memo arguing for the importance of continued Research involvement in MOS technology, claiming that it presented possibilities that CD would not be able to address. By 1970 Research had resumed MOS work. As of this date (2001) there are people in Research, most notably Lew Terman and Bob Dennard, who have spent essentially their entire careers, spanning over thirty-five years, on MOS technology. D. L. Critchlow, "Research Program on Monolithic Memory," 31 December 1968, Box TAR 243, IBM THP.

15. D. L. Critchlow, R. H. Dennard, S. E. Schuster, and E. Y. Rocher, "Design of Insulated-Gate Field-Effect Transistors and Large Scale Integrated Circuit Chips for Memory and Logic Applications," RC 2290, 4 October 1968, IBM Research, Yorktown Heights, New York; Dale Critchlow, note to author, 14 December 1995.

16. Critchlow et al., "Design of Insulated-Gate Field-Effect Transistors"; Dale Critchlow, note to author, 18 December 1995.

17. Dale Critchlow, note to author, 14 December 1995; Dale Critchlow, note to author, 18 December 1995. The manual began by stating: "The main emphasis will be on the more practical design aspects of the devices which are not available in the literature." Critchlow et al., "Design of Insulated-Gate Field-Effect Transistors," 1. Critchlow, who had a long and distinguished career with IBM, said that the most important thing he ever did for the company was to make sure that the design manual was well written and widely distributed throughout the company.

18. Critchlow et al., "Design of Insulated-Gate Field-Effect Transistors," 111–32, 141–70.

19. As an example of how far ahead of its time this program was (and how short institutional memories can be), in the 1980s IBM established joint programs between Research and development organizations as a standard operating procedure to speed technology transfer. IBM's director of research described these joint programs with no apparent awareness of the precedent in the joint MOS program. John Armstrong, "Reinventing Research at IBM," in *Engines of Innovation: U.S. Industrial Research at the End of an Era*, ed. Richard S. Rosenbloom and William J. Spencer (Boston: Harvard Business School Press, 1996), 151–54.

20. Ed Davis, interview by Emerson Pugh, 17 July 1986, 27 April 1987, IBM Technical History Project Oral Histories (*IBM's 360 and Early 370 Systems*), Hagley Museum and Library, Wilmington, Delaware (hereafter these interviews will be cited as IBM THP interviews, with the initial citation for each interview identifying the volume for which the interview was conducted).

21. Emerson W. Pugh, Lyle R. Johnson, and John H. Palmer, *IBM's 360 and Early 370*

Systems (Cambridge, Mass.: MIT Press, 1991), 467–71. In a presentation to the Management Committee in December 1968, Davis asserted that ferrite work was being slowed "because of the high investment required to achieve a relatively small cost improvement compared to monolithics [semiconductor memory]." "Management Committee Minutes," 19 December 1968, Plaintiff exhibit (hereafter PX) 3242, *U.S. v. IBM,* Hagley Museum and Library, Wilmington, Delaware (hereafter *U.S. v. IBM*).

22. "M.C. Presentation," 24 January 1968, Box TBR 113, IBM THP; "Minutes of 1/25 [1968] MRC Meeting," PX 2103, *U.S. v. IBM,* 3; F. T. Cary, "Memory Technology," 15 January 1968, Box TAR 242, IBM THP; Davis, IBM THP interview, 27 April 1987, 39–40. At the January 1968 meeting, Davis reported that he had a backup plan such that even a year and a half later, IBM could shift back to core memories without slipping the delivery schedule of its systems.

23. H. E. Cooley to F. T. Cary, "Memory Review with Corporate Staff," 17 January 1968, Box TAR 242, IBM THP.

24. B. N. Slade, "Report on IGFET [MOS] Study," 23 January 1968, Box TAR 242, IBM THP; Meade, interview.

25. Slade, "Report on IGFET"; Ed Davis's representative on this study recalls that Davis's goals were for the study to keep both bipolar and MOS efforts going. Meade, interview.

26. J. A. Haddad, "NS Memory Program," 24 January 1968, Box TAR 242, IBM THP; J. A. Haddad, "Phase 2I Memory Program," 24 January 1968, Box TAR 242, IBM THP. One of Davis's senior lieutenants recalls being confident early on that MOS memories would be used by IBM if they could be made to work. Robert Elfant, interview by author, 22 August 1997.

27. Pugh et al., *IBM's 360 and Early 370 Systems,* 413–20. Further discussion of the history of cache memory is provided in John L. Hennessy and David A. Patterson, *Computer Architecture: A Quantitative Approach* (San Mateo, Calif.: Morgan Kaufmann, 1990), 486.

28. This is discussed in G. Robert Gunther-Mohr, interview by author, 21 March 1995.

29. Pete Fagg and Ed Davis, "IGFET's," 16 February 1968, Box TAR 242, IBM THP.

30. When Davis (and the managers above and below him) presented the decision to use semiconductor memory to IBM's top executives, one of the points they made was that it would not involve establishing a new "plant location"—that is to say, the semiconductor memory development and manufacturing would be carried on at one of IBM's existing locations. "Minutes of 1/25 MRC Meeting," PX 2103, *U.S. v. IBM.*

31. R. M. Meade, "Monolithic Memory Manufacturing Plans," 13 March 1968, Box TAR 242, IBM THP; Davis, IBM THP interview, 27 April 1987, 38–43.

32. Davis, IBM THP interview, 27 April 1987, 29–30, 35–37.

33. D. E. Rosenheim to E. M. Davis, "Resources for FET Monolithic Memory Program in CD," 7 June 1968, Box TMA 120, IBM THP; E. M. Davis to D. E. Rosenheim, "Resources for FET Monolithic Memory Program in CD," 16 August 1968, Box TMA 120, IBM THP. These positions may have been unappealing to Research people because what might be seen as the most interesting work (designing the architecture of the memory chip and the process) had been done in the joint program. In any case it was a no-lose situation for Davis, for if they accepted, he got a talented group of people, while if they declined, he could claim that he had done his best to advance the MOS technology work. Of those in the Research MOS

program, only two people, Joe Shepard and George Cheroff, moved to MOS work in East Fishkill. (Cheroff remained only a short time.) Both of them did not have Ph.D.'s, a fact that may have circumscribed their career opportunities at Research.

34. Joe Shepard, interview by author, 20 June 1995.

35. William North, interview by author, 7 July 1997; Emerson Pugh, "Fred Neves, FET/CP," 16 May 1979, 22, Box TBR 102, IBM THP. These are notes from a history of IBM's MOS memory effort written for IBM's use in the government's antitrust case. One senior manager responsible for MOS work under Davis who cannot even remember Boeblingen ever having responsibility for MOS memory work asserted that Davis would ultimately never let a program as important as MOS get out of his control. Elfant, interview.

36. International Business Machines Corporation, *Annual Report 1969*, 2–3; Katie Woodruff, *Defining Intel: 25 Years/25 Events* (Santa Clara, Calif.: Intel Corporation, 1993), 6; Katherine Davis Fishman, *The Computer Establishment* (New York: Harper & Row, 1981), 384; Dale L. Critchlow, "The Inside Out World of Silicon," 29 September 1976, Box TAR 243, IBM THP; Elfant, interview. Gordon Moore, Gerhard Parker, and Les Vadasz, interview, 17 October 1983, 91, Intel Museum, Santa Clara, California.

37. Sah, "Evolution of the MOS Transistor," 1295. Bell Labs, which also developed semiconductors for internal markets, also persisted in using the name IGFET. See R. Edwards, "The Anatomy of IGFET's," *Bell Laboratories Record* 46 (September 1968): 246–59.

38. The 1965 International Solid-State Circuits Conference included a panel discussion debating the merits of hybrid circuits (the IBM approach) versus integrated circuits. Ed Davis made the case for hybrids and showed a film of IBM's manufacturing facility. The proponent of integrated circuits began his talk by quipping: "All of you who have one hundred million dollars for automation equipment don't have to hear anymore." The engineer daunted by IBM's approach was not from a small start-up, but Westinghouse, showing that IBM could pour resources into its semiconductor operations that dwarfed those available to even the largest corporations. "Putting Integrated Circuits in Their Place," *Electronics*, 8 March 1965, 126. For the importance of IBM's electromechanical background to its semiconductor work, see Steven W. Usselman, "IBM and Its Imitators: Organizational Capabilities and the Emergence of the International Computer Industry," *Business and Economic History* 22 (winter 1993): 1–35.

39. Shepard, interview. Shepard, a veteran of the Research MOS program who later moved to East Fishkill, said: "We were always a 1/4 inch different in wafer size, the world went 2 inch, we went 2 1/4, the world went 3, we went 3 1/4, the world went 4, we went 5, the world went 6, we went 8."

40. M. C. Duffy and R. P. Gnall, "A Survey of Diffusion Process for Fabricating Integrated Circuits," in *Microelectronic Technology: Selected Articles from Semiconductor Products and Solid State Technology*, ed. Samuel L. Marshall (Cambridge, Mass.: Boston Technical Publishers, 1967), 83–92. Duffy and Gnall, both from East Fishkill, explain the capsule diffusion process in great detail because "it is not as well known as the open tube techniques." William MacGeorge, interview by author, 11 January 1996; Bill Harding, interview by author, 10 December 1995; Shepard, interview. IBM's 1965 document on corporate strategy included a section on IBM's competitive status. It reported IBM's use of capsule diffusions but noted

that the "outside world" used the open tube process. D. T. Doody, "Semiconductor Device, Circuit, and Packaging Strategy," October 1965, 25–28, Box TBR 110, IBM THP.

41. The origin of this work is described in Pugh et al., *IBM's 360 and Early 370 Systems*, 69–78.

42. In discussing the IBM-Intel technical exchanges of the 1980s, Gordon Moore noted that while IBM learned a lot from Intel about the silicon part of the process, Intel learned a lot from IBM about metallization and packaging chips. Gordon Moore, interview by author, 15 May 1996. One measure of the complexity of the C4 can be seen by the articles that were devoted to it in a 1969 issue of the *IBM Journal of Research and Development* and by the technical questions raised in those articles. P. A. Totta and R. P. Sopher, "SLT Device Metallurgy and Its Monolithic Extension," *IBM Journal of Research and Development* 13 (1969): 226–38; L. F. Miller, "Controlled Collapse Reflow Chip Joining," *IBM Journal of Research and Development* 13 (1969): 239–50; L. S. Goldmann, "Geometric Optimization of Controlled Collapse Interconnections," *IBM Journal of Research and Development* 13 (1969): 251–65.

43. A. R. Strube, memo to file, 27 March 1972, PX 6269, *U.S. v. IBM*.

44. Gordon Moore recalls someone from Bell Labs telling an IBM manager that IBM had to be very careful in the technologies it pursued because it had "enough capability to make a bad idea work." Gordon Moore, note to author, 21 April 2001.

45. Throughout the late 1960s and early 1970s, a member of the corporate staff, A.R. DiPietro, wrote a series of perceptive memos critiquing IBM's style of semiconductor development and manufacturing and arguing that it should move closer to industry-standard practices. His pleas fell on deaf ears. His memos, which must have delighted the government lawyers pursuing its antitrust case against IBM, occupy a good portion of Box TAR 242 in the IBM Technical History Project. Among the more comprehensive are A. R. DiPietro, "IBM's Semiconductor Components," 19 May 1971; A. R. DiPietro, "Components," 5 October 1971; A. R. DiPietro, "FET Technology," 23 August 1972.

46. This is consistent with the claim of Alfred Chandler Jr. and John Kenneth Galbraith that the large corporation favored stability over profit maximization. The best example of this for IBM was FS, an enormous program in the mid-1970s to develop a new line of computers to supplant System/360. The program was ultimately cancelled, resulting in a lost generation of computers, which was estimated to have cost IBM billions of dollars in lost revenues. But during the 1970s and early 1980s, IBM had continually increasing profits except for one year. The manager most closely associated with FS, John Opel, went on to become the chief executive. The estimate of the lost revenues from FS comes from Charles H. Ferguson and Charles R. Morris, *Computer Wars: The Fall of IBM and the Future of Global Technology* (New York: Times Books, 1994), 32–37. Chandler, *Visible Hand*, 10; John Kenneth Galbraith, *The New Industrial State* (Boston: Houghton Mifflin, 1967).

47. After IBM's MOS program left Research, innovations were slower to take hold. The silicon gate process, implemented at Intel in 1969, was not generally adopted at IBM until the mid-1980s. Intel and other firms in the semiconductor industry were quicker to use Dennard's one-device cell than IBM was. IBM did not introduce a product using the one-device cell until 1978.

48. Castrucci, interview by author, 7 July 1997. Castrucci, one of IBM's leading semicon-

ductor engineers, gained recognition in the world outside of IBM in the late 1980s when he was appointed to the number-two position in the semiconductor industry's manufacturing consortium, Sematech. He gained even greater recognition in March 1989 when, following clashes with Robert Noyce, the chief executive officer of Sematech, Castrucci resigned after seven months on the job. Darrell Dunn, "Sematech Chief Details Exec Rift," *Electronic News,* 3 April 1989, 1, 8; Larry D. Browning and Judy C. Shetler, *Sematech: Saving the U.S. Semiconductor Industry* (College Station, Tex.: Texas A&M University Press, 2000), 81–96.

49. Harding, interview; Elfant, interview; Castrucci, interview.

50. Castrucci, interview.

51. North, interview; Herb Lehman, interview by author, 6 November 1995; Pugh, "Fred Neves, FET/CP," 15; R. F. Elfant, "T2," 29 September 1970, Box TAR 242, IBM THP.

52. H. R. Koenig and L. I. Maissel, "Application of RF Discharges to Sputtering," *IBM Journal of Research and Development* 14 (March 1970): 168–71. In 1971, A. R. DiPietro wrote that the RF sputtering process was not "a reasonable engineering and manufacturing process" and that an alternate process should be developed. A. R. DiPietro to T. C. Papes, 5 October 1971, Box TAR 242, IBM THP.

53. Elfant, "T2"; Pugh, "Fred Neves, FET/CP," 15–16.

54. T. J. Watson Jr. to T. V. Learson, 6 March 1970, Box TAR 242, IBM THP; T. V. Learson to J. E. Bertram, 18 March 1970, Box TAR 242, IBM THP; "Management Committee Minutes, 8 May 1970," PX 4414, *U.S. v. IBM.*

55. A. R. DiPietro to Erich Bloch, 3 December 1970, Box TAR 242, IBM THP. One of the people DiPietro recommended was Frank Wanlass. Given Wanlass's personality, it is hard to imagine him lasting very long in the IBM environment. (See chapters 4 and 5.)

56. James Winnard, interview by author, 23 August 1997; Pugh, "Fred Neves, FET/CP," 15.

57. "How the High-Fliers Take Off," *Business Week,* 22 November 1969, 112–16; "Former Univac Officials Start Own Company: Mohawk Data," *Electronic News,* 7 September 1964, 24; "Cogar Sets Up Firm in Herkimer," *Electronic News,* 25 March 1968, 52; "Cogar to Build $3 Million Plant," *Electronic News,* 10 March 1969, 43; Robert Markle, interview by author, 9 June 1995; C. J. Bashe, "Interview with Mr. George Cogar, 7/17/74," Box TAR 113, IBM THP. In 1974, after the Technology Division of Cogar failed, Cogar offered to testify on IBM's behalf in the government's antitrust case.

58. Meade, interview; Markle, interview, 9 June 1995; Frank Deverse, interview by author, 24 May 2000.

59. Cogar Corporation, Form S-1 Registration Statement, 10 November 1971, 12–13; "If Cogar Can Keep from Being Suited Up," *Datamation,* October 1969, 147–49.

60. Robert Markle, interview by author, 9 September 1995. Markle's ascent to the position managing the development of SLT suggests some canny political maneuvering. When on Learson's staff, Markle headed up a group auditing the progress of SLT and noted a number of problems with its development. The head of the division in which SLT was developed subsequently hired Markle to head its development, possibly as a way to forestall criticism over its progress. R. E. Markle, "SLT Review," 5 February 1963, Box TBR 110, IBM THP; Component Development Organization Chart, 16 October 1964, Box TBR 112, IBM THP.

61. Standard & Poor's Corp. O-T-C and Regional Exchange Stock Reports, Cogar Corp., 27 January 1971, Box 6, Folder 12, Raytheon Company, Electronics Industry Files, Collection M0661, Department of Special Collections, Stanford University Libraries, Stanford, California (hereafter Raytheon Collection); "Cogar Common Sold Out, Rises to a Sharp Premium," *Wall Street Journal,* 17 June 1969, 30; "Cogar Set to Market Monolithic Memories," *Electronic News,* 29 September 1969, 43; "Supermind," Cogar Advertisement, *Electronic News,* 15 December 1969, 27. Automation was an IBM trait, not at all common in Silicon Valley semiconductor firms.

62. *Dutchess County* (Philadelphia, Pa.: William Penn Association of Philadelphia, 1937), 25; Regional Plan Association, *The Future of Dutchess County* (New York: Regional Plan Association, 1972), 7, 9, 11, 50. The beginnings of IBM's work in Poughkeepsie are described in Emerson W. Pugh, *Building IBM: Shaping an Industry and Its Technology* (Cambridge, Mass.: MIT Press, 1995), 90.

63. "IBM Sues Cogar on Trade Secrets," *Electronic News,* 14 July 1969, 34; Markle, interview, 9 September 1995; Meade, interview; Shepard, interview; Deverse, interview.

64. Markle, interview, 9 September 1995; Cogar Corporation, Form S-1 Registration Statement, 11.

65. Cogar Corporation, *Annual Report, 1970;* W. E. Harding, interview by E. W. Pugh, 24 January 1990, IBM THP Interview (*IBM's 360 and Early 370 Systems*); Bashe, "Interview with Mr. George Cogar," 11; Deverse, interview.

66. Deverse, interview; undated Cogar organization chart in author's possession. Cogar's cafeteria is described in Frank Deverse, note to author, 22 May 2001. Intel's lunchroom is described in Gene Flath, note to author, 7 June 2001

67. Isadore Barmish, "IBM Sues Cogar over Trade Secrets," *New York Times,* 10 July 1969, 47, 55; "IBM Sues Cogar on Trade Secrets," *Electronic News,* 14 July 1969, 34; "Lawsuits: IBM's Unruly Brood," *Newsweek,* 21 July 1969, 85–87; "Goliath Attacks David: IBM Is Suing Cogar: A Fledgling Company," *Wall Street Journal,* 10 July 1969, 6.

68. "Agreement Ends IBM Secrets Suit against Cogar," *Electronic News,* 9 March 1970, 36. From my interviews with former Cogar employees, it is clear that they did take IBM materials with them.

69. R. M. Meade, "On Memory System Design," *Proceedings of the Fall Joint Computer Conference 1970,* 33–43; and "How a Cache Memory Enhances a Computer's Performance," *Electronics,* 17 January 1972, 58–63 (both Meade works offer detailed expositions of the concept of cache memory pioneered by IBM); D. Lund, C. A. Allen, S. R. Anderson, and G. K. Tu, "Design of a Megabit Semiconductor Memory System," *Proceedings of the Fall Joint Computer Conference 1970,* 53–62, which described Cogar's emulation of IBM's overall memory system design; Joe Shepard, note to author, 12 September 1995. Shepard visited Cogar's facilities after it had gone out of business and was being sold at auction.

70. Tom Wolfe, "The Tinkerings of Robert Noyce," *Esquire,* December 1983, 367; Shepard, note.

71. "Where the Action Is in Electronics," *Business Week,* 4 October 1969, 86–98; "An Integrated Circuit That Is Catching Up," *Business Week,* 25 April 1970, 134–36. (The quote is from the latter article, 134.)

72. Quantum Science, "The Impact of MOS/LSI," Box 6, Folder 1, Raytheon Collection;

R. T. Jackson to B. S. Goldberg, "Cogar Capabilities," 20 October 1970, PX 4352, *U.S. vs. IBM* ("IBM trained"); W. E. Burdick to C. E. Branscomb, "Memorex/Cogar," 17 November 1970, PX 4299, *U.S. vs. IBM* ("grow as a business").

73. Cogar Corporation, Form S-1 Registration Statement, 7.

74. Pugh et al., *IBM's 360 and Early 370 Systems*, 458–76; "Special Report: Memory Technology," *Electronics*, 28 October 1968, 103; "For Memory Men the Sweetest Word Is 'Semiconductor,'" *Electronics*, 24 November 1969, 33; "How Will You Sell against New Systems?" Cogar Advertisement, *Electronic News*, 27 April 1970, 41; Meade, interview; Markle, interview, 9 June 1995, 9 September 1995.

75. Markle, interview, 9 September 1995.

76. Gordon Moore, interview by Allen Chen, 6 January 1993, 20–21, Intel Museum, Santa Clara, CA.

77. For the classic statement of the modern industrial corporation's avoidance of market forces, see Galbraith, *The New Industrial State*, 33–45.

78. Cogar Corporation, *Annual Report, 1970*; "Cogar Out to Replace 360/65 Core Memory with MOS Modules," *Electronics*, 18 January 1971, 25; "Cogar Corp. Forecasts Switch to Profitability," *Wall Street Journal*, 21 January 1971, 23; "Cogar, Potter in Add-on Memory Dispute," *Electronic News*, 27 March 1972, 1, 14.

79. Denis Arvay, "Cogar Seeks $9 Million in Financing," *Electronic News*, 16 August 1971, 1, 10; "Cogar, Potter in Add-on Memory Dispute," 1, 14; Joe McLean, "SEMI Chief Joins Potter: Eye Memories," *Electronic News*, 10 April 1972, 1, 21; Joe McLean, "Cogar Seeks 11th Hour $, Cuts Work Force Again," *Electronic News*, 24 April 1972, 1, 2; "SEC Suspends Trading in Cogar for 10 More Days," *Electronic News*, 1 May 1972, 6; Meade, interview; Deverse, interview.

80. Markle, interview, 9 June 1995; Meade, interview; Deverse, interview; "The Antenna," *Electronic News*, 12 June 1972, 14.

81. Ted Jenkins, interview by author, 18 February 1997; Cliff Forbes, "Memory Firm Prime Bidder," *Electronic News*, 31 July 1972, 28. *Electronic News* described the auctioneer as disappointed with the "lackadaisical bidding." Jenkins and Gerhard Parker, the Intel employees, ended up buying several microscopes but thought much of the rest of the production equipment was impractical. The star-crossed facility was later taken over by Fairchild, which used it for MOS production. John Rhea and Neil Kelley, "Motorola, FC&I Set MOS Sites," *Electronic News*, 23 April 1972, 1, 6; "FC&I Names Deardorf Plant Mgr.," *Electronic News*, 6 August 1973, 6.

82. *IBM and Vermont* (Essex Junction, Vt.: IBM, n.d.), 32. On Watson's ski lodge in Vermont, see Thomas J. Watson Jr. and Peter Petre, *Father, Son & Co.: My Life at IBM and Beyond* (New York: Bantam Books, 1990), 318–20. In 1947 John Gunther called Vermont "the only state in the union that, in a way, the industrial revolution never hit." *Inside U.S.A.*, 50th anniv. ed. (New York: New Press, 1997), 492–93.

83. MacGeorge, interview, 11 January 1996.

84. Paul Low, interview by author, 8 August 1997. Documents related to the $\phi2i$ ramp-up problem include "Management Committee Minutes," 8 May 1970, PX 4414A, *U.S. vs. IBM*; C. E. Branscomb, "Imperfect Phase 2I Memory Chips," 8 April 1970, PX 4547, *U.S. vs.*

IBM; A. R. DiPietro, "Phase 2I Memory Use in NS1," 30 March 1970, Box TAR 242, IBM THP. IBM's audacity can be seen in an analysis of the problems facing φ2i by a member of the corporate staff in the May 1970 Management Committee meeting cited above. In a statement that would have been self-evident everywhere but IBM, he stated: "Hindsight shows we should not have planned on building production up to 7,000 wafer starts per day."

85. "IBM Component History," 20, IBM Archives, Somers, New York; Low, interview; MacGeorge, interview, 23 January 1996, 30 January 1996.

86. MacGeorge, interview, 23 January 1996, 30 January 1996.

87. MacGeorge, interview, 23 January 1996, 30 January 1996; Bill Rowe, interview by author, 7 July 1997.

88. Linton, interview; Gladu, interview; MacGeorge, interview, 25 August 1997; Bill Rowe, interview. The Burlington organization at this time is shown in "IBM Organization Directory," 31 August 1971, 69, Box TMB 147, IBM THP.

89. Low, interview; MacGeorge, interview, 23 January 1996, 30 January 1996.

90. MacGeorge, interview, 23 January 1996, 30 January 1996; C. H. Stapper, P. P. Castrucci, R. A. Maeder, W. E. Rowe, and R. A. Verhelst, "Evolution and Accomplishments of VLSI Yield Management at IBM," *IBM Journal of Research and Development* 26 (September 1982): 532–45. When IBM invested in Intel in the early 1980s, the two companies had technical exchanges in areas where one company believed it could learn from the other. Yield management was one area where IBM gave information to Intel. Lehman, interview; Bill Rowe, interview.

91. MacGeorge, interview, 23 January 1996, 30 January 1996. The comparison in sizes for an SLT transistor and a 2,048-bit memory chip are from Pugh et al., *IBM's 360 and Early 370 Systems,* 77; and Rolf Remshardt and Utz G. Baittinger, "A High Performance Low Power 2,048-Bit Memory Chip in MOSFET Technology and Its Application," *IEEE Journal of Solid-State Circuits,* 11 (June 1976): 356. For comparisons of the costs of IBM memory chips against the prices of chips available commercially, see, for example, Lewis M. Branscomb to T. C. Papes, 17 December 1975, PX 6306, *U.S. v. IBM.*

92. MacGeorge, interview, 23 January 1996, 30 January 1996. An internal publication detailing IBM's MOS work gives some suggestions about the extensive infrastructure IBM put in place to manufacture MOS memories. *IBM Burlington, This Week,* 27 July 1972, IBM Archives, Somers, New York.

93. MacGeorge, interview, 23 January 1996.

94. Linton, interview; MacGeorge, interview, 30 January 1996, 25 August 1997. Low, interview; Lehman, interview; Dale Critchlow, note to author, 29 January 1996. The differences in site cultures between Burlington and East Fishkill would warrant a lengthy study of its own. (Most of these people worked at both locations in their careers.)

95. C. H. Stapper et al., "Evolution and Accomplishments of VLSI Yield Management at IBM," 539–40. MacGeorge, interview, 23 January 1996; Remshardt and Baitinger, "A High Performance Low Power 2048-Bit Memory Chip," 352–59.

96. MacGeorge, interview, 23 January 1996. Even in a mature technology, problems could strike that would drive yield down dramatically and take months to fully recover from. In 1980 Burlington ran into difficulties manufacturing the 2,048-bit memory chip that had

been in production for seven years. The yield went down by 10 percentage points, and it took six months to get back to the previous yield levels. "Riesling: Final Test/Month," graph from 1978–1982, in author's possession. I thank Bill Rowe for sending me this document.

97. "IBM Component History," 22–23; John Rhea, "N/MOS Makers Hail IBM Move," *Electronics News*, 7 August 1972, 28.

98. Martin Gold, "MOS, Bipolar Memories Boosted by New IBM Machines," *Electronic News*, 5 February 1973, 44.

99. Low, interview; "IBM Component History," 22; "IBM Organization Directory," 3 June 1974, Box TMB 148, IBM THP.

100. D. P. Phypers, "Corporate Organization Manual," 26 October 1967, Box TAR 242, IBM THP. DiPietro describes the meeting where SDD said it had no need for MOS products in A. R. DiPietro, "N-Channel vs. P Channel FET's," 7 May 1971, Box TAR 242, IBM THP. In early 1968 one of the engineers in a systems development group that was interested in MOS technology but lacked large production requirements to get it off the ground wrote to DiPietro, stating: "If the charter has been changed, I doubt if anyone in CD is aware of it since I see no alteration of their plan." R. A. Vaughn, "Introduction of New Technologies," 18 March 1968, Box TAR 242, IBM THP.

101. Cheroff, interview.

102. E. R. Piore to F. T. Cary, 13 June 1969, Box TAR 242, IBM THP.

103. "Minutes of the Corporate Technical Committee Meeting, July 12, 1971," 1–2, Box TAR 132, IBM THP.

104. Ibid.; "Minutes of the Corporate Technical Committee Meeting, March 27, 1972," Box TAR 132, IBM THP.

105. "IBM Component History," 21–28.

106. FS is discussed at greater length in Pugh et al., *IBM's 360 and Early 370 Systems*, 538–53; Ferguson and Morris, *Computer Wars*, 31–37.

107. Triebwasser, Dennard, and Terman, interview; Carl Caricari, interview by author, 24 August 1997.

108. The fact that the Beacon group was among the first in the industry to discover the hot electron problem suggests how aggressive it was. Shakir A. Abbas and Robert C. Dockerty, "N-Channel IGFET Design Limitations Due to Hot Electron Trapping," *1975 International Electron Device Meeting Digest of Technical Papers*. The Beacon chip had eighteen hundred circuits and average delays of 4.3 nanoseconds per circuit. "Minutes of the CTC Meeting—March 20, 1972," Box TAR 132, IBM THP; Caricari, interview; Lehman, interview.

109. "IBM Burlington: More Than Just Memory, A Look at the Years 1975–1979," *Burlington Closeup*, February 1991, 12. (This is an internal publication produced by IBM Burlington for IBM Burlington employees.)

EIGHT · THE LOGIC OF MOS

1. E. A. Sack, R. C. Lyman, and G. Y. Chang, "Evolution of the Concept of a Computer on a Slice," *Proceedings of the IEEE* 52 (December 1964): 1713.

2. Integrated Circuit Engineering Corporation, *Integrated Circuit Engineering: Basic Technology*, 4th ed. (Cambridge, Mass.: Boston Technical Publishers, 1966), iii.

3. D. L. Critchlow, R. H. Dennard, S. E. Schuster, and E. Y. Rocher, "Design of Insulated-Gate Field-Effect Transistors and Large Scale Integrated Circuit Chips for Memory and Logic Applications," RC 2290, 4 October 1968, 218, IBM Research, Yorktown Heights, New York.

4. "Integrator on a Chip," *Electronics,* 22 August 1966, 38–39; "General Instrument to Show Monolithic Digital Analyzer," *Electronic News,* 22 August 1966, 92; Lee Boysel, interview by author, 23 January 1996.

5. Computer Research Bureau, "CST Broker Special," 21 October 1968, 1 (in author's possession). UCC is University Computing Corporation. I thank Lee Boysel for providing me with the Computer Research Bureau newsletter and other documents describing the work of Viatron and Four-Phase Systems.

6. "Can EDP Work on a Shoestring?" *Business Week,* 19 October 1968, 182. On Viatron's early plans, see also James Brinton, "Computer with Peripherals Will Rent for $40 a Month," *Electronics,* 14 October 1968, 193–95.

7. Computer Research Bureau, "CST Broker Special," 1; "Computer with Peripherals," 193.

8. Viatron, "Viatron System 21, System Description," n.d. (in author's possession).

9. More about Viatron can be found in W. David Gardner, "The Rise and Fall of Viatron," *Datamation,* 15 May 1971, 38–40; W. David Gardner, "The Rise and Fall of Viatron—Part II," *Datamation,* 1 July 1971, 44–47; Paul E. Ceruzzi, *A History of Modern Computing* (Cambridge, Mass.: MIT Press, 1998), 252–54.

10. Computer Research Bureau, "CST Broker Special," 1–3; "Can EDP Work on a Shoestring?" 182–86; James Brinton, "Viatron—Vibrant and Probably Viable," *Electronics,* 23 June 1969, 141–47. Laurence C. Drew, Viatron's manager of development engineering, had worked on MOS at RCA.

11. Bennett is quoted in "Viatron—Vibrant and Probably Viable," 141. Gardner, "The Rise and Fall of Viatron," 38, 40, provides the figures for letters of intent, MOS contracts, and funds Viatron raised.

12. W. David Gardner, "Viatron Loses $9.4 Million as Sales Rise Almost 10-Fold," *Electronic News,* 16 March 1970, 18; James Poulos, "Activity, Silence at Viatron," *Electronic News,* 19 October 1970, 24; James Poulos, "Bondholders of Viatron to Hold Meeting," *Electronic News,* 28 December 1970, 44; "Viatron: A Collector's Item," *Electronic News,* 15 March 1971, 1, 62. Viatron lost $9.4 million in the year ending October 31, 1969, and $7.3 million in the next six months.

13. A report made for investors in Viatron by the consulting firm Integrated Circuit Engineering noted that due to the increases in yields over time, "a system design which may not be practical for production in 1969 may become feasible by 1971." Integrated Circuit Engineering, "A Review of the Engineering Feasability of the MOS Electronics Proposed for the Viatron System 21," Box 4, File 2, Series 16, Integrated Circuit Engineering Collection, National Museum of American History, Archives Center, Washington, D.C.

14. Gardner, "The Rise and Fall of Viatron—Part II," 47. Shortly after he was removed from Viatron, Bennett gave a talk to an engineering conference on his principles of entrepreneurship. One attendee called it "Machiavelli with a briefcase." One of Bennett's many quotable lines was "It's more fun to spend someone else's money than your own." James Poulos, "Nerem Group Hears Dr. Bennett's Bible," *Electronic News,* 9 November 1970, 32.

15. Lee Boysel, interview by author, 28 February 1997. Boysel's departure from Fairchild is documented in "Five More Resign Posts at Fairchild," *Electronic News*, 18 November 1968, 48. In an October 1968 memo, a Fairchild manager was making detailed plans for the MOS circuits Fairchild would design in the upcoming year. He put Boysel's group down for 50 designs. Within weeks Boysel would be gone, and in 1969 many other members of his group would join him. Harry Neil, "1969 Custom Design Forecast," 9 October 1968, Box 6, Rob Walker Papers, Collection M0742, Department of Special Collections, Stanford University Libraries, Stanford, California.

16. Blyth & Co., "Confidential Report on Four-Phase Systems, Inc." 30 April 1971 (in author's possession); Boysel, interview, 28 February 1997. Corning had an agreement with Boysel that if his company failed, he and his engineers would join Signetics. Comparing the names Intel and Four-Phase Systems suggests how Boysel's vision differed from Moore and Noyce's. Four-phase was a specific type of MOS circuitry, while "Intel" was short for "integrated electronics," suggesting Moore and Noyce saw their charter as much broader, encompassing all of integrated electronics.

17. Boysel, interview, 28 February 1997.

18. Ibid. Four-Phase's strategy is described in Blyth & Co., "Confidential Report on Four-Phase Systems, Inc," 1–2. An early report on Four-Phase and its plans is Marge Scandling, "A Way to Cut Computer Costs," *Palo Alto Times*, 28 April 1969, 9.

19. Boysel, interview, 28 February 1997.

20. Cartesian, "Prospectus," n.d. (in author's possession); "New MOS Firm Bows, Eyes Custom Wafers," *Electronic News*, 19 May 1969, 43. While they were both at Fairchild, Cole quoted Boysel prices per wafer processed, which provided the basis for Four-Phase's early plans. These prices were written on the back of a sheet of yellow paper. Bob Cole, "Price per Wafer per Design per Order," n.d. (in author's possession).

21. Lee Boysel, letter to shareholders, 1 April 1970 (in author's possession); Jim Vincler, "New Four Phase Computer Uses Only Semicon Memories," *Electronic News*, 9 February 1970, 10; "This System Starts with Integration," *Business Week*, 12 September 1970, 27; Boysel, interview, 28 February 1997.

22. Michael Merritt, "'Intelligent' IV-70 Outguns IBM's 2260s and 3270s," *Computerworld*, 23 June 1971, 26; Four-Phase Systems, "Preliminary Prospectus," 30 May 1973, 2 (in author's possession). As a measure of how optimistic Viatron's plans were, in two years of production Four-Phase had placed fewer terminals than Viatron had planned to produce every month. The one caveat raised about the Four-Phase system in the *Computerworld* article was from a representative of Eastern Airlines, who said that it was not possible to use the Four-Phase system as a general-purpose computer (suggesting that the claim that the Four-Phase System was equivalent in power to an IBM 360/30 may have been exaggerated).

23. Lee L. Boysel and Joseph P. Murphy, "Four-Phase LSI Logic Offers New Approach to Computer Designer," *Computer Design*, April 1970, 141–46; L. Boysel, "History of the Microprocessor," 11 August 1995 (in author's possession).

24. Boysel and Murphy, "Four-Phase LSI Logic Offers New Approach to Computer Designer," 141, 144, 146.

25. In the 1990s, in conjunction with a group of personal computer manufacturers trying to challenge a Texas Instruments microprocessor patent, Boysel put together a dem-

onstration system showing that the AL1 could have operated as a microprocessor according to a modern definition of the term, where the AL1 coupled with RAM, ROM, and I/O formed a complete computing system. Lee Boysel, "Court Room Demonstration System 1969 AL1 Microprocessor," 3 April 1995 (in author's possession).

26. Quoted in Elizabeth de Atley, "LSI Poses Dilemma for Systems Designers," *Electronic Design*, 1 February 1970, 44.

27. Boysel, interview, 28 February 1997.

28. Boysel, interview, 28 February 1997. Four-Phase's concern with integrated circuit packaging is consistent with IBM's; IBM also put more effort into packaging than firms in the semiconductor industry. For IBM's efforts in semiconductor packaging, see chapter 7.

29. Lee Boysel, "The Way It Really Was," 24 May 1974 (in author's possession). For a discussion of the effect of computer leasing on the computer industry, see Katherine Davis Fishman, *The Computer Establishment* (New York: Harper & Row, 1981), 15–18.

30. On Four-Phase's stockpiling of chips, see Boysel, interview, 28 February 1997. "LSI Poses Dilemma for Systems Designers," 45, includes a picture of a wafer of AL1 chips with the bad chips marked, suggesting a yield of roughly 50 percent.

31. On the cover of its 1976 annual report, Four-Phase had a Social Security operations center using its systems and asserted that the agency had acquired thirteen hundred Four-Phase terminals. Four-Phase Systems, *Annual Report, 1976,* 1–2; Four-Phase Systems, *Annual Report, 1980,* 1; Thomas J. Lueck, "Motorola Moves Further into Technology with Its Risky Decision to Buy Four-Phase," *Wall Street Journal,* 4 January 1982, 33; "Stockholder Meeting Briefs," *Wall Street Journal,* 3 March 1982, 44.

32. Robert N. Noyce, "A Look at Future Costs of Large Integrated Arrays," *Proceedings of the Fall Joint Computer Conference, 1966,* 113. Another former Fairchild manager held the same position. Charles Sporck, the president of National Semiconductor, stated: "Custom work kills a company when it's trying to grow. Henry Ford's Model T approach is just as valid now as it was then." Quoted in Don C. Hoefler, "Nat'l Semicon's Sporck Sees Firm Gaining Ground," *Electronic News,* 16 December 1968, 49.

33. Noyce, "A Look at Future Costs," 113.

34. Quoted in Elizabeth de Atley, "Can You Build a System with Off-the-Shelf LSI?" *Electronic Design,* 1 March 1970, 50. On Intel's interaction with Honeywell in designing the 1103, see chapter 6. On the 3101, see Gordon Moore, Gerhard Parker, and Les Vadasz, interview, 17 October 1983, 20, Intel Museum, Santa Clara, California.

35. Quoted in de Atley, "Can You Build a System," 47–48.

36. Quoted in ibid., 48. Each participant was interviewed separately, and then responses were put together (apparently after getting participants to respond to each other's comments) to simulate a panel discussion.

37. Ibid.

38. M. E. Hoff Jr. and Stanley Mazor, "Standard LSI for a Microprogrammed Processor," *1970 NEREM Conference Record,* 92; Stanley Mazor, interview by author, 6 June 1997.

39. Ted Hoff, interview by author, 13 March 1997; Tekla S. Perry, "Marcian E. Hoff," *IEEE Spectrum,* February 1994, 46–49.

40. Mazor, interview; SYMBOL is discussed in William R. Smith et al., "SYMBOL: A Large Experimental System Exploring Major Hardware Replacement of Software," *Proceedings of*

the Spring Joint Computer Conference, 1971, 601–16; Gordon Moore, interview by Rob Walker, 3 March 1995, Silicon Genesis: Oral History Interviews with Silicon Valley Scientists, Collection M0741, Department of Special Collections, Stanford University Libraries, Stanford, California (hereafter any reference to an interview in this collection will be cited as Silicon Genesis interviews); Myron Calhoun, note to author, 16 January 2001.

41. Joel Karp, interview by author, 29 January 1997; Marcian E. Hoff, "Solid-State Memory Offers New Organizations," *EDN,* 15 October 1969, 55–61.

42. "Desk Calculators Booming in Japan," *Electronic News,* 3 March 1969, 39; Kenneth Flamm, *Mismanaged Trade? Strategic Policy and the Semiconductor Industry* (Washington, D.C.: Brookings Institution, 1996), 70–77.

43. All accounts of Intel's early microprocessor work have to rely primarily on the participants. Unless otherwise noted, the remainder of this section is based on the following accounts: Federico Faggin, Marcian E. Hoff Jr., Stanley Mazor, and Masatoshi Shima, "The History of the 4004," *IEEE Micro,* December 1996, 10–20; Tekla S. Perry, "Marcian E. Hoff," *IEEE Spectrum,* February 1994, 46–49; Robert N. Noyce and Marcian E. Hoff Jr., "A History of Microprocessor Development at Intel," *IEEE Micro,* February 1981, 8–21; Federico Faggin, "The Birth of the Microprocessor," *Byte,* March 1992, 145–50; Stanley Mazor, "The History of the Microcomputer: Invention and Evolution," *Proceedings of the IEEE* 83 (December 1995): 1601–8. Faggin, Hoff, Mazor, and Shima give the number of chips requested by Busicom as seven, while Noyce and Hoff state that it was twelve.

44. Federico Faggin, interview by author, 6 June 1997.

45. Hal Feeney, interview by author, 23 April 1997.

46. "CPU Chip Turns Terminal into Stand Alone Machine," *Electronics,* 7 June 1971, 36–37; H. Smith, "Impact of LSI on Micro Computer and Calculator Chips," *1972 NEREM Technical Papers,* 143–46. Smith, an Intel engineer, compares Intel's silicon gate 8008 with a "Company T Equivalent Metal-Gate CPU Chip." Pages 27 and 28 of the same issue of *Electronics* carry an advertisement by Texas Instruments that mentions its CPU chip.

47. Although Hoff sometimes presents himself as a heroic lone inventor battling to have the part marketed generally, Gordon Moore asserts that Hoff's emphasis on opposition within Intel to marketing the 4004 broadly is based on a misapprehension about who the key decision makers were within the company—Intel's top managers quickly understood the importance of the 4004 and the 8008. Ted Hoff, interview by Rob Walker, 3 March 1995, Silicon Genesis interview; Gordon Moore, interview by Rob Walker, 18 September 1995, Silicon Genesis interview; Gordon Moore, Gerhard Parker, and Les Vadasz, interview, 17 October 1983, 60, Intel Museum, Intel Corporation, Santa Clara, California. Moore's statements are consistent with Noyce's earlier claims about standard LSI products.

48. The initial ad is reprinted in Noyce and Hoff, "History of Microprocessor Development at Intel," 9. For the confusion between microprocessor and microcomputer, see Rebecca Smith, "What Does 'Microcomputer' Mean?" *San Jose Mercury News,* 13 January 1994, 1E–2E.

49. William Aspray, "The Intel 4004 Microprocessor: What Constituted Invention?" *IEEE Annals of the History of Computing* 19 (1997): 4–15, makes a similar point.

50. C. A. Mead, "The Tunnel-Emission Amplifier," *Proceedings of the IRE* 48 (March 1960): 359–61. George Gilder provides biographical information on Mead in *Microcosm:*

The Quantum Revolution in Economics and Technology (New York: Simon and Schuster, 1989), 38–42.

51. Ted Jenkins, interview by author, 18 February 1997; Carver Mead, "Computers That Put the Power Where It Belongs," *Engineering and Science*, February 1972, 9. Moore credits Mead with coining the term "Moore's Law." Moore, Silicon Genesis interview.

52. Mead, "Computers That Put the Power," 6. In a November 1973 article in *Fortune*, Noyce also used the small electric motor analogy. Gene Bylinsky, "How Intel Won Its Bet on Memory Chips," *Fortune*, November 1973, 143.

53. Mead, "Computers That Put the Power," 9.

54. In its 1970 report on the semiconductor industry, the consulting firm Integrated Circuit Engineering said: "Silicon monolithic integrated circuits for logic applications constitute the largest segment of the integrated circuit market," estimating that bipolar logic accounted for 61% of the total market for integrated circuits. Integrated Circuit Engineering, *Status of Integrated Circuits 1970* (Phoenix, Ariz.: Integrated Circuit Engineering, 1970), 6, 23.

55. "Replace All This Random Logic with a One-Chip Computer," Intel Advertisement, *Electronic News*, 7 August 1972, 24.

56. Feeney, interview; Michael S. Malone, *The Microprocessor: A Biography* (New York: Springer-Verlag, 1995), 129–30; Noyce and Hoff, "History of Microprocessor Development at Intel," 13–14; Regis McKenna, interview by Rob Walker, 22 August 1995, Silicon Genesis interview. This process seems analogous to the one described by Nathan Rosenberg for the machine tool industry: "Technological Change in the Machine Tool Industry, 1840–1910," in *Perspectives on Technology* (Cambridge: Cambridge University Press, 1976), 9–31.

57. Intel Advertisements, *Electronic News*, 20 August 1973, 33, 35, 37, 39, 41; Andrew Grove, Ed Gelbach, and Ted Jenkins, interview, 59, 24 October 1983, Intel Museum, Santa Clara, California.

58. George Sideris, "Microcomputers Muscle In," *Electronics*, 1 March 1973, 63. Ed Gelbach recalled that for every dollar's worth of microprocessor sold, Intel would sell ten dollars' worth of memories. Gelbach, Grove, and Jenkins, interview, 63.

59. Hank Smith, "Marketing the Early Microprocessors," *IEEE Micro*, December 1996, 18; "Intel Makes It Easy with Unprecedented Design Support," Intel Advertisement, *Electronic News*, 7 August 1972, 25.

60. "Davidow Will Manage Intel Group," *Electronic News*, 27 August 1973, 8; William H. Davidow, *Marketing High Technology: An Insider's View* (New York: Free Press, 1986).

61. Ed Gelbach says Intel did not know who the software people were and had no basis for judging competency. Kildall later went on to write the popular CP/M operating system for personal computers. Gelbach, Grove, and Jenkins, interview, 47–48; Robert Slater, *Portraits in Silicon* (Cambridge, Mass.: MIT Press, 1987), 251–61; Ceruzzi, *History of Modern Computing*, 222–24; Feeney, interview.

62. Gelbach, Grove, and Jenkins, interview, 48–50; Mimi Real, *A Revolution in Progress: A History of Intel to Date* (Santa Clara, Calif.: Intel, 1984), 33; Davidow, *Marketing High Technology*, 15–17.

63. Feeney, interview.

64. Noyce and Hoff, "History of Microprocessor Development at Intel," 15; Faggin, "Birth of the Microprocessor," 150.

65. Tom Rowe, interview by author, 19 June 1997.

66. Masatoshi Shima, Federico Faggin, and Stanley Mazor, "An N-channel 8-bit Single Chip Microprocessor," *1974 International Solid-State Circuits Conference: Digest of Technical Papers*, 56; John Rhea, "May 2d-Source Intel Microprocessor," *Electronic News*, 15 April 1974, 34. Mazor provides a detailed discussion of the technical changes made to the 8080 in "The History of the Microcomputer," 1604. Other accounts of the development of the 8080 include Noyce and Hoff, "History of Microprocessor Development," 15, and Faggin, "Birth of the Microprocessor," 150.

67. "From CPU to Software, the 8080 Microcomputer Is Here," Intel Advertisement, *Electronic News*, 15 April 1974, 44–45; Hal Feeney, "A New Microcomputer Family," *1974 Wescon Technical Papers*, Session 15/1.

68. John Day, "Unveil Microprocessors, Low-End PDP-8 at DEC," *Electronic News*, 18 March 1974, 65. Noyce and Hoff call the Digital announcement a "coming of age for the microprocessor." Noyce and Hoff, "History of Microprocessor Development at Intel," 15. In the *Electronic News* article, a Digital vice president acknowledged that with Intel's integration upward into the sales of computers on a board, Digital was buying the 8008 from a competitor.

69. "Wescon Is Fully Booked Despite Industry Softness," *Electronic News*, 9 September 1974, 1; *1974 Wescon*, Sessions 11, 15, 19, 23.

70. "Unicom Acquisition to Move AMI into Desk Calculators," *Electronic News*, 16 August 1971, 2; "TI's Radical Move into Calculators," *Business Week*, 23 September 1972, 28; John Rhea, "New Unit Markets Nat'l Semi Calculators," *Electronic News*, 7 May 1973, 52; Walter Mathews, "Mostek Subsidiary Sets Calculator Introduction," *Electronic News*, 23 July 1973, 26; "$895 SCM Calculator Due," *Electronic News*, 28 April 1969, 27; "Mostek Puts Calculator on One Chip," *Electronic News*, 1 February 1971, 25; "Version from TI Expected by June," *Electronic News*, 1 February 1971, 25; "TI Introduces Chip Calculator," *Electronic News*, 20 September 1971, 32; "Sanyo Launches Calculator Line for Sale in U.S.," *Electronic News*, 1 February 1971, 25; "Mini Calculators Proliferate as More Producers Go after the Consumer Market," *Wall Street Journal*, 21 September 1972, 1; Raymond A. Joseph, "Calculator Competition Helps Consumer, but Many Manufacturers Are Troubled," *Wall Street Journal*, 14 January 1975, 38; "Rescuing Rockwell's Calculators," *Business Week*, 18 August 1975, 25; Integrated Circuit Engineering, *Status 1978: A Report on the Integrated Circuit Industry* (Scottsdale, Ariz.: Integrated Circuit Engineering, 1978), 2–5. For more on the trouble the calculator brought to the semiconductor industry, see Braun and Macdonald, *Revolution in Miniature*, 189–90.

71. Robert W. Wilson, Peter K. Ashton, and Thomas P. Egan, *Innovation, Competition, and Government Policy in the Semiconductor Industry* (Lexington, Mass.: D.C. Heath, 1980), 99–100, provides further details on the semiconductor industry and calculators in the early 1970s. The authors assert that twenty-nine American and Japanese companies exited the calculator business in 1973 and 1974. A description of the HP-80 and a statement of company president William Hewlett's disavowal of interest in the four-function calculator business can be found in "H-P Business Unit Bows," *Electronic News*, 22 January 1973, 36. Further details of Hewlett-Packard's calculators are provided in Chuck House, "Hewlett-Packard and

Personal Computing Systems," in *A History of Personal Workstations,* ed. Adele Goldberg (New York: ACM Press, 1988), 403–32.

72. Accounts of MITS and the Altair include Ceruzzi, *History of Modern Computing,* 221–41; Paul Freiberger and Michael Swaine, *Fire in the Valley: The Making of the Personal Computer,* 2d ed. (New York: McGraw-Hill, 2000), 33–73; Steven Levy, *Hackers: Heroes of the Computer Revolution* (Garden City, N.Y.: Anchor Press/Doubleday, 1984), 181–200; Martin Campbell-Kelly and William Aspray, *Computer: A History of the Information Machine* (New York: Basic Books, 1996), 233–44; and Richard N. Langlois, "External Economies and Economic Progress: The Case of the Microcomputer Industry," *Business History Review* 66 (spring 1992): 1–50.

73. Ceruzzi, *History of Modern Computing,* 205–41.

74. On Moore's initial response to the Altair, see Moore, Parker, and Vadasz, interview, 88–89.

75. A 1977 report by the consulting firm Integrated Circuit Engineering asserted that the microprocessor market was growing 50 percent per year and had been $36 million in 1975. Integrated Circuit Engineering, *Status 1977: A Report on the Integrated Circuit Industry* (Scottsdale, Ariz.: Integrated Circuit Engineering, 1977), 2-10–2-11.

76. Gene Bylinsky, "Here Comes the Second Computer Revolution," *Fortune,* November 1975, 135.

77. Ibid., 138.

78. Gordon E. Moore, "Microprocessors and Integrated Electronic Technology," *Proceedings of the IEEE* 64 (June 1976): 837.

79. Dennis A. Roberson, "A Microprocessor-Based Portable Computer: The IBM 5100," *Proceedings of the IEEE* 64 (June 1976): 994–99. IBM consistently defined a microprocessor as a microprogrammed processor, which meant that the number of chips was irrelevant. IBM used this definition of microprocessor as late as 1992. J. R. Flanagan, T. A. Gregg, and D. F. Casper, "The IBM Enterprise Systems Connection (ESCON) Channel—A Versatile Building Block," *IBM Journal of Research and Development* 36 (July 1992): 619. On the earlier definition of computers, see chapter 1, titled, "When Computers Were People," of Campbell-Kelly and Aspray, *Computer: A History of the Information Machine,* 9–28. On page 9 they cite the *Oxford English Dictionary*'s early definition of computer as "one who computes; a calculator, reckoner; specifically a person employed to make calculations in an observatory, in surveying, etc."

80. Bylinsky, "Here Comes the Second Computer Revolution," 137.

81. Ibid., 184.

CONCLUSION/EPILOGUE

1. Integrated Circuit Engineering, *Status 1977: A Report on the Integrated Circuit Industry* (Scottsdale, Ariz.: Integrated Circuit Engineering 1977), 3–2.

2. Integrated Circuit Engineering Corporation, *Status 1979: A Report on the Integrated Circuit Industry* (Scottsdale, Ariz.: Integrated Circuit Engineering, 1979), 3–1. In 1977 General Instrument was advertising a line of microprocessors that could be used for such func-

tions as controlling a microwave oven, a television, a music system, or a telephone. *Electronic News,* 10 January 1977, 30.

3. Robert H. Dennard, Fritz H. Gaensslen, Hwa-Nien Yu, V. Leo Rideout, Ernest Bassous, and Andre R. LeBlanc, "Design of Ion-Implanted MOSFET's with Very Small Physical Dimensions," *IEEE Journal of Solid-State Circuits* 9 (October 1974): 256–68. For previous knowledge of the principles of scaling, see Thomas O. Stanley, "The Validity of Scaling Principles for Field Effect Transistors," Princeton Technical Report 1282, 13 August 1962, RCA Collection, Hagley Museum and Library, Wilmington, Delaware, and Gordon Moore, interview by author, 15 May 1996. The first widely publicized effort to apply scaling to MOS transistors in production came from Intel, with its H-MOS family, which scaled its transistors down from 6 micron gate lengths to 3.5 micron gate lengths. Richard Pashley et al., "H-MOS Scales Traditional Devices to Higher Performance Level," *Electronics,* 18 August 1977, 94–99.

4. The relationship between science and technology has been one of the major interests of historians of technology. David A. Hounshell, "The Evolution of Industrial Research in the United States," in *Engines of Innovation: U.S. Industrial Research at the End of an Era,* ed. Richard S. Rosenbloom and William J. Spencer (Boston: Harvard Business School Press, 1996), 13–85; George Wise, "Science and Technology," *Osiris* 2nd ser., 1 (1985): 229–46; Donald E. Stokes, *Pasteur's Quadrant: Basic Science and Technological Innovation* (Washington, D.C.: Brookings Institution, 1997).

5. Another study that makes this point is Ann Johnson, "Engineering Culture and the Production of Knowledge: An Intellectual History of Anti-Lock Braking Systems" (Ph.D. diss., Princeton University, 2000), 185–237.

6. Alan B. Fowler, "A Semicentury of Semiconductors," *Physics Today,* October 1993, 61; R. W. Keyes and M. I. Nathan, "Semiconductors at IBM: Physics, Novel Devices, and Materials Science," *IBM Journal of Research and Development* 25 (September 1981): 782.

7. After a pessimistic discussion about the possibilities of competing technologies to the MOS transistor, Alan Fowler, a member of the IBM Research MOS effort, concludes: "If there is no attempt to find alternatives, they will never be found. Most of the large electronics companies have grown so sophisticated in finding reasons not to work on new technologies, and so bottom-line conscious, that they are unwilling to risk significant money in looking at or for alternatives." "On Some Modern Uses of the Electron in Logic and Memory," *Physics Today,* October 1997, 54. Fowler's article was subtitled "How Silicon MOSFET Technology Came to Dominate the Ways in Which Electrons Are Used in Logic and Memory Devices; Will This Dominance Continue?" His implicit answer was yes.

8. Although the issue of *Technology Review* considering technologies "Beyond Silicon" contained some cautions in its articles on molecular computing, quantum computing, biological computing, and DNA computing, its overall tone was very positive. David Rotman, "Molecular Computing," *Technology Review,* May/June 2000, 52–58; M. Mitchell Waldrop, "Quantum Computing," *Technology Review,* May/June 2000, 60–66; Simson L. Garfinkel, "Biological Computing," *Technology Review,* May/June 2000, 70–77; Antonio Regalado, "DNA Computing," *Technology Review,* May/June 2000, 80–84. Science and technology writers have a challenging job reporting on these technologies and have a strong incentive to overstate their case, for not many people want to read about a technology that has no

chance to make it out of the laboratory. An IBM Research manager makes the case for the longevity of CMOS technology in R. D. Isaac, "The Future of CMOS Technology," *IBM Journal of Research and Development* 44 (May 2000): 369–78. Sunlin Chou, as of this writing an Intel senior vice president and the general manager of the technology and manufacturing group, told me (in a note that he said reflected his personal opinions) that rather than being worried about competitors to MOS technology, he is concerned that when MOS technology runs out of gas "no effective alternative will be available for a long time." Sunlin Chou, note to author, 26 January 2001.

9. Quoted in George Johnson, "Rolf Landauer, Pioneer in Computer Theory, Dies at 72," *New York Times*, 30 April 1999, 21. Further details about Landauer's skepticism about technologies based on new physical phenomena can be found in Alan Fowler et al., "Rolf William Landauer," *Physics Today*, October 1999, 104–5; Rolf Landauer, "Advanced Technology and Truth in Advertising," *Physica A* 168 (1990): 75–87; Rolf Landauer, "Need for Critical Assessment," *IEEE Transactions on Electron Devices* 43 (October 1996): 1637–39.

10. Gordon Moore, interview by Rob Walker, 3 March 1995, Silicon Genesis: Oral History Interviews with Silicon Valley Scientists, Collection M0741, Department of Special Collections, Stanford University Libraries, Stanford, California (hereafter any reference to an interview in this collection will be cited as Silicon Genesis interviews).

11. Rob Walker, *Silicon Destiny: The Story of Application Specific Integrated Circuits and LSI Logic Corporation* (Milpitas, Calif.: CMC Publications, 1992), 13–65. While Gordon Moore had incomparable judgment as a manager of semiconductor research, he was less effective at Fairchild as a manager of computer research. Fairchild's major computer research effort proved to be barren, with respect to both Fairchild and Silicon Valley as a whole. The project, the SYMBOL computer, was conducted in a spirit of academic research that Moore would have been unlikely to tolerate in semiconductors. When the project was finally cancelled (years after Moore had left to found Intel), the researchers primarily went to work at large, established companies or at universities. The most famous alumni of SYMBOL was Stan Mazor, who worked at Intel on the microprocessor.

12. Nathan Rosenberg, *Exploring the Black Box: Technology, Economics, and History* (Cambridge: Cambridge University Press, 1994), 92–94.

13. "RCA Develops Revolutionary Solid-State Element Combining Best Properties of Transistors and Vacuum Tubes," 11 February 1963, David Sarnoff Library, Princeton, New Jersey. I thank Alex Magoun of the David Sarnoff Library for providing me with this document. RCA also played a key role in developing thin-film transistors and liquid crystal displays that form the fundamental display technology for portable computers. T. Peter Brody, "The Thin Film Transistor—A Late Flowering Bloom," *IEEE Transactions on Electron Devices* 31 (November 1984): 1614–28; Hedrick Smith, *Rethinking America* (New York: Random House, 1995), 6–18.

14. D. L. Critchlow, R. H. Dennard, S. E. Schuster, and E. Y. Rocher, "Design of Insulated-Gate Field-Effect Transistors and Large Scale Integrated Circuit Chips for Memory and Logic Applications," RC 2290, 4 October 1968, 218, IBM Research, Yorktown Heights, New York. One might imagine a thought experiment where one takes Intel and IBM engineers from 1969 to the world of 1989 and shows them the Intel 486 microprocessor. If one were to ask them if this was something they would see value in, they would all say yes. Intel was able to

get the 486 microprocessor because it started with the 4004, whose value was less clear. IBM passed on the four-bit microprocessor and had a difficult time jumping back into the fray.

15. F. E. Terman to Nicholas Reznick, 9 May 1961, Box 18, Folder 4, Frederick Terman Papers, Collection SC160, Stanford University Archives, Stanford University Libraries, Stanford, California.

16. Progress Report, Digital Systems Research, 8, 1 May 1965, Box 11, Folder 7, Fairchild Camera and Instrument Technical Reports and Progress Reports, Collection M1055, Department of Special Collections, Stanford University Libraries, Stanford, California.

17. Gordon Moore's assessment of Stanford's role in the semiconductor industry of Silicon Valley is very similar to this. Gordon Moore, interview by Ross Bassett and Christophe Lécuyer, 18 February 1997; James F. Gibbons, "The Role of Stanford University: A Dean's Reflections," in Chong-Moon Lee, William F. Miller, Marguerite Gong Hancock, and Henry S. Rowen, eds., *The Silicon Valley Edge: A Habitat for Innovation and Entrepreneurship* (Stanford, Calif.: Stanford University Press, 2000), 200–217.

18. IBM devoted an entire issue of its *Journal of Research and Development* to describing the technology in the 3081. *IBM Journal of Research and Development* 26 (January 1982). For IBM's personal computer, see James Chposky and Ted Leonsis, *Blue Magic: The People, Power and Politics behind the IBM Personal Computer* (New York: Facts on File, 1988).

19. Andrew Pollack, "In Unusual Step, IBM Buys Stake in Big Supplier of Parts," *New York Times*, 23 December 1982, A1, D3; Thomas J. Lueck, "For Intel, a Vote of Confidence," *New York Times*, 23 December 1982, D1, D3; Allen R. Myerson, "IBM Aims to Cut 25,000 More Jobs; Sees Record Loss," *New York Times*, 16 December 1992, A1, D4; John Markoff, "Shifting Role in Technology," *New York Times*, 16 December 1992, A1, D4.

20. On estimates of the mainframe's contribution to IBM's profits, see Michael W. Miller, "IBM Forecasts That Core Business Will Post Revenue Growth in 1991," *Wall Street Journal*, 12 September 1991, B4. This figure was confirmed in Paul Rizzo, interview by author, 7 November 2000.

21. A discussion of standardization in the semiconductor industry is provided in Nathan Rosenberg and W. Edward Steinmueller, "The Economic Implications of the VLSI Revolution," in Nathan Rosenberg, *Inside the Black Box: Technology and Economics* (Cambridge: Cambridge University Press, 1982), 182–83.

22. Regis McKenna described Intel's process of mobilizing the company behind Operation Crush in two weeks to a Motorola executive, who said it would have taken him that long to get an airplane ticket approved to visit a customer. Regis McKenna, interview by Rob Walker, 22 August 1995, Silicon Genesis interviews. For Galbraith's claim that large corporations do not lose money, see John Kenneth Galbraith, *The New Industrial State* (Boston: Houghton Mifflin, 1967), 81–82. On the 125% Solution, see Mimi Real, *A Revolution in Progress: A History of Intel to Date* (Santa Clara, Calif.: Intel, 1984), 47. Americo DiPietro, a member of the corporate staff, assessed IBM's semiconductor program in 1971 by noting that IBM produced only things where large volumes could be ensured. He went on to say this: "We have not been as versatile as the typical vendor. We do not develop as many different components nor do we produce our components as early or as easily as the typical vendor." A. R. DiPietro to J. E. Bertram, 19 May 1971, Box TAR 242, IBM Technical History Project, IBM Archives, Somers, New York (hereafter IBM THP).

23. SAMOS is described in Richard A. Larsen, "A Silicon and Aluminum Dynamic Memory Technology," *IBM Journal of Research and Development* 24 (1980): 268–82. In a typically IBM obfuscatory naming practice, SAMOS stood for silicon aluminum metal oxide semiconductor, in some ways implying that the technology was similar to silicon gate, which it was not.

24. Dale L. Critchlow, "Burlington Visit," 22 June 1976, Box TAR 243, IBM THP. Critchlow was the manager of IBM Research's Solid-State Engineering Department and a veteran of the original MOS program.

25. J. Riseman, "Some Memory Product Concerns," 15 December 1975, Plaintiff Exhibit 6306, *U.S. v. IBM,* Hagley Museum and Library, Wilmington, Delaware.

26. The deep problems at IBM that required it to license Intel's technology were never publicly acknowledged. "IBM Component History," 37, IBM Archives, Somers, New York.

27. Other perspectives on the transfer have been provided by participants at IBM and Intel. Ken Moyle, interview by author, 13 December 2000; Bill Rowe, interview by author, 7 July 1997; Paul Castrucci, interview by author, 7 July 1997; Ken Moyle, note to author, 6 January 2001.

28. IBM's missed opportunities in the personal computer are detailed in Charles H. Ferguson and Charles R. Morris, *Computer Wars: The Fall of IBM and the Future of Global Technology* (New York: Times Books, 1994), 37–83. Ironically, in the late 1970s, IBM created two separate semiconductor organizations, one focused on mainframe computers, the other focused on lower-end systems, apparently so that if the government's antitrust suit ended in a breakup of IBM into two separate companies, each would have its own semiconductor organization. In 1980, when it looked less likely that the government would win the case, IBM went back to a single centralized semiconductor organization. Had IBM maintained two separate semiconductor organizations in the 1980s, it might have been more successful in developing semiconductor technologies for its personal computers and lower-end systems. "IBM Component History," 27–33.

29. Ferguson and Morris cite Jack Kuehler's comments in *Computer Wars*, 60. For Moore's statement about the revolutionary nature of the microprocessor, see Gordon E. Moore, "Microprocessors and Integrated Electronic Technology," *Proceedings of the IEEE*, 64 (June 1976): 837.

30. This analogy is based on one made by Paul Ceruzzi in *A History of Modern Computing* (Cambridge, Mass.: MIT Press, 1998), 177–78.

31. The price for an IBM 3081 is given in James A. White, "IBM Introduces Its Fastest Computer, Cuts Prices and Upgrades Other Lines," *Wall Street Journal*, 22 October 1981, 10. The memory size for the 3081 is from A. Padegs, "System/360 and Beyond," *IBM Journal of Research and Development* 25 (September 1981): 389. The figures for the IBM personal computer are from Chposky and Leonsis, *Blue Magic*, 220.

32. The figure for the number of chips in a mid-1970s mainframe is from Lewis M. Terman, "The Role of Microelectronics in Data Processing," *Scientific American*, September 1977, 174. The figure for a 1990s mainframe is from Isaac, "Future of CMOS Technology," 376. The complexity of the mainframe can be seen by looking at any of the articles on the 3081 in the *IBM Journal of Research and Development* 26 (January 1982).

33. Ferguson and Morris, *Computer Wars*, 211–12; Kenneth Flamm, *Mismanaged Trade?*

Strategic Policy and the Semiconductor Industry (Washington, DC: Brookings Institution, 1996): 216–19; George W. Cogan and Robert A. Burgelman, "Intel Corporation (A): The DRAM Decision," Stanford Graduate School of Business Case S-BP-256, 22 January 1991; Bruce Graham and Robert Burgelman, "Intel Corporation (B): Implementing the DRAM Decision," Stanford Graduate School of Business Case BP-256B, 20 December 1990.

34. The differences in the two logic technologies can be seen in M. S. Pittler, D. M. Powers, and D. L. Schnabel, "System Development and Technology Aspects of the IBM 3081 Processor Complex," *IBM Journal of Research and Development* 26 (January 1982): 6, and Charles L. Cohen, "Hitachi CPU Challenges IBM," *Electronics*, 18 March 1985, 16–17.

35. The comparisons between the 486 and the 4004 are from "Intel Microprocessors," provided by the Intel Museum, Santa Clara, California. The introduction of the 486 is covered in Robert Ristelhueber, "Intel i486 Introduced; Integrates MMU, FPU," *Electronic News*, 17 April 1989, 1, 28. See also Richard Brandt and Otis Port, "Intel: The Next Revolution," *Business Week*, 26 September 1988, 74–79.

36. K. H. Brown, D. A. Grose, T. H. Ning, and P. A. Totta, "Advancing the State of the Art in High-Performance Logic and Array Technology," *IBM Journal of Research and Development* 36 (September 1992): 826. IBM's system is described in "IBM Introduces 18 New Systems Including Two Summit Models," *Electronic News*, 10 September 1990, 12, 18.

37. Steve Lohr, "Mainframes Aren't All That Dead," *New York Times*, 9 February 1993, D1, D6; David R. Brousell, "What's Really Wrong with IBM," *Datamation*, 1 March 1993, 142; Brandt and Port, "Intel: The Next Revolution," 74–79; Otis Port, "Put Enough PCs Together and You Get a Supercomputer," *Business Week*, 26 September 1988, 80.

38. The forty salespeople assigned to one account was described to me by Rizzo, interview.

39. The best case for the advantages of CMOS over bipolar technology is given by IBM. G. S. Rao et al., "IBM S/390 Parallel Enterprise Servers G3 and G4," *IBM Journal of Research and Development* 41 (1997): 397–403.

40. Laurence Zuckerman, "Adding Power, Hitachi Becomes No. 2 to IBM in Mainframes," *New York Times*, 7 July 1997, D1, D4.

41. Chad Fasca, "Copper Strikes Gold for IBM," *Electronic News*, 29 September 1997, 6; Gale Bradley, "IBM Debuts Alliance Chip," *Electronic News*, 16 June 1997, 1; Rao et al., "IBM S/390 Parallel Enterprise Servers G3 and G4," 397–403.

A large body of work exists on the history of semiconductor technology, but it is uneven, reflecting the writers' diverse backgrounds as historians, journalists, participants, and economists. This is reflected in the only two works that deal with the history of the MOS transistor at any length. George Gilder's *Microcosm: The Quantum Revolution in Economics and Technology* (New York: Simon and Schuster, 1989) is alternately exhilarating and exasperating. Gilder is an enthusiastic student of technology drawn to the big picture—he understood the implications of MOS technology well before they became clear to most of the rest of the world. But his work is burdened by his own philosophical and political predilections, and one senses that in his eagerness to look at the big picture, he gets bored with details, where he is frequently unreliable. (Gilder's work is reviewed by two Silicon Valley executives in T. J. Rodgers, "Landmark Messages from the Microcosm," *Harvard Business Review*, January–February 1990, 24–30; and Robert N. Noyce, "False Hopes and High-Tech Fiction," *Harvard Business Review*, January–February 1990, 31–36.) Chih-Tang Sah's "Evolution of MOS Transistor—From Conception to VLSI," *Proceedings of the IEEE* 76 (October 1988): 1280–1326, provides the authoritative knowledge of a leading researcher in the field and a comprehensive list of technical citations but is less successful at presenting the historical development of the technology.

Of the broader histories of semiconductor technology, Michael Riordan and Lillian Hoddeson, *Crystal Fire: The Birth of the Information Age* (New York: W.W. Norton, 1997), provides the definitive history of the invention and early development of the transistor but stops before MOS work begins in earnest. Their research also led to the articles "Origins of the p-n Junction," *IEEE Spectrum*, June 1997, 46–51; and "The Moses of Silicon Valley," *Physics Today*, December

1997, 42–47. Previously, Hoddeson had written a number of articles that had established her position as the preeminent scholar of the history of solid-state physics. These include "The Discovery of the Point-Contact Transistor," *Historical Studies in the Physical Sciences* 12 (1981): 41–76; "The Entry of the Quantum Theory of Solids into the Bell Telephone Laboratories, 1925–1940: A Case Study of the Industrial Application of Fundamental Science," *Minerva* 18 (1980): 422–47; and "Research on Crystal Rectifiers during World War II and the Invention of the Transistor," *History and Technology* 11 (1994): 121–30. Hoddeson, along with Ernest Braun, Jürgen Teichmann, and Spencer Weart, also edited *Out of the Crystal Maze: Chapters from the History of Solid-State Physics* (New York: Oxford University Press, 1992), which concentrates on the scientific developments in solid-state physics prior to the development of the transistor. Ernest Braun and Stuart MacDonald produced one of the earliest histories of semiconductor technology in *Revolution in Miniature: The History and Impact of Semiconductor Electronics*, 2d ed. (Cambridge, England: Cambridge University Press, 1982), and although it gives only cursory attention to MOS technology, it introduced a number of important themes. Stan Augarten's *State of the Art: A Photographic History of the Integrated Circuit* (New Haven, Conn.: Ticknor and Fields, 1983) is not a conventional history of semiconductor technology, but much can be gleaned from his collection of pictures of transistors and integrated circuits spanning a thirty-five-year period. Raymond Warner provides an insightful participant's account of the history of semiconductor technology up to the early 1970s based on his experiences at Bell Labs, Motorola, and Texas Instruments in the introduction to his engineering text, R. M. Warner Jr. and B. L. Grung, *Transistors: Fundamentals for the Integrated Circuit Engineer* (New York: Wiley, 1983), 1–91. Frederick Seitz and Norman G. Einspruch's *Electronic Genie: The Tangled History of Silicon* (Urbana, Ill.: University of Illinois Press, 1998) provides a history of silicon science and technology that is much broader in its coverage than other work. While the special issue of *Electronic Engineering Times* on the history of the integrated circuit, "Thirty Who Made a Difference: The 30th Anniversary of the Integrated Circuit," *Electronic Engineering Times,* September 1988, introduces many personalities, it is highly idiosyncratic and not completely reliable. Bruce E. Deal and James M. Early provide an authoritative technical review in "The Evolution of Silicon Semiconductor Technology, 1952–1977," *Journal of the Electrochemical Society* 126 (January 1979): 20C–32C.

A good deal of historical writing on semiconductor technology has clustered around the invention of the integrated circuit and the microprocessor. Herbert S. Kleiman, "The Integrated Circuit: A Case Study of Product Innovation in the Electronics Industry" (D.B.A. diss., George Washington University, 1966), makes the argument that government funding did not directly contribute to the development of the integrated circuit. Daniel U. Holbrook, "Technical Diversity and Technological Change in the American Semiconductor Industry, 1952–1965" (Ph.D. diss., Carnegie-Mellon University, 1999), explores how a diversity of approaches contributed to the development of the integrated circuit. One of the most detailed histories (with copies of pages from laboratory notebooks) is Michael F. Wolff, "The Genesis of the Integrated Circuit," *IEEE Spectrum,* August 1976, 45–53. Jack Kilby gives his account of the early history of the integrated circuit in "Invention of the Integrated Circuit," *IEEE Transactions on Electron Devices* 23 (July 1976): 648–54. Lester Hogan provides an account of the invention of the integrated circuit that gives more attention to work prior to Kilby and Noyce in "Reflections on the Past and Thoughts about the Future of Semiconductor Technology," *Interface Age,* March 1977, 24–36. T. R. Reid presents a history of the integrated circuit and the microprocessor based on interviews with the participants in *The Chip: How Two Americans Invented the Microchip and Launched a Revolution* (New York: Simon and Schuster, 1985). Michael S. Malone presents a history of the microprocessor that goes beyond Intel's work in *The Microprocessor: A Biography* (New York: Springer-Verlag, 1995). William Aspray gives a history of Intel's 4004 that pays more attention to the Japanese side of the story in "The Intel 4004 Microprocessor: What Constituted Invention?" *IEEE Annals of the History of Computing* 19 (1997): 4–15. The history of the microprocessor is particularly rich in participants' accounts. Among them are Robert N. Noyce and Marcian E. Hoff Jr., "A History of Microprocessor Development at Intel," *IEEE Micro,* February 1981, 8–21; Stanley Mazor, "The History of the Microcomputer: Invention and Evolution," *Proceedings of the IEEE* 83 (December 1995): 1601–8; Federico Faggin, "The Birth of the Microprocessor," *Byte,* March 1992, 145–50. The key participants make a joint statement in Federico Faggin, Marcian E. Hoff Jr., Stanley Mazor, and Masatoshi Shima, "The History of the 4004," *IEEE Micro,* December 1996, 10–20.

Economists were among the first group to study the history of semiconductor technology. Useful studies include Richard R. Nelson, "The Link between Science and Invention: The Case of the Transistor," in *The Rate and Direction of Inventive Activity: Economic and Social Factors; A Conference of the Universities-National Bureau Committee for Economic Research and the Committee on Economic Growth of the Social Science Research Council* (Princeton, N.J.: Princeton University Press, 1962), 549–83; John E. Tilton, *The International Diffusion of Technology: The Case of Semiconductors* (Washington, D.C.: Brookings Institution, 1971); Richard Levin, "The Semiconductor Industry," in *Government and Technical Progress: A Cross Industry Analysis*, ed. Richard R. Nelson (New York: Pergamon Press, 1982), 9–100; Robert W. Wilson, Peter K. Ashton, and Thomas P. Egan, *Innovation, Competition, and Government Policy in the Semiconductor Industry* (Lexington, Mass.: D.C. Heath, 1980); A. M. Golding, "The Semiconductor Industry in Britain and the United States: A Case Study in Innovation, Growth, and the Diffusion of Technology (D. Phil. thesis, University of Sussex, 1971).

Unfortunately, there has of yet been no full-scale history of Silicon Valley. Very insightful analysis, only partially grounded in history, is provided in AnnaLee Saxenian, *Regional Advantage: Culture and Competition in Silicon Valley and Route 128* (Cambridge, Mass.: Harvard University Press, 1994). The diversity of approaches to studying Silicon Valley can be seen by the pieces in two recent volumes: Martin Kenney, ed., *Understanding Silicon Valley: The Anatomy of an Entrepreneurial Region* (Stanford, Calif.: Stanford University Press, 2000); and Chong-Moon Lee, William F. Miller, Marguerite Gong Hancock, and Henry S. Rowen, ed., *The Silicon Valley Edge: A Habitat for Innovation and Entrepreneurship* (Stanford, Calif.: Stanford University Press, 2000). Stuart Leslie and his colleagues at Johns Hopkins University have written a series of articles about Stanford's role in Silicon Valley. Stuart W. Leslie and Bruce Hevly, "Steeple Building at Stanford: Electrical Engineering, Physics, and Microwave Research," *Proceedings of the IEEE* 73 (July 1985): 1169–80; Robert Kargon, Stuart W. Leslie, and Erica Schoenberger, "Far Beyond Big Science: Science Regions and the Organization of Research and Development," in *Big Science: The Growth of Large Scale Research*, ed. Peter Galison and Bruce Hevly (Stanford, Calif.: Stanford University Press, 1992, 334–54); Robert Kargon and Stuart Leslie, "Imagined Geographies: Princeton, Stanford, and the Boundaries of Useful Knowledge in Postwar America," *Minerva* 32 (summer 1994): 121–43; and Stuart W. Leslie and Robert H. Kargon, "Selling Silicon Valley: Frederick Terman's Model for Regional Advantage," *Business History Review* 70 (winter 1996): 435–72. Journalistic studies include Michael S. Malone's *The Big Score: The Billion-Dollar Story of Silicon Valley* (Garden City, N.Y.: Doubleday, 1985); Everett M. Rogers and Judith K. Larsen, *Silicon Valley Fever: Growth of High-Technology Culture* (New York: Basic Books, 1984); and Dirk Hansen's *The New Alchemists: Silicon Valley and the Microelectronics Revolution* (Boston: Little, Brown, 1982). Henry Lowood provides a revisionist perspective on Stanford and the development of the high-technology industry in *From Steeples of Excellence to Silicon Valley: The Story of Varian Associates and Stanford Industrial Park* (Palo Alto, Calif.: Varian Associates, 1988).

Christophe Lécuyer explores the early history of Fairchild in his dissertation, "Making Silicon Valley: Engineering Culture, Innovation, and Industrial Growth, 1930–1970" (Ph.D. diss., Stanford University, 1999); his article "Silicon for Industry: Component Design, Mass Production and the Move to Commercial Markets at Fairchild Semiconductor, 1960–1967," *History and Technology* 16 (1999): 179–216; and his chapter "Fairchild Semiconductor and Its Influence" in *The Silicon Valley Edge*. Rob Walker gives a participant's account of Fairchild's R&D work on designing MOS integrated circuits using computer-aided-design techniques in *Silicon Destiny: The Story of Application Specific Integrated Circuits and LSI Logic Corporation* (Milpitas, Calif.: CMC Publications, 1992).

Whatever Stanford's role in the development of the semiconductor industry in Silicon Valley, the Stanford University Library's Department of Special Collections and University Archives has one of the richest collections of archival material relating to semiconductor history. Particularly important are the Fairchild Camera and Instrument Technical Reports and Progress Reports (M1055), which include progress reports of the R&D laboratory. Other important collections of Fairchild material are the Bruce Deal Papers (M1051), which include published papers and technical reports; and the Rob Walker Papers (M0742), which include material on Fairchild's work on computer-aided design, the photographic files of Steve Allen (M1038), and several other collections of images (M0985, MISC 587). Important interview collections at Stanford include Silicon Genesis: Oral History Interviews with Silicon Valley Scientists (M0741), videotaped interviews with both senior statesmen of Silicon Valley and lesser-known figures, often in their own homes; the George Rostky collection (M0851), interviews that are chock full of anecdotes but must be used with some caution, done for the special issue of *Electronic Engineering Times* cited above; and the Herb Kleiman collection (M0827), interviews done for his 1966 doctoral dissertation. The Raytheon Collection (M0661) includes materials tracking developments throughout the semiconductor industry. The Frederick Terman papers (SC 160) contain materials that provide some idea of the relations between Stanford and the electronics industry. Although not consulted for this study, the William Shockley papers (SC 222) are also housed at Stanford.

The only book-length treatment of Intel's history, Tim Jackson's *Inside Intel: Andy Grove and the Rise of the World's Most Powerful Chip Company* (New York: Dutton, 1997), contains some interesting information but on the whole is a disappointment. It was quickly written, based largely on interviews with ex-Intel employees, and while Jackson tried to write it as an exposé, he ultimately did not find much to expose. Intel has put out three anniversary brochures that are good sources for its corporate history: Mimi Real, *A Revolution in Progress: A History of Intel to Date* (Santa Clara, Calif.: Intel, 1984) and *Intel: Architect of the Microcomputer Revolution* (Santa Clara, Calif.: Intel, 1988); as well as Katie Woodruff, *Defining Intel: 25 Years/25 Events* (Santa Clara, Calif.: Intel, 1993). The closest thing to a company-sponsored academic history of Intel is the collection of Stanford Business School case studies written by Robert Burgelman and his colleagues. Robert Burgelman, Dennis L. Carter, and Raymond Bamford, "Intel Corporation: The Evolution of an Adaptive Organization," Stanford Graduate School of Business, Case SM-65, 22 July 1999; George B. Cogan and Robert Burgelman, "Intel Corporation (A): The DRAM Decision," Stanford Graduate School of Business, Case BP-256A, 22 January 1991; Bruce Graham and Robert Burgelman, "Intel Corporation (B): Implementing the DRAM Decision," Stanford Graduate School of Business, Case BP-256B, 20 December 1990. Several present and former Intel employees have written books that provide small pieces of historical information about the company. Andrew S. Grove, *Only the Paranoid Survive: How to Survive the Crisis Points That Challenge Every Company and Career* (New York: Doubleday, 1996), provides some discussion of Intel's early semiconductor memory work and Intel's decision to leave the semiconductor memory business. Albert Yu, *Creating the Digital Future: The Secrets of Consistent Innovation at Intel* (New York: The Free Press, 1998), contains some history of Intel's microprocessor business. William H. Davidow, *Marketing High Technology: An Insider's View* (New York: Free Press, 1986), has insights about Intel's microprocessor work. Intel's Museum has several useful interviews with senior Intel employees. *Time*'s 1997 man of the year portrait of Andrew Grove is Joshua Cooper Ramo, "A Survivor's Tale," *Time*, 29 December 1997/5 January 1998, 54–72. The *IEEE Spectrum* profiled Grove in Linda Geppert, "Profile: Andy Grove," *IEEE Spectrum*, June 2000, 34–38. Gordon Moore has written several insightful articles covering aspects of both Fairchild's history and Intel's history: "Intel—Memories and the Microprocessor," *Daedalus* 125 (spring 1996): 55–80; "Some Personal Perspectives on Research in the Semiconductor Industry," in *Engines of Innovation: U.S. Industrial Research at the End of an Era*, ed. Richard S. Rosenbloom and William J. Spencer (Boston: Harvard Business School Press, 1996), 165–74; and "The Role of Fairchild in Silicon Technology in the Early Days of 'Silicon Valley,'" *Proceedings of the IEEE* 86 (January 1998): 53–62. The history of Intel's 1103 is described in Brian Santo, "1K-Bit DRAM," *IEEE Spectrum*, 25 November 1988, 108–12. While Tom Wolfe's "The Tinkerings of Robert Noyce," *Esquire*, December 1983, 346–74, is not the place to go for technical or business details, it is still a classic description of Intel's culture. Sematech published a volume in memory of Robert Noyce, *Robert Noyce, 1927–1990* (Austin, Tex.: Sematach, n.d.). Leslie Berlin has examined Noyce in "Robert Noyce and Fairchild Semiconductor, 1957–1968," *Business History Review* 75 (spring 2001): 63–101; and "Robert Noyce and the Rise of Silicon Valley, 1956–1990" (Ph.D. diss., Stanford University, 2001). Public documents such as annual reports and Securities and Exchange Commission filings are a vital source of information about Intel and other semiconductor firms.

IBM has sponsored two histories that provide a good perspective on its computer and semiconductor development efforts: Charles J. Bashe, Lyle R. Johnson, John H. Palmer, and Emerson W. Pugh, *IBM's Early Computers* (Cambridge, Mass.: MIT Press, 1986); and Emerson W. Pugh, Lyle R. Johnson, and John H. Palmer, *IBM's 360 and Early 370 Systems* (Cambridge, Mass.: MIT Press, 1991). Emerson W. Pugh himself then wrote *Building IBM: Shaping an Industry and Its Technology* (Cambridge, Mass.: MIT Press, 1995), covering the entire history of IBM; it is not in the same detail as the earlier works, but it still displays an authoritative knowledge of the company. In the course of writing these books the authors assembled a very rich archive, which the company has in the past made available to qualified scholars. The interviews conducted for these volumes are on deposit at the Hagley Library in Wilmington, Delaware, and the Charles Babbage Institute in Minneapolis, Minnesota. Steven Usselman's "IBM and Its Imitators: Organizational Capabilities and the Emergence of the International Computer Industry," *Business and Economic History* 22 (winter 1993): 1–35, provides an understanding of how IBM made the transition from punched cards to computing. One of the few positive outcomes of the government's antitrust case against IBM was that its interminable proceedings provided many important documents for historians. Materials from the antitrust trial are available at the Hagley Library and the Babbage Institute. Two dramatically different guides to these materials are provided by Franklin M. Fisher and Richard Thomas DeLamarter. Fisher, an MIT economist and expert witness for IBM, and his colleagues find that the material exonerates IBM in *IBM and the U.S. Data Processing Industry: An Economic History* (New York: Praeger, 1983) and *Folded, Spindled and Mutilated: Economic Analysis and U.S. v. IBM* (Cambridge, Mass.: MIT Press, 1983). DeLamarter, an economist who worked for the government, comes to the opposite conclusion in *Big Blue: IBM's Use and Abuse of Power* (New York: Dodd, Mead, 1986). Useful accounts of IBM's semiconductor work by participants include Robert Dennard, "Evolution of the MOSFET Dynamic RAM—A Personal View," *IEEE Transactions on Electron Devices* 31 (November 1984): 1549–55, as well as three articles published in the twenty-fifth anniversary issue of the *IBM Journal of Research and Development* 25 (September 1981): E. W, Pugh, D. L. Critchlow, R. A. Henle, and L. A. Russell, "Solid State Memory Development in IBM," 585–602; E. J. Rymaszewski, J. L. Walsh, and G. W. Leehan, "Semiconductor Logic Technology in IBM," 603–16; and William E. Harding, "Semiconductor Manufacturing in IBM, 1957 to the Present: A Perspective," 647–58. Alan Fowler, one of IBM's leading semiconductor physicists, presents his perspectives on semiconductor history in "A Semicentury of Semiconductors," *Physics Today,* October 1993, 59–62; and "On Some Modern Uses of the Electron in Logic and Memory," *Physics Today,* October 1997, 50–54. The best book describing IBM's difficulties in the early 1990s (in spite of its hyperbolic title and its now-dated emphasis on the Japanese threat) is Charles H. Ferguson and Charles R. Morris, *Computer Wars: The Fall of IBM and the Future of Global Technology* (New York: Times Books, 1994). Less successful by failing to understand the technological changes facing IBM are Paul Carroll, *Big Blues:*

The Unmaking of IBM (New York: Crown Publishers, 1993); and D. Quinn Mills and G. Bruce Friesen, *Broken Promises: An Unconventional View of What Went Wrong at IBM* (Boston: Harvard Business School Press, 1996). An article that attempts to come to grips with how IBM executives themselves managed information about technological change is Martin Fransman, "Information, Knowledge, Vision, and Theories of the Firm," *Industrial and Corporate Change* 3 (1994): 713–57.

AT&T has the richest corporate archive available for scholarly use. It has an abundance of material related to Bell Labs' invention and development of the transistor. One particularly important document, originally written as part of a corporate history but never published, is A. E. Anderson and R. M. Ryder, "Development History of the Transistor in Bell Laboratories and Western Electric (1947–1975)." It provides a very detailed history of the early period of transistor development by two participants as well as more candor than is commonly found in corporate histories. Most of the pieces written by Bell Labs engineers on the original MOS work at Bell Labs are not very enlightening. Unfortunately, Dawon Kahng's "A Historical Perspective on the Development of MOS Transistors and Related Devices," *IEEE Transactions on Electron Devices* 23 (July 1976): 655–57, falls into this category, but Ian M. Ross's "The Invention of the Transistor," *Proceedings of the IEEE* 86 (January 1998): 7–28, is a notable exception. The AT&T-sponsored series *A History of Engineering and Science in the Bell System* contains two volumes that discuss AT&T's semiconductor work in detail: *Electronics Technology (1925–1975)* (n.p.: AT&T Bell Laboratories, 1985) and *Physical Sciences (1925–1980)* (n.p.: AT&T Bell Laboratories, 1983). William Shockley wrote two detailed articles on his invention of the junction transistor, "The Path to the Conception of the Junction Transistor," *IEEE Transactions on Electron Devices* 23 (July 1976): 597–620, and "The Invention of the Transistor: An Example of Creative-Failure Methodology," *National Bureau of Standards Special Publication 388: Proceedings of Conference on the Public Need and the Role of the Inventor (11–14 June 1973), Monterey, Calif.,* 47–89. Thomas J. Misa's "Military Needs, Commercial Realities, and the Development of the Transistor, 1948–1948," in *Military Enterprise and Technological Change: Perspectives on the American Experience,* ed. Merritt Roe Smith (Cambridge, Mass.: MIT Press, 1985), 253–87, is a history of Bell's early transistor development effort that tends to overplay the role of the military.

RCA's work in semiconductor technology has often been neglected, due to its lack of success commercially and its place in the shadow of Bell Labs. G. B. Herzog gives a history of RCA's early commitment to complementary MOS in "COS/MOS from Concept to Manufactured Product," *RCA Engineer,* Dec.–Jan. 1975–76, 36–38. The David Sarnoff Library in Princeton, New Jersey, has a rich collection of reports and laboratory notebooks. The Hagley Library's RCA collection supplements the Sarnoff holdings. The *RCA Engineer* provides some insight into the technical details of RCA's semiconductor work. The "RCA Engineers Interviews" at the IEEE History Center in New Brunswick, New Jersey, includes oral history interviews with several key figures in RCA's semiconductor work.

One of the more intriguing questions behind the transistor is the work preceding it by Julius Lilienfeld and Oskar Heil. The best introduction to this work is Robert G. Arns, "The Other Transistor: Early History of the Metal-Oxide-Semiconductor Field-Effect Transistor," *Engineering Science and Education Journal* 7 (October 1998): 223–40.

Industrial research in post–World War II America has received a good deal of attention from historians of technology recently. The best overview is David A. Hounshell's "The Evolution of Industrial Research in the United States," in *Engines of Innovation,* 13–85. The volume as a whole provides perspectives from both industry and the academy on the changing nature of industrial research in the late twentieth century. Hounshell and John Kenly Smith Jr.'s *Science and Corporate Strategy: Du Pont R&D, 1902–1980* (Cambridge, England: Cambridge University Press, 1988) shows much about the promise and perils of research for one leading corporation, while Margaret B. W. Graham's *The Business of Research: RCA and the VideoDisc* (Cambridge, England: Cambridge University Press, 1986) explores similar themes in a case study. David C. Mowery and Nathan Rosenberg, *Technology and the Pursuit of Economic Growth* (Cambridge, England: Cambridge University Press, 1989), is invaluable for understanding the history of R&D in America, as is their later work *Paths of Innovation: Technological Change in 20th-Century America* (Cambridge, England: Cambridge University Press, 1998). Robert Buderi carries the story of R&D labs into the 1990s in his *Engines of Tomorrow: How the World's Best Companies Are Using Their Research Labs to Win the Future* (New York: Simon & Schuster, 2000), although it is impossible to know if his optimism is justified.

The history of computing has matured as a field recently. Paul E. Ceruzzi, *A History of Modern Computing* (Cambridge, Mass.: MIT Press, 1998), and Martin Campbell-Kelly and Bill Aspray, *Computer: A History of the Information Machine* (New York: Basic Books, 1996), are surveys that provide good accounts of the development of the personal computer. Paul Freiberger and Michael Swaine's *Fire in the Valley: The Making of the Personal Computer*, 2d ed. (New York: McGraw-Hill, 2000), and Steven Levy, *Hackers: Heroes of the Computer Revolution* (Garden City, N.Y.: Anchor Press/Doubleday, 1984), provide a good sense of the milieu in which the personal computer developed. Richard N. Langlois, "External Economies and Economic Progress: The Case of the Microcomputer Industry," *Business History Review* 66 (spring 1992): 1–50, provides an important perspective on the development of the personal computer industry.

The sheer volume of primary source material presented in the journals and conferences of professional societies is overwhelming. S. M. Sze has edited *Semiconductor Devices: Pioneering Papers* (Singapore: World Scientific, 1991); its two dozen or so MOS-related papers in the collection of 141 papers provide a good introduction to the field. Important publications for following purely technical developments in semiconductor technology are the *International Solid-State Circuits Conference Digest of Technical Papers,* which often presents work just prior to its commercial introduction. Its sister publication is the *IEEE Journal of Solid-State Circuits.* Other important publications, although often presenting work that is still in the research stage, are the *IEEE Transactions on Electron Devices* and the *IEDM Technical Digest.* Professional publications in semiconductor technology must be used with caution. Just because an advance is described in a journal does not mean that it will ever see the light of day, and many advances are never described in articles. The trade publications *Electronic News* and *Electronics* (the latter no longer published) are essential resources for providing both a technical and a business perspective. They contain valuable information about companies' products, strategies, and personnel. Since 1967 the consulting firm Integrated Circuit Engineering (ICE) has published an annual report on the semiconductor industry, providing statistics and evaluations of the positions of major firms.

One of the richest archival collections covering the industry as a whole is the ICE Collection at the National Museum of American History (NMAH). ICE was one of the leading semiconductor consulting firms and had contacts with many corporations. It also, as noted above, published an annual report on the state of the industry, providing statistics and an evaluation of the positions of firms both big and small. The ICE Collection includes its evaluations of integrated circuits as well as a large number of integrated circuits themselves. The NMAH also holds hard-to-find copies of Don Hoefler's *Microelectronics News,* an important source of news and gossip in Silicon Valley in the 1970s and 1980s.

There has been little written on the history of the semiconductor industry in the 1980s and 1990s. A starting point is provided by Richard N. Langlois and W. Edward Steinmueller, "The Evolution of Competitive Advantage in the World-wide Semiconductor Industry, 1947–1996," in *Sources of Industrial Leadership: Studies of Seven Industries,* ed. David C. Mowery and Richard R. Nelson (Cambridge, England: Cambridge University Press, 1999), 19–78; and Jeffery T. Macher, David C. Mowery, and David A. Hodges, "Semiconductors," in *U.S. Industry in 2000: Studies in Competitive Performance,* ed. David C. Mowery (Washington, D.C.: National Academy Press, 1999), 245–86. Kenneth Flamm covers the trade dispute between the United States and Japan in *Mismanaged Trade? Strategic Policy and the Semiconductor Industry* (Washington, D.C.: Brookings Institution Press, 1996). Some of America's anxiety over the Japanese threat in the 1980s is shown in Fred Warshofsky, *The Chip War: The Battle for the World of Tomorrow* (New York: Charles Scribner's Sons, 1989). Larry D. Browning and Judy C. Shetler, *Sematech: Saving the U.S. Semiconductor Industry* (College Station, Tex.: Texas A&M University Press, 2000), is an officially sponsored history of the semiconductor manufacturing consortium.

The literature of business and economic history has played an important role in helping me frame the issues involved in this study. Any study of big business in America has to begin with Alfred Chandler's classic works *The Visible Hand: The Managerial Revolution in American Business* (Cambridge, Mass.: Harvard University Press, 1977), *Strategy and Structure: Chapters in the History of the American Industrial Enterprise* (Cambridge, Mass.: MIT Press, 1962), and *Scale and Scope: The Dynamics of Industrial Capitalism* (Cambridge, Mass.: Harvard University Press, 1990). Nathan Rosenberg's works, particularly the essays in *Perspectives on Technology* (Cambridge, England: Cambridge University Press, 1976), *Inside the Black Box: Technology and Economics* (Cambridge, England: Cambridge University Press, 1982), and *Exploring the Black Box: Technology, Economics, and History* (Cambridge, England: Cambridge University Press, 1994) are indispensable for their understanding of technology and economics. His essay "Economic Experiments" in *Exploring the Black Box* has been particularly important in shaping my understanding of economic and institutional factors in technological change.

Clayton Christensen has developed a powerful method of understanding technological change in a corporate environment. He lays out his ideas very accessibly in *The Innovator's Dilemma: When New Technologies Cause Great Firms to Fail* (Boston: Harvard Business School Press, 1997). This should be supplemented with his writings in scholarly journals: "The Rigid Disk Drive Industry: A History of Commercial and Technological Turbulence," *Business History Review* 67 (winter 1993): 531–88; Richard S. Rosenbloom and Clayton M. Christensen, "Technological Discontinuities, Organizational Capabilities, and Strategic Commitments," *Industrial and Corporate Change* 3 (1994): 655–85; Clayton M. Christensen and Richard S. Rosenbloom, "Explaining the Attacker's Advantage: Technological Paradigms, Organizational Dynamics, and the Value Network," *Research Policy* 24 (1995): 233–57. One of the foundations of Christensen's work is Giovanni Dosi's article on technological trajectories, "Technological Paradigms and Technological Trajectories: A Suggested Interpretation of the Determinants and Directions of Technical Change," *Research Policy* 11 (1982): 147–62. An excellent historical study using Christensen's framework is Albert J. Churella, *From Steam to Diesel: Managerial Customs and Organizational Capabilities in the Twentieth-Century American Locomotive Industry* (Princeton, N.J.: Princeton University Press, 1998). Churella provides a briefer version of his argument in "Corporate Culture and Marketing in the America Railway Locomotive Industry: American Locomotive and Electro-Motive Despond to Dieselization," *Business History Review* 69 (summer 1995): 191–229.

INDEX

Page numbers in italics refer to illustrations or tables.

IBM (*cont.*)

Burlington facility, 239–46, 250, 385n.94; cache memory at, 103, 220; CMOS development effort at, 248–49, 386n.108; corporate staff at, 92, 95–96, 103–4, 218–19, 230–31, 246–247; Corporate Technical Board at, 77–78, 343n.62, 348n.44; Corporate Technical Committee at, 104; Data Processing Group, 103–4; development cycle at, 212; early transistor work at, 59, 66, 338n.4; East Fishkill facility, 5, 69, 228–29, 320n.8; exchanges with Intel, 297–98, 385n.90; Federal Systems Division, 93; Future System (FS), 248–49, 381n.46; focus on performance at, 75–76, 77, 87, 101, 347n.32; formation of Components Division at, 67; Future Manufacturing System at, 320n.8; Huntsville, Lee Boysel and, 124, 126; integrated circuits and, 69–70, 71–72, 221, 341n.37; Intel compared with, 210–11, 222, 227–28, 230, 294–96, 301, 381nn.42, 47; introduction of MOS memories into products at, 245; joint MOS development program at, 105–6, 211–17, 378n.19; LSI and, 91–92; mainframe computer at, 299–307; microprocessor and, 280, 393n.79; MOS logic development program at, 246–48, 386n.100; MOS manufacturing at, 239–46; MOS memory development at, 228–31; 1956 reorganization of, 60; in 1990s, 4–5, 303–7, 320n.8; organization of semiconductor operations at, 397n.28; papers from, at 1965 SSDRC, 141; personal computer and, 298–99; SAMOS, development of, at, 296–98, 397n.23; semiconductor operations of, compared to semiconductor industry, 91–92, 94–95, 222–28, 230–31, 247, 347n.31, 348n.42, 350n.67; 380nn.38, 39, 40, 381nn.42, 45, 47; Sindelfingen plant, 244; Stanford and, 293; tensions between bipolar and MOS at, 72–73, 218–22, 304–6; yield management system at, 241, 242–44, 385n.90. *See also* IBM Research; IBM System/360; IBM System/370

IBM Research: advocacy of MOS within IBM by, 76, 91, 100–102; audit of MOS program at, 92, 96; background of MOS personnel at, *80*, 80–81, 87; CD, and, 70, 73– 74, 81, 88–89, 92, 104–6, 157, 228–29, 363n.37; commitment to MOS transistor at, 78, 86; compared to Bell Labs, 59, 74, 337n.3; criticism of, by Thomas Watson Jr., 75, 343n.55; design automation work at, 86–87, 88–91; early wariness toward MOS transistor at, 345n.14; efforts to halt MOS work at, 77–78, 86, 215–16, 344n.66, 378n.14; emphasis of engineering considerations in MOS work by, 74–75, 88–89, 99, 346nn.19, 24; formation and early history of, 60–62, 64–67, 339n.9; germanium, and, 76, 148–49, 361n.20; IBM Burlington and, 241; integrated circuit, response to, 71; interfirm communications of, 98–99, 156–61, 363nn.36, 39; linear model of R&D and, 61–62, 284–85; magnetic memory programs at, 100; microprocessor and, 252; MOS design manual at, 216–17, 378n.17; MOS memory work of, 101–2, 104–6, 214–15, 377n.11; MOS scaling and, 283–84; MOS stability work at, 84–85, 97–100; n-channel MOS and, 83–85; in 1990s, 307; organization of, compared to Bell Labs, 74; organization of, in 1950, 58–59; organization of MOS work at, 86, *311*; other (nonMOS) semiconductor programs at, 65, 75, 81, 147–49, 342n.40, 344n.1; proposal for military funding of MOS work by, 76–77, 89; SAMOS and, 296–97; scientific work of, on MOS structures, 85, 285

IBM System/360: Cogar and, 237, 238; control stores and, 102; at Fairchild, 133; Intel design of memory systems for, 201; Lee Boysel's study of, 125–26; use of SLT in, 68–69, 94–95

IBM System/370, 201, 245, 257

IEEE: awards, 327n.43, 344n.1, 363n.35, 369n.27; conferences of, 82, 119, 137, 140, 142, 205, 276–77, 380n.38; publications of, 143–46, 252. *See also* Institute of Radio Engineers; International Electron Device Meeting; International Solid State Circuits Conference; Solid State Device Research Conference

IEEE *Spectrum*, 189–90

IGFET, usage of term for MOS transistor, 223, 346n.15, 380n.37

RELATED BOOKS IN THE SERIES

William M. McBride, *Technological Change and the United States Navy, 1865–1945*

Arthur L. Norberg and Judy E. O'Neill, with contributions by Kerry Freedman, *Transforming Computer Technology: Information Processing for the Pentagon, 1962–1986*

Walter G. Vincenti, *What Engineers Know and How They Know It: Analytical Studies from Aeronautical History*

Printed in the United States
72141LV00002B/265-285